Sandwich Composites

Sandwich Composites
Fabrication and Characterization

Edited by
Senthilkumar Krishnasamy
Chandrasekar Muthukumar
Senthil Muthu Kumar Thiagamani
Sanjay Mavinkere Rangappa
Suchart Siengchin

CRC Press
Taylor & Francis Group
Boca Raton London New York

CRC Press is an imprint of the
Taylor & Francis Group, an **Informa** business

First edition published 2022
by CRC Press
6000 Broken Sound Parkway NW, Suite 300, Boca Raton, FL 33487-2742

and by CRC Press
4 Park Square, Milton Park, Abingdon, Oxon, OX14 4RN

© 2022 Taylor & Francis Group, LLC

CRC Press is an imprint of Taylor & Francis Group, LLC

Library of Congress Cataloging-in-Publication Data
Names: Krishnasamy, Senthilkumar, editor. | Muthukumar, Chandrasekar, editor. | Rangappa, Sanjay Mavinkere, editor. | Thiagamani, Senthil Muthu Kumar, editor. | Siengchin, Suchart, editor.
Title: Sandwich composites : fabrication and characterization / edited by Senthilkumar Krishnasamy, Chandrasekar Muthukumar, Senthil Muthu Kumar Thiagamani, Sanjay Mavinkere Rangappa, and Suchart Siengchin.
Description: First edition. | Boca Raton, FL : CRC Press, 2022. | Includes bibliographical references and index. | Summary: "This book provides insight into composite sandwich panels based on the material aspects, mechanical properties, defect characterization, and secondary process after fabrication, such as drilling and repair. It helps readers select the right material for the right application. Due to the wide scope of the topics covered, this book is suitable for researchers and scholars involved in the research and development of composite sandwich panels. This book can also be used as a reference book by professionals and engineers interested in understanding the factors governing the material properties, material response, and the failure behavior under various mechanical loads"— Provided by publisher.
Identifiers: LCCN 2021047641 (print) | LCCN 2021047642 (ebook) | ISBN 9780367697273 (hbk) | ISBN 9780367697327 (pbk) | ISBN 9781003143031 (ebk)
Subjects: LCSH: Laminated materials. | Sandwich construction.
Classification: LCC TA418.9.L3 S348 2022 (print) | LCC TA418.9.L3 (ebook) | DDC 620.1/18—dc23/eng/20211115
LC record available at https://lccn.loc.gov/2021047641
LC ebook record available at https://lccn.loc.gov/2021047642

ISBN: 978-0-367-69727-3 (hbk)
ISBN: 978-0-367-69732-7 (pbk)
ISBN: 978-1-003-14303-1 (ebk)

DOI: 10.1201/9781003143031

Typeset in Times
by codeMantra

Contents

Preface..ix

Editors...xi

Contributors xv

Chapter 1   Introduction to Sandwich Composite Panels and Their
Fabrication Methods.. 1

                   *Athul Joseph, Vinyas Mahesh, Vishwas Mahesh, and
Dineshkumar Harursampath*

Chapter 2   Corrugated Core- and Fold Core-Based Sandwich Panels 27

                   A. Farrokhabadi and M. Naghdi Nasab

Chapter 3   Metallic Core- and Truss Core-Based Composite Sandwich
Panels... 45

                   Jin-Shui Yang and Shuang Li

Chapter 4   Composite Sandwich Panels with the Metallic Facesheets........... 61

                   Lin Feng Ng and Kathiravan Subramaniam

Chapter 5   Failure Behavior and Residual Strength of the
Composite Sandwich Panels Subjected to Compression
after Impact Testing 75

                   *Dhaneshwar Mishra, Charanjeet Singh Tumrate, and
Anoop Kumar Mukhopadhyay*

Chapter 6   Low-Velocity Impact Response of the Composite Sandwich
Panels... 99

                   *Vishwas Mahesh, Vinyas Mahesh, and
Dineshkumar Harursampath*

Chapter 7   High-Velocity Impact Properties of the Composite
Sandwich Panels..115

                   Hossein Ebrahimnezhad-Khaljiri

Chapter 8   Investigation of Blast Loading Response of the Composite
                Sandwich Panels..131

                Hukum Chand Dewangan, Subrata Kumar Panda,
                Nitin Sharma, and Chetan Kumar Hirwani

Chapter 9   Flexural Behavior of Reinforced Concrete Sandwich
                Wall Panels Enabled by Fiber-Reinforced Polymer (FRP)
                Connectors...147

                Junqi Huang, Sushil Kumar, and Jian-Guo Dai

Chapter 10 Axial Behavior of Reinforced Concrete Sandwich Wall
                Panels Enabled by Fiber-Reinforced Polymer (FRP)
                Connectors... 167

                Jian-Guo Dai, Sushil Kumar, and Junqi Huang

Chapter 11 Enhanced Failure Behavior for Sandwiches with Hybrid
                Wire Mesh/FRP Face Sheets 191

                Çağrı Uzay and Necdet Geren

Chapter 12 Low-Velocity Impact Behaviour of Textile-Reinforced
                Composite Sandwich Panels... 213

                B.K. Behera, Manya Jain, Lekhani Tripathi, and
                Soumya Chowdhury

Chapter 13 Drilling and Repair of the Composite Sandwich Panels............. 261

                Shashikant Verma, Lalit Ranakoti, Brijesh Gangil, and
                Manoj Kumar Gupta

Chapter 14 Composite Sandwich Structures in the Marine Applications 277

                Thuy Thi Thu Nguyen, Tuan Anh Le, and Quang Huy Tran

Chapter 15 Composite Sandwich Structures in Aerospace Applications 293

                Erik Vargas-Rojas

Chapter 16 Crashworthiness Applications of the Composite Sandwich
                Structures 321

                S. Hassouna, M. Janane Allah, and A. Timesli

Chapter 17 Role of 3D Printing in the Fabrication of Composite
Sandwich Structures.. 349

*Athul Joseph, Vinyas Mahesh, Vishwas Mahesh,
Dineshkumar Harursampath, and Vasu Mallesha*

Index... 377

Preface

A composite sandwich panel is a hybrid material made up of face sheet, core, and adhesive film to bond the face sheet and core together. Advancements in materials have provided the designers with several choices for developing sandwich structures with advanced functionalities. The choice of selecting a material in the sandwich construction is decided based on the cost, availability, strength requirements, ease of manufacturing, machinability, and post-manufacturing process requirements.

This book provides a good insight into the composite sandwich panels based on the material aspects, mechanical properties, applications, and secondary processes after the fabrication, such as drilling and repair.

Due to the broad scope of the topics covered, this book is suitable for research scholars, researchers, and faculties doing research on composite sandwich panels. This book can also be used as a reference book by the professionals and engineers interested in understanding the factors governing the material properties, material response to mechanical loads, and the failure behavior to various mechanical loads.

This book consists of 17 chapters, which are structured as follows: Chapter 1 presents the history of sandwich structures and outlines the existing fabrication methods. Chapters 2–4 are devoted to the composite sandwich panels made of different face sheets and core materials. Chapters 5–12 covers the response of composite sandwich panels to the static and dynamic loads such as (i) compression after impact testing, (ii) low-velocity and high-velocity impact loads, (iii) blast loading response, and (iv) flexural and axial behaviors of sandwich structures. Chapters 13–17 discuss the various applications and challenges in drilling the sandwich composite panels. Chapter 13 describes the parameters governing the drilling process and repair procedures in the composite sandwich panels. Chapters 14–16 discuss the applications of the composite sandwich panels in marine, aerospace, and crashworthiness. Chapter 17 provides an overview of the role of 3D printing in the fabrication of composite sandwich panels.

All the authors wish to acknowledge the publisher's help, editorial manager, and associates for their support in the publishing process of this book.

Editors

Senthilkumar Krishnasamy, PhD, is an Associate Professor in the Department of Mechanical Engineering at Francis Xavier Engineering College, Tirunelveli, Tamilnadu, India. He earned a bachelor's degree in mechanical engineering at Anna University, Chennai, India, in 2005. He earned a master's degree in CAD/CAM at Anna University, Tirunelveli, India, in 2009. He earned a PhD in mechanical engineering at Kalasalingam University, Krishnankoil, Tamilnadu, India in 2016. From 2010 to 2018, Dr. Krishnasamy worked in the Department of Mechanical Engineering at Kalasalingam Academy of Research and Education (KARE), India. He completed a postdoctoral fellowship at Universiti Putra Malaysia, Serdang, Selangor, Malaysia and King Mongkut's University of Technology North Bangkok (KMUTNB) under the research topics of experimental investigations on mechanical, morphological, thermal, and structural properties of kenaf fiber/mat epoxy composites and sisal composites and fabrication of eco-friendly hybrid green composites on tribological properties in a medium-scale application. His area of research includes the modification and treatment of natural fibers, nanocomposites, 3D printing, and hybrid reinforced polymer composites. He has published research papers in international journals, book chapters, and at conferences in the field of natural fiber composites. He also edits books with different publishers.

Chandrasekar Muthukumar, PhD, is an Assistant Professor in the Department of Aeronautical Engineering, Hindustan Institute of Technology and Science, Chennai, India. He earned a bachelor's degree in Aeronautical engineering from the Kumaraguru College of Technology, Coimbatore, India. His master's degree is in aerospace engineering from the Nanyang Technological University – TUM Asia, Singapore and his PhD in aerospace engineering is from the Universiti Putra Malaysia (UPM), Malaysia. His PhD was funded through a research grant from the Ministry of Education, Malaysia. During his association with the UPM, he obtained internal research funding from the university worth 16,000 and 20,000 MYR, respectively. He has 5 years of teaching and academic research experience. His interests include fiber metal laminates (FMLs), natural fibers, biocomposites, additive manufacturing and non-destructive testing. His publications are based on the fabrication and characterization techniques of biocomposites. He has authored and co-authored research publications in SCIE journals, book chapters, and articles in the conference proceedings. He is a peer reviewer for the *Journal of Industrial Textiles, Polymer Composites, Materials Research Express, Journal of Natural Fibres, etc.*

Senthil Muthu Kumar Thiagamani, PhD, is an Associate Professor in the Department of Mechanical Engineering at Kalasalingam Academy of Research and Education (KARE), Tamil Nadu, India. In 2004, he earned a Diploma in Mechanical Engineering at the State Board of Technical Education and Training, Tamil Nadu. In 2007, he graduated from Anna University, Chennai with a Bachelor's degree in Mechanical Engineering. In 2009, he obtained his Master's degree in Automotive Engineering at Vellore Institute of Technology, Vellore. In 2018, he earned a PhD in Mechanical Engineering, specializing in Biocomposites, at KARE. He completed postdoctoral research in the Materials and Production Engineering Department at the Sirindhorn International Thai-German Graduate School of Engineering (TGGS), KMUTNB, Thailand in the year 2019. He started his academic career in 2010 as Lecturer in Mechanical Engineering at KARE. He has around 12 years of teaching and research experience. He is a visiting researcher at KMUTNB, Thailand. He is a member of the Society of Automotive Engineers and the International Association of Advanced Materials. His research interests include biodegradable polymer composites and their characterization. He has authored several articles in peer-reviewed international journals and book chapters and in conference proceedings. He has edited books on the different themes of Biocomposites published by CRC Press, John Wiley and Springer. He also serves as a reviewer for various journals to name few are *Journal of Industrial Textiles,* Journal of Cleaner Production, Materials Today Communications, *Journal of Polymers and the Environment, SN Applied Sciences, Mechanics of Composite Materials,* and *International Journal of Polymer Science.*

Sanjay Mavinkere Rangappa, PhD, is a Senior Research Scientist and an advisor within the Office of the President for University Promotion and Development toward international goals at King Mongkut's University of Technology, Bangkok, Thailand. In 2010, he earned a BE in mechanical engineering. In 2013, he earned an MTech in computational analysis in mechanical sciences. In 2018, he earned a PhD in mechanical engineering at Visvesvaraya Technological University, Belagavi, India. In 2019, Dr. Rangappa did postdoctoral work at King Mongkut's University of Technology North Bangkok, Thailand. He is a Lifetime Member of the Indian Society for Technical Education (ISTE) and an Associate Member of the Institute of Engineers (India). He is a board member of various international journals in the fields of materials science and composites. He is a reviewer for more than 85 international journals (Nature, Elsevier, Springer, Sage, Taylor & Francis, Wiley, American Society for Testing and Materials, American Society of Agricultural and Biological Engineers, IOP, Hindawi, North Carolina State University – USA, ASM International, Emerald Group, Bentham Science Publishers, and Universiti Putra Malaysia) and a reviewer for book proposals and international conferences. He has published more than 140 articles in international peer-reviewed journals indexed by SCI/Scopus, 5 editorial corners, 40 book chapters, 1 book, and 15 books as an editor (Elsevier, Springer, Taylor & Francis, and Wiley) and also presented papers at national and international conferences. In 2021, his 17 articles received top-cited status in various journals (*Journal of Cleaner Production, Carbohydrate Polymers, International Journal of Biological Macromolecules, Journal of Natural Fibers, Journal of Industrial Textiles*). He is a lead editor of special issues on artificial intelligence and machine learning in composites and metamaterials, *Frontiers in Materials* (ISSN 2296–8016) indexed in Web of Science, and also trends and developments in natural fiber composites, *Applied Science and Engineering Progress* (ASEP) indexed in Scopus. He has delivered keynote and invited talks at various international conferences and workshops. His research areas include natural fiber composites, polymer composites, and advanced material technology. He has received the DAAD Academic Exchange – PPP Programme between Thailand and Germany to the Institute of Composite Materials, University of Kaiserslautern, Germany. He has received a Top Peer Reviewer 2019 award, Global Peer Review Awards, Powered by Publons, Web of Science Group. KMUTNB awarded him the Outstanding Young Researcher Award 2020. In 2019, he was recognized by Stanford University's list of the world's top 2% of the most-cited scientists in single year.

Suchart Siengchin, PhD, is the President of King Mongkut's University of Technology North Bangkok (KMUTNB), Thailand. He earned a Dipl-Ing in mechanical engineering at the University of Applied Sciences Giessen/Friedberg, Hessen, Germany, in 1999; an MSc in polymer technology at the University of Applied Sciences Aalen, Baden-Wuerttemberg, Germany, in 2002; an MSc in materials science at the Erlangen-Nürnberg University, Bayern, Germany, in 2004; a PhD in engineering (Dr-Ing) at the Institute for Composite Materials, University of Kaiserslautern, Rheinland-Pfalz, Germany, in 2008; and a postdoctoral research at Kaiserslautern University and the School of Materials Engineering, Purdue University, USA. In 2016, he received the habilitation at the Chemnitz University in Sachsen, Germany. He worked as a Lecturer for Production and Material Engineering Department at the Sirindhorn International Thai-German Graduate School of Engineering (TGGS), KMUTNB. He has been a full-time professor at KMUTNB and became the President of KMUTNB. He received the Outstanding Researcher Award in 2010, 2012, and 2013 at KMUTNB. His research interests include polymer processing and composite materials. He is Editor-in-Chief of *KMUTNB International Journal of Applied Science and Technology* and the author of more than 150 peer-reviewed journal articles. He has participated with presentations at more than 39 international and national conferences with respect to materials science and engineering topics.

Contributors

B.K. Behera
Department of Textile and Fiber
  Engineering
Indian Institute of Technology
New Delhi, India

Soumya Chowdhury
Department of Textile and Fiber
  Engineering
Indian Institute of Technology
New Delhi, India

Jian-Guo Dai
Department of Civil and
  Environmental Engineering
The Hong Kong Polytechnic
  University
Hong Kong, China

Hukum Chand Dewangan
Department of Mechanical Engineering
National Institute of Technology
Rourkela, India

Hossein Ebrahimnezhad-Khaljiri
Faculty of Materials Science and
  Engineering
K.N. Toosi University of Technology
Tehran, Iran

A. Farrokhabadi
Department of Mechanical
  Engineering
Tarbiat Modares University
Tehran, Iran

Brijesh Gangil
Department of Mechanical
  Engineering
H.N.B. Garhwal University
Srinagar Garhwal, India

Necdet Geren
Department of Mechanical
  Engineering
Çukurova University
Adana, Turkey

Manoj Kumar Gupta
Department of Mechanical
  Engineering
H.N.B. Garhwal University
Srinagar Garhwal, India

Dineshkumar Harursampath
Department of Aerospace Engineering
Indian Institute of Science
Bangalore, India

S. Hassouna
Artificial Intelligence and Complex
  Systems Engineering Laboratory
National Higher School of Arts and
  Crafts of Casablanca (ENSAM
  Casablanca)
Hassan II University of Casablanca
and
National Higher School of Arts and
  Crafts of Casablanca (ENSAM
  Casablanca)
Casablanca, Morocco

Chetan Kumar Hirwani
Department of Mechanical
  Engineering
National Institute of Technology Patna
Patna, India

Junqi Huang
School of Civil and Hydraulic
  Engineering
Hefei University of Technology
Anhui, China

Manya Jain
Department of Textile and Fiber
  Engineering
Indian Institute of Technology
New Delhi, India

M. Janane Allah
Artificial Intelligence and Complex
  Systems Engineering Laboratory
National Higher School of Arts and
  Crafts of Casablanca (ENSAM
  Casablanca)
Hassan II University of Casablanca
and
National Higher School of Arts and
  Crafts of Casablanca (ENSAM
  Casablanca)
Casablanca, Morocco

Athul Joseph
Department of Aerospace
  Engineering
Indian Institute of Science
Bangalore, India

Sushil Kumar
Department of Civil and
  Environmental Engineering
The Hong Kong Polytechnic
  University
Hong Kong, China

Tuan Anh Le
Phenikaa University Nano Institute
Phenikaa University
Hanoi, Vietnam

Shuang Li
Key Laboratory of Advanced Ship
  Materials and Mechanics
College of Aerospace and Civil
  Engineering
Harbin Engineering University
Harbin, China

Vinyas Mahesh
Department of Aerospace Engineering
Indian Institute of Science
Bangalore, India
and
Department of Mechanical
  Engineering
Nitte Meenakshi Institute of
  Technology
Bangalore, India
and
Department of Mechanical
  Engineering
National Institute of Technology
  Silchar
Assam, India

Vishwas Mahesh
Department of Industrial Engineering
  and Management
Siddaganga Institute of Technology
Tumkur, India
and
Department of Aerospace Engineering
Indian Institute of Science
Bangalore, India

Vasu Mallesha
Department of Production
  Engineering
National Institute of Technology
Tiruchirappalli, India

Dhaneshwar Mishra
Multiscale Simulation Research
  Center (MSRC)
Department of Mechanical
  Engineering
Faculty of Engineering
Manipal University Jaipur
Jaipur, India

Anoop Kumar Mukhopadhyay
Department of Physics
School of Basic Sciences
Faculty of Science
Manipal University Jaipur
Jaipur, India

M. Naghdi Nasab
Department of Mechanical
Engineering
Rowan University
Glassboro, New Jersey

Lin Feng Ng
Centre for Advanced Composite
Materials
School of Mechanical Engineering
Universiti Teknologi Malaysia
Johor Bahru, Malaysia

Thuy Thi Thu Nguyen
Phenikaa University Nano Institute
Phenikaa University
Hanoi, Vietnam

Subrata Kumar Panda
Department of Mechanical
Engineering
National Institute of Technology
Rourkela, India

Lalit Ranakoti
Mechanical Engineering Department
National Institute of Technology
Uttarakhand, India

Nitin Sharma
School of Mechanical Engineering
KIIT University
Bhubaneswar, India

Kathiravan Subramaniam
Centre for Advanced Research on
Energy
Faculty of Mechanical Engineering
Universiti Teknikal Malaysia Melaka
Durian Tunggal, Malaysia

A. Timesli
Artificial Intelligence and Complex
Systems Engineering Laboratory
National Higher School of Arts and
Crafts of Casablanca (ENSAM
Casablanca)
Hassan II University of Casablanca
and
National Higher School of Arts and
Crafts of Casablanca (ENSAM
Casablanca)
Casablanca, Morocco

Quang Huy Tran
Phenikaa University Nano Institute
Phenikaa University
Hanoi, Vietnam

Lekhani Tripathi
Department of Textile and Fiber
Engineering
Indian Institute of Technology
New Delhi, India

Charanjeet Singh Tumrate
Multiscale Simulation Research
Center (MSRC)
Department of Civil Engineering
Faculty of Engineering
Manipal University Jaipur
Jaipur, India

Çağrı Uzay
Department of Mechanical
Engineering
Kahramanmaraş Sütçü İmam
University
Kahramanmaraş, Turkey

Erik Vargas-Rojas
National Polytechnic Institute
Mexico City, Mexico
and
Franche-Comté University
Besançon, France

Shashikant Verma
Department of Mechanical
    Engineering
Institute of Engineering and
    Technology
Bundelkhand University
Jhansi, India

Jin-Shui Yang
Key Laboratory of Advanced Ship
    Materials and Mechanics
College of Aerospace and Civil
    Engineering
Harbin Engineering University
Harbin, China

1 Introduction to Sandwich Composite Panels and Their Fabrication Methods

Athul Joseph
Indian Institute of Science

Vinyas Mahesh
Indian Institute of Science
Nitte Meenakshi Institute of Technology
National Institute of Technology Silchar

Vishwas Mahesh
Siddaganga Institute of Technology
Indian Institute of Science

Dineshkumar Harursampath
Indian Institute of Science

CONTENTS

1.1 What Are Sandwich Composite Structures? 2
1.2 Brief History of Sandwich Composite Structures 4
1.3 Common Materials Used in Composite Sandwich Structures and
Their Properties 4
   1.3.1 Face Materials.. 4
   1.3.2 Core Materials 5
      1.3.2.1 Foams or Solid Cores.. 6
      1.3.2.2 Corrugated or Truss Cores.. 7
      1.3.2.3 Honeycomb Structures.. 8
   1.3.3 Adhesive Materials 8
      1.3.3.1 Epoxy Resins and Toughened Epoxies 8
      1.3.3.2 Phenolics 9
      1.3.3.3 Polyurethanes.. 9

DOI: 10.1201/9781003143031-1

1

          1.3.3.4   Urethane Acrylates 9
          1.3.3.5   Polyester and Vinyl Ester Resins 10
1.4   Fabrication Methods for Composite Sandwich Laminates....................... 10
     1.4.1   Hand Layup Process... 10
     1.4.2   Compression Moulding Process 11
     1.4.3   Filament Winding Process.. 11
     1.4.4   Vacuum Bagging and Autoclave Moulding Process 13
     1.4.5   Pultrusion Process 14
     1.4.6   Resin Transfer Moulding (RTM).. 15
     1.4.7   Additive Manufacturing Processes.. 15
1.5   Conclusions... 19
References.. 19

1.1   WHAT ARE SANDWICH COMPOSITE STRUCTURES?

Sandwich composite structures are a special class of laminated composites in which different forms of materials are bonded to each other to obtain a desired functionality. The overall behaviour of these composite structure is dependent on the properties of the constituent layers [1]. Conventionally, a sandwich composite structure consists of two main parts: the face sheets or faces and the core structure. Usually, the face sheets are adhesively bonded to the core on either side. Also, the face sheets are much thinner than the core as seen in Figure 1.1. The adhesive used to bond these two entities must have enough strength to withstand the stresses set up between the face and the core [2]. The basic principle of design for a sandwich composite structure is inspired from an I-beam (shown in Figure 1.2). In view of sandwich composite structures, the face sheets of the structure resemble the flanges of the I-beam and the core represents the web of the I-beam that connects both the flanges. The face sheets form a stress couple to counteract the bending stresses applied on the structure with one face under compression and the other under tension as seen in Figure 1.3 [3]. The core resists the shear loads and improves the stiffness of structure, providing adequate support to the face sheets. A continuous support is provided by the core structure so that a uniformly stiffened structure is obtained. Some of the primary advantages of sandwich composites are their very high stiffness-to-weight ratio and high bending strength-to-weight ratio coupled with good thermal insulation properties. Furthermore, some sandwich composite structures have excellent mechanical

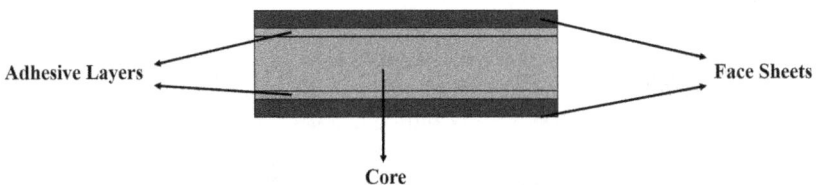

Adhesive Layers   **Face Sheets**

Core

FIGURE 1.1   Depiction of a typical sandwich composite panel.

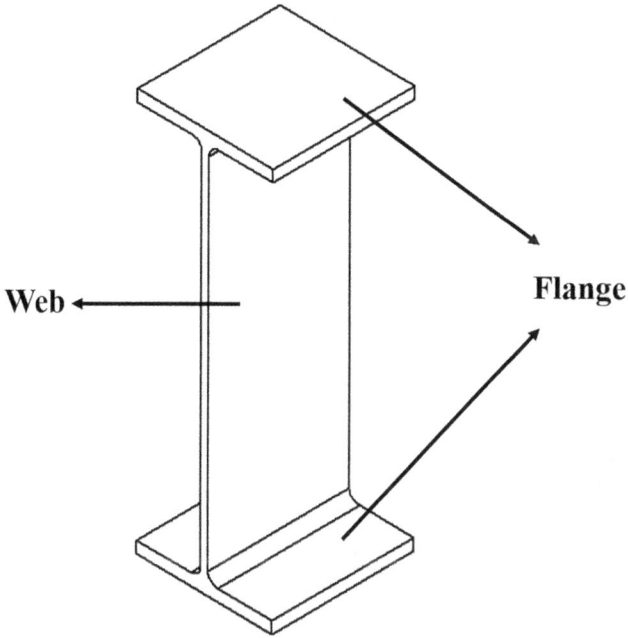

FIGURE 1.2 Structure of an I-beam.

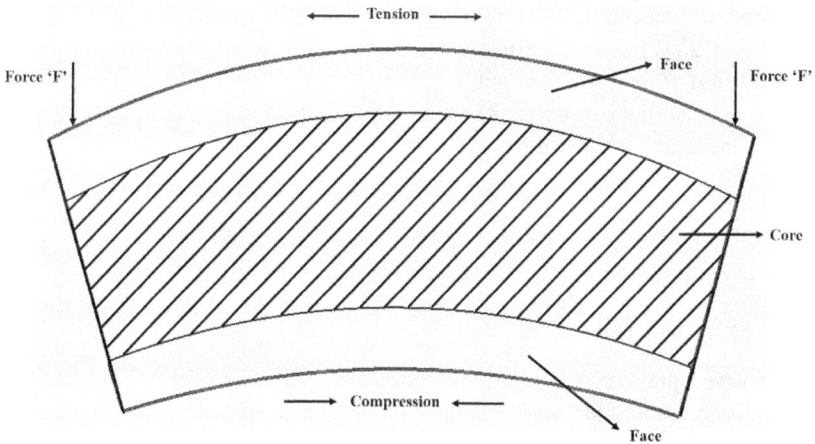

FIGURE 1.3 Behaviour of sandwich composite laminates under bending forces.

energy absorption characteristics, fatigue life and acoustical insulation. These characteristics of sandwich structures make them desirable in a lot of applications such as marine structures, aerospace components, automotive parts and civil engineering structures.

1.2   BRIEF HISTORY OF SANDWICH COMPOSITE STRUCTURES

Sandwich constructions date back to 1849 when Noor, Burton and Bert stated the first use of sandwich construction back to Fairbairn in England in 1849 [4]. From there, a major development was seen when the Wright-Patterson Air Force Base designed and fabricated the Vultee BT-15 fuselage using fibreglass-reinforced polyester as the face material and both a glass-fabric honeycomb and a balsa core in 1943. However, the first documented research paper concerning sandwich construction was written by Marguerre [5] in Germany in 1944 to determine the in-plane compressive properties of sandwich panels. Hoff [6] put forth a set of differential equations and boundary conditions to analyse the bending and buckling of sandwich plates. The equations were derived using the principle of virtual displacements, which was primarily suited for buckling problem only. Shortly, a small deflection theory for sandwich plates was put forth by Libove and Batdorf [7]. Later in the year 1949, a structural optimization theory for sandwich panels was put forth by Flügge [8]. The study presented nomograms for the solution of several problems related to composites structures.

In 1956, Gerard [9] discussed different sandwich plate optimization routes in his book *Minimum Weight Analysis of Compression Structures.* Shortly, Kaechele [10] expounded on the minimum weight design of sandwich panels in a report submitted to the United States Forest Products Laboratory (USFPL). In 1966, Plantema [11], in the Netherlands, published the first book on sandwich structures, followed by another book by Allen in England in 1969. These books remained as the standard reference for composite structures until the mid-1990s. The first work on the finite element analysis of sandwich constructions was presented by Ha [12] in 1990. Since then, composite structures have evolved to be included in a wide variety of applications from aerospace structures to marine structures to automotive parts and many others, where effective weight reduction with uncompromised strength and stiffness was desired [13].

1.3   COMMON MATERIALS USED IN COMPOSITE SANDWICH STRUCTURES AND THEIR PROPERTIES

1.3.1   FACE MATERIALS

As discussed previously, the primary purpose of the face sheet is to resist the bending stresses applied on the sandwich laminate structure. In addition to this, properties such as high tensile and compressive strength, good impact resistance, good surface finish, environmental resistance and wear resistance are desirable to obtain strong and reliable structures [2]. The materials used for the face sheets are usually in the form of thin sheets. It is common to use both metals and non-metallic materials as face sheets, although the latter are widely preferred due to their excellent weight reduction and stiffness enhancement. Metallic materials vastly include steel, stainless steel and aluminium alloys. There are other alloys with better strength properties. However, they find limited applications due to their higher material density and limited stiffness variation. Non-metallic materials are the most common materials used for face sheets. Materials such as plywood, cement, plastics and fibre-reinforced plastics are the major ones used as face sheets in most of the applications. Engineering plastics such as polypropylene, polyamide and high-performance polymers such as polyetherimide (PEI) and polyether ether ketone (PEEK) have widely been used. In addition to this, the use of fibre-reinforced plastics has been effective in significantly reducing weight while providing sufficient mechanical capabilities for the structure. Some high-performance polymers such as PEI are used in sophisticated applications such as aerospace components.

1.3.2 Core Materials

Core materials should primarily possess low density in order to reduce the total weight of the sandwich structure as much as possible. Core materials are primarily subjected to shear stresses which produce global deformation as a result of the strains produced within the core material. Therefore, one of the primary characteristics that a core material should possess is high shear modulus to give the required shear stiffness so that the structure does not fail under transverse loads. However, the critical wrinkling load that the structure can bear is dependent on the shear and Young's moduli of the material. Based on the type of support offered, the core materials used in sandwich composite laminates may be classified as shown in Figure 1.4. Foam cores or homogenous core materials are used in low-cost applications. They exhibit good out-of-plane compression and shear performance. However, they are comparatively lower than most honeycomb cores. Some of the common foam materials used for homogenous core materials are listed below with their fundamental characteristics and recent developments. Typically, there are three types of sandwich cores: (1) foam or solid cores, (2) corrugated or truss cores and (3) honeycomb cores with different unit cell structures.

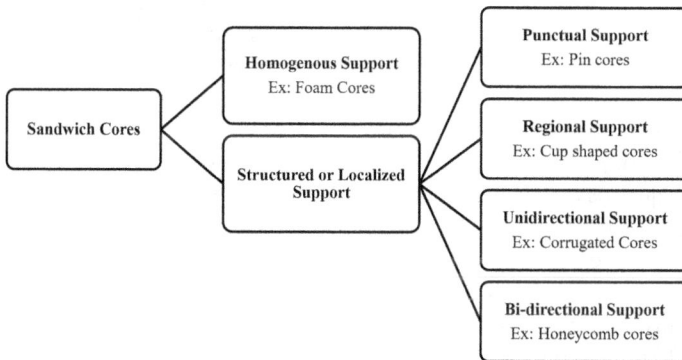

FIGURE 1.4   Classification of core structures for sandwich composite panels.

1.3.2.1 Foams or Solid Cores

Foam is a solid material that is used in the preparation of core materials for sandwich structures. These materials have a low density and are generally realized by conventional manufacturing processes. Foam cores generally have a perforated geometry which consequently reduces the weight of the overall composite. However, once cured, these structures have good shear properties and stiffness to resist considerable loads. Some of the most common foam materials include polyurethane foams, polystyrene foams, polyvinylchloride foams and polymethacrylimide (PMI) foams. A description of the general properties of these foams is given below.

Polyurethane foams are produced by the reaction of isocyanate and polyol in which either trichlorofluoromethane or carbon dioxide is used as the blowing agent. Polyurethane foams come in many variations such as soft foams with more open-cell designs to rigid foams with a predominant closed-cell structure. The densities of the foams range from 30 to 500 kg/m^3 [14]. Polyurethane foams are widely used as core foam materials because of their dependable mechanical properties such as high shear strength and stiffness. However, one of the primary disadvantages of these materials is the environmental pollution caused when disposed. Traditionally, since the material is composed of chlorine and fluorine compounds, they release toxic gases upon incineration. Therefore, modern research has focused on obtaining bio-based polyurethane alternatives and polyurethane foams with enhanced mechanical, thermal and acoustical properties [15–19]. Additions of nanoparticles and fibrous reinforcements have improved the load-bearing capacities of the composites [20–24]. Moreover, modifications in the manufacturing techniques have been helpful in producing structures with improved flame retardancy, reduced moisture uptake and improved thermal stability.

Polystyrene foams are prepared by a similar method as polyurethane foams, in closed moulds by either extrusion or expansion. Polystyrene foams exhibit good mechanical and thermal insulation properties while also being comparatively cheaper [25]. The typical densities exhibited by polystyrene foams usually lie in between 15 and 300 kg/m^3. However, the use of chlorofluorocarbons (CFCs) has been highly condemned due to their adverse environmental effects. Additionally, the high sensitivity of these foams to solvents restricts the use of common ester-based matrices as adhesives. Numerical and experimental investigations on the mechanical properties of polystyrene foams have shown promising results for their application in practical scenarios [26]. In addition to this, new manufacturing techniques have aided in improving the thermal responses of these foams with better cell distribution and properties [25,27,28]. Additionally, efforts are made to reuse these foams to extract viable compounds for further manufacture and subsequently reduce the ill effects on the environment [25,26,29–31].

Polyvinylchloride (PVC) exists as thermoplastic entities and cross-linked isocyanate-modified entities. The thermoplastic PVC has a linear molecular configuration with good mechanical properties. However, its application to sandwich core materials is limited by its tendency to soften at elevated temperatures [32–35].

On the other hand, cross-linked PVC is less sensitive to heat, but is more brittle. Cross-linked PVC foams have higher strength and stiffness, and are more rigid. Even though the foam is rigid, it can elongate to up to 10% of its initial length in tension which is much higher than polyurethane and polystyrene foams. PVC foams usually have densities between 30 and 400 kg/m³ [36–39].

Expanded imide-modified polyacrylates are used to synthesize acrylimide cellular plastics. PMI foams exhibit the best properties among all the previously discussed core sandwich materials [40–43]. However, they are fairly brittle with an ultimate elongation of 3% in tension. One of the major advantages of PMI foams is their temperature resistance, making it possible to use these foams in autoclave manufacturing to about 180°C–200°C along with epoxy prepregs. These foams have very fine cell structures with closed cells, and their density usually lies in between 30 and 300 kg/m³.

1.3.2.2 Corrugated or Truss Cores

Corrugated composite sandwich panels are widely used in engineering applications where extreme anisotropic stiffness properties are desired. These composites are stiff along the transverse direction and flexible across the longitudinal direction [44]. These structures also possess high strength-to-weight ratio as a result of the lower requirement of materials. As opposed to soft honeycomb structures, corrugated cores opposed bending, twisting and vertical shear forces while effectively separating the two faces and thus functioning as a single entity. One of the most profound benefits of this behaviour is the use of these composites in morphing wings of an aircraft [45]. The stiffness in the transverse direction is essential to withstanding the bending due to aerodynamic forces, and the longitudinal flexibility is needed to deform the structure to match the flight regime efficiently. These composites also reduce the number of parts used in the wing structure, which results in reduced manufacturing costs and improved production speeds. Additionally, these structures also possess good fatigue resistance and energy absorption capabilities as added advantages for the aforementioned application [46–50]. When compared to honeycomb structures, corrugated cores have minimal humidity retention. A number of materials ranging from metallic materials to polymer composite materials are used for corrugated structures. Metallic corrugated cores are used for industrial and transportation applications using dissimilar core structures [45,51]. Typical corrugated structures are depicted in Figure 1.5.

FIGURE 1.5 Cross section of a typical corrugated composite panel.

1.3.2.3   Honeycomb Structures

Honeycomb structures are one of the most widely adapted core configurations besides foam-based cores. Honeycomb structures have great energy absorption capabilities coupled with good mechanical strength. Some of the typical core materials used in honeycomb core structures are Nomex, fibreglass and aluminium [52]. While hexagonal honeycombs are the most common type of honeycomb structures, a number of structures such as over-expanded structures, under-expanded structures, square-cell structures, reinforced hexagonal structures and flexible core structures are used in modern materials [53,54]. Honeycomb structures are usually prepared by methods such as adhesive bonding, welding, brazing, diffusion bonding and thermal fusion. Of these, the adhesive bonding method is the most commonly used one. When foams are combined with honeycomb structures, there is a phenomenal improvement in the behaviour of these structures. Such composites have superior debonding resistance combined with enhanced acoustic absorption and heat resistance [55–58]. The combination of foams and honeycomb structures increases the adhesive area of the core structures, which in turn enhances the dynamic properties of the honeycomb structures. In addition to these benefits, the overall natural frequencies of the structure see a considerable reduction in its magnitude.

1.3.3   Adhesive Materials

1.3.3.1   Epoxy Resins and Toughened Epoxies

Epoxy resins are one of the most commonly used materials in the fabrication of composite sandwich laminates. They are typically low-temperature-curing resins, but some variants are apt for higher-temperature curing at 130°C–220°C as well. They are characterized by low-volume shrinkage due to the absence of the use of any kind of solvent, which prevents the formation of volatile by-products. This property broadens the application spectrum of these materials to be used in any core materials. They come in different forms and typically have shear strengths ranging between 20 and 25 MPa [59–61]. With the addition of nanoparticles, the adhesiveness of these adhesives is seen to improve and they also exhibit multifunctional behaviour [62,63]. However, the properties may be hindered if the nanoparticles are not dispersed evenly across the matrix. In some cases, they can inhibit the formation of 3D cross-linking networks of epoxy by interfering with the curing time.

Toughened epoxies are very much similar to regular epoxy resins, but the addition of synthetic rubber compounds such as polysulfide elastomers improves their peel resistance [64]. However, as the content of the elastomers increases, the creep resistance and heat resistance decrease. Nevertheless, these adhesives are not utilized to their full potential due to the lack of modelling studies that quantitatively predict the engineering performance of these structures. There is a growing need to realize the properties and capabilities of toughened epoxy adhesives as structures and components made of engineering plastics, lightweight glass-reinforced plastics (GRPs) and aluminium alloys are becoming increasingly popular. For such structures, spot welding techniques cannot be readily used and adhesives such as toughened epoxies can effectively assume the role of the primary joining method.

1.3.3.2 Phenolics

Phenolic adhesives are known for their excellent mechanical properties, high-temperature performance and durability. However, phenolic resins are seen to give out water when cured. Thus, venting is an essential aspect when it comes to the use of phenolic resins [65,66]. This has limited the use of phenolic resins to structures such as the honeycomb cores where venting is not an issue. It is seen that the addition of both organic and inorganic fillers to the adhesives did not affect the curing process [67–69]. Additionally, these resins showed differential thermal activity: they were highly exothermic at lower temperatures and endothermic at higher temperatures. However, the addition of ingredients such as walnut shell flour and amorphous silica 200 fillers improved the fracture toughness and the brittle nature of the adhesives. In a novel attempt, Kalami et al. [70] replaced the phenolic compounds with different lignins in regular phenol-formaldehyde adhesives to obtain highly renewable and environmentally safe phenolic resins. They even found application in bone fracture fixations when environmentally friendly and biocompatible materials are used to prepare the resin [71].

1.3.3.3 Polyurethanes

Polyurethanes provide excellent adhesion to most materials. These adhesives also possess high chemical resistivity, excellent drying, low-temperature flexibility and good abrasion resistance. The adhesive bonding strength of these compounds is dependent on the structure of the polyol and isocyanate used in its manufacture. Moreover, there is a substantial decrease in the cross-linking time with the increase in the molecular weight of the polyurethane adhesive. In addition to this, additions and modifications in the manufacturing process can improve their fire retardancy and water-resistant behaviour [72,73]. Like epoxy resins, these adhesives do not contain any solvent and are the least toxic of all major adhesives. It is seen that different types of polyols extracted from castor oils were useful in the adhesion of wood-to-wood composites. Some polyol variants had shear strengths which are about ten times those of the commercially available wood adhesives, while some polyols had better bonding strength compared to the others. Similarly, the addition of nanosilica particles is seen to improve the thermal, rheological, mechanical and adhesion properties of thermoplastic polyurethane adhesives [74].

1.3.3.4 Urethane Acrylates

These adhesives are most compatible with polyesters and vinyl esters. These adhesives are extremely tough and exhibit very minimal volume shrinkage during curing, and they can be used in wet polymer laminates as well. These adhesives can enhance the bonding of GRP sandwich structures when they are used for the first reinforcing layer that is closest to the core. Even if the subsequent layers are bonded with other adhesives, the laminate can offer perfect interlaminar bonding. These adhesives can further be enhanced to improve their flexibility and reduce their viscosity to be used in different demanding applications. Recent studies have focused on obtaining fluorinated urethane acrylate-based UV-curable adhesives that possess lesser surface energy compared to the urethane adhesive. These materials have improved surface scratch resistance and hardness values along with good chemical resistance [75–77].

1.3.3.5   Polyester and Vinyl Ester Resins

Polyester and vinyl ester are commonly used thermosets to obtain reinforced plastic components. The resin can be used to bond prefabricated laminates onto foams or balsa cores. Usually, the first layer of the laminate is bonded wet to the core in such a manner that all the surface cells are filled. This is particularly important because they undergo considerable volume shrinkage while curing. This can be reduced by priming the core with a thin layer of the resin which is allowed to cure before the rest of the laminate is applied [78–80].

1.4   FABRICATION METHODS FOR COMPOSITE SANDWICH LAMINATES

Sandwich composites can be manufactured by a number of techniques. Conventionally, moulding techniques are used to fabricate such structures. Methods such as vacuum bagging, autoclave moulding, adhesive bonding, compression moulding and layup techniques are the most prominent ones [81,82]. However, modern additive manufacturing techniques have been used to fabricate complex core structures with multifunctional capabilities. Fused deposition modelling (FDM), stereolithography (SLA), selective laser melting (SLM) and selective laser sintering (SLS) are a few examples of such methods. The upcoming sections describe these fabrication methods in sufficient detail.

1.4.1   Hand Layup Process

Hand layup is one of the oldest, easiest and most commonly used methods to fabricate composite laminate structures [83–85]. This method utilizes continuous fibres which are usually available as unidirectional, woven, knitted or stitched fabrics. Different fibre orientations and laminar layers are incorporated as per the requirement of the intended application. The matrix is distributed over the fibre layers with the help of a roller to ensure uniform distribution and removal of additional resin. A releasing agent is placed on the surface of the mould so that the material does not stick to the mould and facilitates easy removal. The same process is repeated for the subsequent layers and is set for curing at standard atmospheric temperature. A schematic representation of the process is depicted in Figure 1.6. The quality of the laminates prepared by this method is influenced

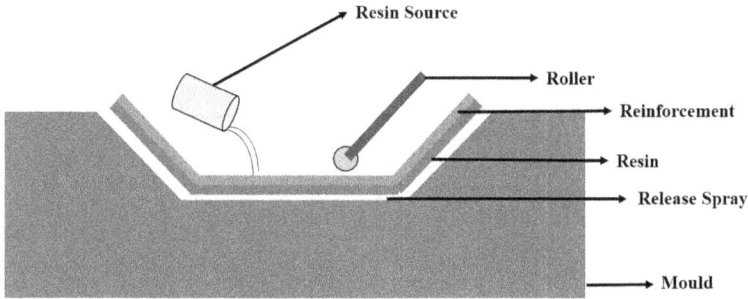

FIGURE 1.6 Illustration of hand layup process.

by a lot of factors such as fibre reinforcement type, resin material properties, curing temperature and time, and pressure. However, the process is fairly inexpensive and facilitates the production of large structures as well. However, skilled labour is strictly required to minimize faults in the structures and to obtain uniform properties. This method is not preferred for the production of large structures and is often restricted to structures such as boat hulls and vessels. Further enhancements in the properties of these composites can be made by the inclusion of nanoparticles such as graphene.

1.4.2 COMPRESSION MOULDING PROCESS

Compression moulding is another conventional method used in the manufacture of composite laminates. This method is capable of rapid mass production as compared to other methods used in the manufacture of sandwich laminates. It follows a two-step approach to fabricate laminate parts. Initially, the charge is introduced into the cavity of the matched mould while it is in the open position. Once the charge is filled, the mould is closed, and pressure is applied to squeeze the resin in order to fill the mould. During the application of pressure, the materials are also cured by the application of heat. Upon the completion of the curing process, the pressure is released, and the part is removed. Figure 1.7 gives an overview of the compression moulding process. Compression-moulded parts have extensively been researched over the years to clearly define the capability of the process and the resultant parts. This method can produce parts with lower-dimensional variations and better finish as compared to the hand layup process. The parts are seen to have good mechanical properties along with improved fatigue life. The inclusion and orientation of different fibres and nanoparticles can result in improved mechanical properties of the moulded parts [86–88].

1.4.3 FILAMENT WINDING PROCESS

Filament winding process involves the continuous rotation of fibres onto a rotating mandrel in a specific orientation for the desired functionality. It is primarily used to produce parts that have a symmetric cylindrical or oval-shaped geometry.

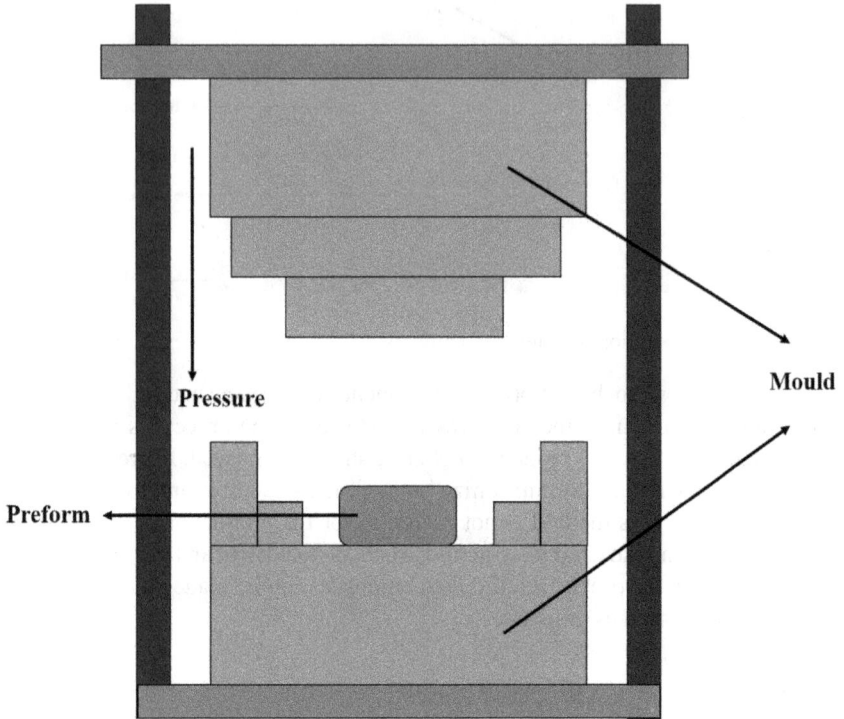

FIGURE 1.7 Schematic representation of compression moulding process.

The fibres are passed through a resin bath to wet them before winding them onto the mandrel structure. The thickness of the winding and the number of layers determine the final properties of the structure. The compaction of the fibres onto the mandrel is achieved by giving the appropriate tension to the filaments. Once the winding is complete, the parts are cured either at the room temperature or at an elevated temperature. Variations can be achieved with respect to the winding pattern to ensure different properties. Some of the common winding patterns are polar winding, hoop winding and helical winding. Moreover, the mandrel used in this process defines the final geometry of the composite part. The mandrel used can be removable or non-removable. A visual representation of the entire process is illustrated in Figure 1.8.

Filament winding is a quick and fairly automated process that requires minimal manual intervention. The composites produced by this method are seen to have excellent mechanical properties with better thickness control, proper fibre orientation and good internal finish [89]. However, the process is complicated for winding complex shapes. Moreover, the poor external finish of the parts, high costs of the mandrel and demand for low-viscosity resins limit its application to a few structural and material combinations. Research in the filament winding process has focused on obtaining multifunctional composites through the addition

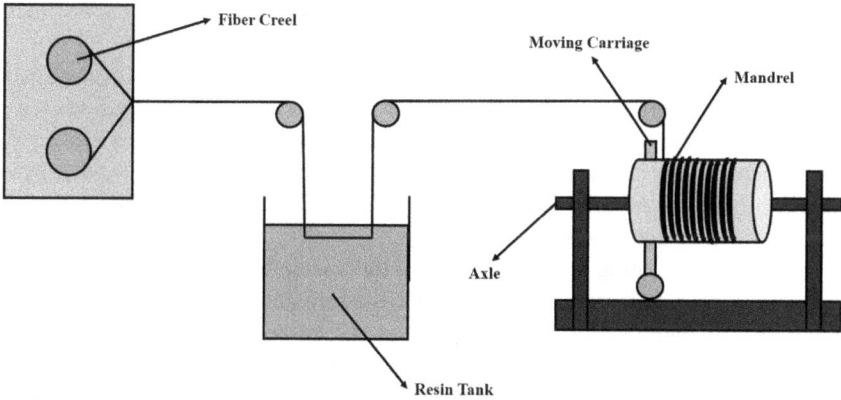

FIGURE 1.8 Schematic representation of filament winding process.

of various fibres to improve the mechanical properties of the composite laminate structures [90].

1.4.4 VACUUM BAGGING AND AUTOCLAVE MOULDING PROCESS

Vacuum-bagging process gets its name due to the application of vacuum pressure to realize the final composite parts during the curing cycle of the process. This process utilizes uniform atmospheric pressure to hold the resin and fibres in the desired place to consolidate the layers in the laminate. The laminate is sealed in the airtight bag, and the air inside the bag is evacuated using a vacuum pump to maintain uniform air pressure over the laminate. In addition to the resin and fibres, a number of components are used in this process as represented in Figure 1.9. A peel ply creates a clean moulding surface for perfect bonding, and a releasing agent prevents sticking of the resin to the mould. The releasing agent is usually perforated to allow entrapped gases and other volatile compounds to escape. Sealant tapes are attached to both the sides of the bag to ensure a vacuum-tight seal, and a bleeder layer removes the extra resin from the laminate. Uniform pressure is applied to evenly distribute the resin and to ensure proper bonding

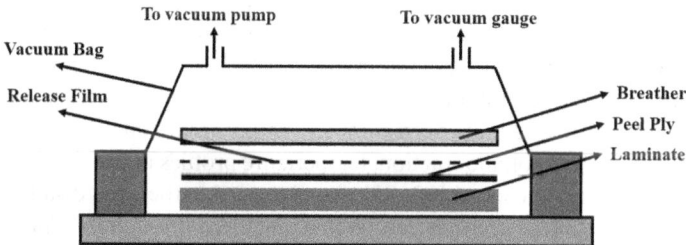

FIGURE 1.9 Representation of vacuum-bagging process.

between the resin and the fibres. The fabricated composite laminates are then cured at a pre-defined temperature to complete the fabrication process.

This process offers substantial advantages in terms of the quality of the parts produced. The parts show good layer adhesion even at high fibre loadings with very minimal emissions as the process is carried out in a sealed bag. However, this process is not suitable for high-volume production and requires very expensive curing ovens. A variety of fibre reinforcements have been experimented with to achieve enhanced properties [91].

Autoclave moulding is an extension of the vacuum-bagging process to produce more precise and high-quality parts that are generally not achievable by the vacuum-bagging process. This process is advanced in terms of the quality of the parts fabricated. This method imparts high pressure and temperature with precise control to compact and make the composite void-free. The method primarily uses composite prepregs as the feed material. These prepregs are lined with the fibres to obtain the necessary thickness. This arrangement is pressurized onto the moulding plate and is later hardened to obtain the desired shape. The peel plies and breather cloths perform a similar function as seen in the vacuum-bagging process. Later, this whole assembly is vacuum bagged to remove any air present in the layers. Finally, the vacuumed parts are later moved to an autoclave for curing. The process is expensive and requires a lot of effort and time to obtain quality parts. Often, the application of this process is restricted to the aerospace and military parts. The quality of the fabricated parts depends on the pressure applied, curing temperature, curing time and the constituent materials used [92].

1.4.5 Pultrusion Process

The pultrusion process is a high-volume composites fabrication process that utilizes a continuous fibre lamination route to achieve parts of a uniform cross section. This process is quite similar to extrusion. The main difference is that the parts are pulled out here as opposed to being pushed in extrusion. The process begins with the impregnation of the fibres with the resin. The excess resin is then removed to ensure proper bonding. This arrangement is then sent to the preform die which shapes the composites. Finally, the composites are cured in the heating dies before they are sent to the cut-off saw where they are cut to the desired dimensions. Pultrusion process is highly automated and facilitates quick production. Figure 1.10 illustrates a typical pultrusion process. The method can produce composite laminates containing high volume fraction of fibres, minimal labour and better structural and surface properties. The quality of the parts produced by this method is influenced by the resin properties such as viscosity and polymerization, fibre volume fraction, temperature profiles maintained during the process, and pultrusion speed [93–95]. As discussed earlier, the disadvantage is that only constant cross sections can be manufactured through this method. The addition of fibres such as kenaf, glass and jute can enhance the properties of the pultruded composites [96].

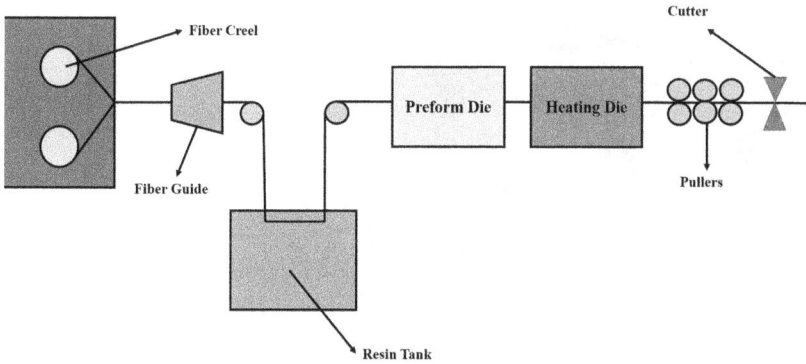

FIGURE 1.10 Schematic representation of pultrusion process.

1.4.6 Resin Transfer Moulding (RTM)

Liquid-moulding processes use a liquid resin that infuses into the reinforcement fabric through a pressure gradient. Some of the most common liquid-moulding processes include resin transfer moulding (RTM), vacuum injection moulding and structural reaction injection moulding. RTM is widely used for fabrication of composite laminates made of thermosetting polymer compounds. Initially, the reinforcement structure is placed in the lower mould and the assembly is closed with the upper mould. Later, the catalysed low-viscosity resin is introduced into the mould at high pressure and temperature such that the reinforcements are well impregnated with the resin. The high-pressure injection causes the gases inside the mould to escape through the vents created in the mould. Once the injection is complete, the part is cured at the room temperature or elevated temperature within the mould. Figure 1.11 represents the process of RTM. Later, the mould is opened and the laminate is removed. RTM can be extensively used for the rapid production of complex and high-performance composite structures that have an excellent surface finish on both sides. This method is also capable of producing laminates with high fibre-to-resin ratio and outstanding specific strength. However, this process is limited to low-viscosity resins and is often expensive.

These costs can be minimized by using vacuum-assisted resin transfer moulding (VARTM). The primary difference between VARTM and RTM is the introduction of the resin into the mould. In this method, the resin is drawn to the fibre preform through the vacuum. This mitigates the need for high power and heat requirements, thereby reducing the tooling costs. As seen for RTM, this method is useful for rapid fabrication of complex parts with varying fibre architectures for high-performance structural applications.

1.4.7 Additive Manufacturing Processes

Additive manufacturing (AM) techniques are relatively new to the fabrication of sandwich composite laminates. Conventionally, AM techniques follow a

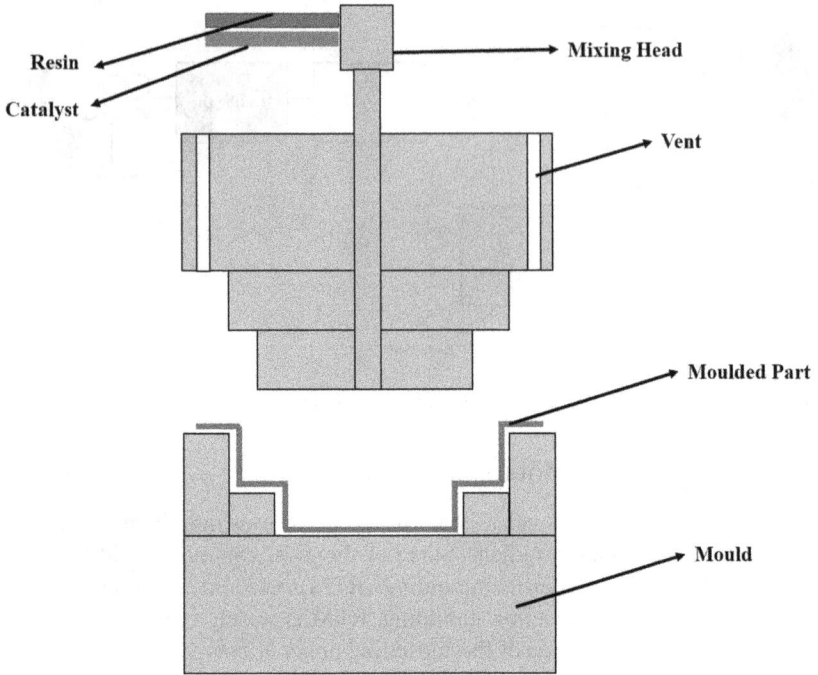

FIGURE 1.11   Schematic representation of RTM.

layer-wise manufacturing approach where structures of complex geometry and arrangements can be manufactured in a relatively easy manner [97]. AM offers the much needed versatility for realizing structures with multifunctional capabilities. A lot of methods such as FDM, SLS, SLM and SLA are used to fabricate composite structures [98]. In the modern context, AM techniques are primarily used as an assistive technique in combination with the aforementioned fabrication methods to realize complex laminates. In this regard, AM is primarily used in the manufacture of the sandwich core structure of different geometries. In addition to providing the required stiffness and support to face structures, AM core structures can provide multifunctional capabilities such as negative Poisson's ratio, high-impact energy absorption and vibration damping without the increasing cost for manual labour and extensive tooling as seen for other conventional methods [99].

The AM methods are chosen based on the build material of the structure and the required resolution. FDM and SLS can produce composite parts made of polymers and polymer composite materials, while SLM deals with the fabrication of metallic parts [100]. SLS and SLM are powder manufacturing process, and FDM follows a material deposition route as stated by the name. SLA is used to realize core structures through a photopolymeric reaction with the incident laser beam. Hence, the method is limited to photosensitive materials only. Nevertheless, SLA offers the best resolution and quality as compared to other AM processes. FDM is the cheapest and the most readily available AM technique among these. There are several other techniques such as electron beam melting and direct laser writing which are much more sophisticated. However, it must be noted that AM processes are a new venture and they are currently limited by the availability of materials and the high initial costs [101]. Therefore, only after sufficient research and development can AM be exploited for the quick, versatile and cost-effective fabrication of composite laminate structures. Table 1.1 provides a consolidated understanding of the different processes discussed here along with their advantages, disadvantages and applications for a comparative study.

TABLE 1.1

Comparison of Different Fabrication Methods along with Their Advantages, Limitations and Applications

Fabrication Method	Advantages	Limitations	Applications
Hand layup	• The process is relatively simple and versatile. • Cost-effective.	• Time-consuming. • Disorientation of fibres. • Inconsistent part geometries. • Need for skilled labour. • Low- to medium-volume production.	• Boat hulls. • Storage tanks. • Furniture components. • Bathtubs. • Swimming pools.
Compression moulding	• Can obtain parts of different geometries. • Consistent with respect to the geometrical tolerances of the structure. • Simple fabrication. • The parts have good structural stability.	• High cost of machinery and moulds. • Not possible to fabricate intricate geometries. • Large volume of products cannot be produced.	• Automobile panels. • Wing fences. • Missile components. • Brackets and clip structures. • Ground exploration hardware.
Vacuum bagging and autoclave moulding	• Simple process design. • A large number of fibre–matrix combinations are possible. • Cost-efficient method.	• Highly time-consuming. • Low pressure and temperature tolerances. • Inconsistent parts. • Breather cloth needs frequent replacement.	• Turbine blades.

(Continued)

TABLE 1.1 (*Continued*)

Comparison of Different Fabrication Methods along with Their Advantages, Limitations and Applications

Fabrication Method	Advantages	Limitations	Applications
Filament winding	• Large parts can be fabricated. • Existing textile processes can be used. • The process is quick and easy to handle.	• Limited spinning speed. • Frequent yarn breakage. • Curing cannot be easily done. • Only cylindrical geometries can be fabricated.	• Light poles. • Compressed natural gas masks. • Sailboat masts. • Rescue air tanks. • Water softener systems. • Aerospace components.
Pultrusion	• Minimal manual intervention. • High processing speeds. • Versatile cross-sectional shapes can be realized. • Parts can have continuous reinforcements.	• Mainly used for thermoset polymers only. • The dies are expensive. • The die can undergo frequent repairs due to excessive heat generation.	• Composite bars and rods. • Pipes and tubing. • Ladder rails and rungs. • Structural supports of various kinds.
Resin transfer moulding	• Products have good surface finish with selective fibre reinforcement with up to 65% loading. • Lower tooling costs. • Can produce very thin cross section. • Highly automated with higher production rates. • Lower material voids. • Versatile parts can be fabricated with thickness ranging from 0.5 to 90 mm.	• Longer curing periods. • Intricate geometries are at times tough to fabricate. • There is a loss of material due to resin spill when it is transferred to the mould.	• Turbine blades. • Boat hulls. • Truck panels. • Bathroom fixtures. • Aerospace components. • Safety helmets. • Automotive structural parts.
Additive manufacturing	• Quick production. • Versatile part production. • Faster fabrication. • Cost-effective when produced at large. • Realization of multifunctional capabilities is easy.	• Lesser number of readily available materials for AM. • Higher initial cost. • Parts produced have properties lower than that of moulded parts. • High levels of anisotropy.	• Secondary aerospace structures. • Automotive interior components. • Multifunctional test structures. • Surgical tools. • Medical implants.

1.5   CONCLUSIONS

As seen from the above studies, a variety of manufacturing processes and technologies are capable of realizing sandwich composite panels. These advances have promoted the use of sandwich panels from secondary structures to primary structures. A lot of industries such as aerospace and automotive industries have adopted the use of sandwich panels for effective weight reduction and improved performance of structural elements. However, the selection of a feasible process is dependent on a number of factors such as the quality, size, cost and properties required in the end product. Numerous sophistications in the highly capable processes such as VARTM and filament winding can produce parts of extremely good quality at lower production costs. Furthermore, the increasing demand for composite panels in aerospace parts has led to further enhancements in processes such as vacuum bagging and autoclave moulding. AM techniques are also being promoted considerably to realize composite core structure with multifunctional applications. The use of AM can considerably reduce the time and costs involved in the fabrication of sandwich composite panels with increasing development and demand for these processes despite the current cost levels. Hence, it can be firmly assured that, with significant technological advancements and improved material characteristics, sandwich panels can be extensively used for most engineering applications.

Acknowledgement: The author Vinyas Mahesh acknowledges the support of Indian Institute of Science, Bangalore, through C.V. Raman Post-doctoral Fellowship R(IA)/CVR-PDF/2019/1630, under Institution of Eminence scheme.

REFERENCES

1. Birman, V.; Kardomatea, G. A. Review of Current Trends in Research and Applications of Sandwich Structures. *Compos. Part B* **2018**, *142*, 221–240. Doi: 10.1016/j.compositesb.2018.01.027.
2. Ramnath, B. V.; Alagarraja, K.; Elanchezhian, C. Review on Sandwich Composite and Their Applications. *Mater. Today Proc.* **2019**, *16*, 859–864. Doi: 10.1016/j.matpr.2019.05.169.
3. Mohammed, J.; Moheb, A.; Al-ameen, E. S. Study of the Bending Characteristics in Composite Sandwich Structures – A Review Study of the Bending Characteristics in Composite Sandwich Structures – A Review. In 2nd International Scientific Conference of Al-Ayen University (ISCAU-2020): IOP Conference Series: Materials Science and Engineering; 2020. Doi: 10.1088/1757-899X/928/2/022147.
4. Noor, A. K.; Burton, W. S.; Bert, C. W. Computational Models for Sandwich Panels and Shells. *Appl. Mech. Rev.* **1996**, *49* (3), 155–199.
5. Bernard, Lisa A. *Low Velocity Impact Testing of Sandwich Panels with Polymeric cores*; Thesis submitted to the University of New South Wales at the Australian Defence Force Academy, 2011.
6. Hoff, N. J. *Bending and Bucking of Rectangular Sandwich Plates*; National Advisory Committee for Aeronautics, Polytechnic Institute of Brooklyn, Washington, 1950.
7. Libove, C.; Batdorf, S. B. *A General Small-Deflection Theory for Flat Sandwich Plates*; National Advisory Committee for Aeronautics, Report No. 899, Washington 1947.

8. Flügge, W. The Optimum Problem of the Sandwich Plate. *J. Appl. Mech.* **1952**, *19* (1), 104–108. Doi: 10.1016/B978-0-08-012870-2.50007-9.

9. Gerard, G. Minimum Weight Analysis of Compression Structures. New York University Press, New York, **1956**.

10. Kaechele, L. E. *Minimum-Weight Design of Sandwich Panels*; United States Air Force and RAND Corporation, RM-1895, 1957.

11. Plantema, F. J. *Sandwich Construction - The Bending and Buckling of Sandwich Beams, Plates and Shells*; John Wiley and Sons, London, 1966.

12. Ha, K. H. Finite Element Analysis of Sandwich: An Overview. *Comput. Struct.* **1990**, *37* (4), 397–403.

13. Vinson, J. R. Sandwich Structures. *Appl. Mech. Rev.* **2001**, *54* (3), 201–214.

14. Taghipoor, H.; Eyvazian, A.; Musharavati, F.; Sebaey, T. A.; Ghiaskar, A. Experimental Investigation of the Three-Point Bending Properties of Sandwich Beams with Polyurethane Foam-Filled Lattice Cores. *Structures* **2020**, *28* (April), 424–432. Doi: 10.1016/j.istruc.2020.08.082.

15. Keller, M.; Ambrosio, E.; de Oliveira, V. M.; Góes, M. M.; de Carvalho, G. M.; Batistela, V. R.; Garcia, J. C. Polyurethane Foams Synthesis with Cassava Waste for Biodiesel Removal from Water Bodies. *Bioresour. Technol. Reports* **2020**, *10* (December 2019), 100396. Doi: 10.1016/j.biteb.2020.100396.

16. Kecerdasan, I.; Ikep, P. Polyetherols and Polyurethane Foams from Starch. *Polym. Test.* **2021**, *93*, 106884.

17. Andersons, J.; Kirpluks, M.; Cabulis, P.; Kalnins, K.; Cabulis, U. Bio-Based Rigid High-Density Polyurethane Foams as a Structural Thermal Break Material. *Constr. Build. Mater.* **2020**, *260*, 120471. Doi: 10.1016/j.conbuildmat.2020.120471.

18. Kanchanapiya, P.; Intaranon, N.; Tantisattayakul, T. Assessment of the Economic Recycling Potential of a Glycolysis Treatment of Rigid Polyurethane Foam Waste: A Case Study from Thailand. *J. Environ. Manage.* **2020**, (June), 111638. Doi: 10.1016/j.jenvman.2020.111638.

19. Contreras, J.; Valdés, O.; Mirabal-Gallardo, Y.; de la Torre, A. F.; Navarrete, J.; Lisperguer, J.; Durán-Lara, E. F.; Santos, L. S.; Nachtigall, F. M.; Cabrera-Barjas, G.; Abril, D. Development of Eco-Friendly Polyurethane Foams Based on Lesquerella Fendleri (A. Grey) Oil-Based Polyol. *Eur. Polym. J.* **2020**, *128* (January), 109606. Doi: 10.1016/j.eurpolymj.2020.109606.

20. Verdolotti, L.; Di Caprio, M. R.; Lavorgna, M.; Buonocore, G. G. *Polyurethane Nanocomposite Foams: Correlation between Nanofillers, Porous Morphology, and Structural and Functional Properties*; Elsevier Inc., 2017. Doi: 10.1016/ B978-0-12-804065-2.00009-7.

21. Kim, M. S.; Kim, J. D.; Kim, J. H.; Lee, J. M. Mechanical Performance Degradation of Glass Fiber-Reinforced Polyurethane Foam Subjected to Repetitive Low-Energy Impact. *Int. J. Mech. Sci.* **2021**, *194*, 106188. Doi: 10.1016/j.ijmecsci.2020.106188.

22. Yu, C. T.; Lai, C. C.; Wang, F. M.; Liu, L. C.; Liang, W. C.; Wu, C. L.; Chiu, J. C.; Liu, H. C.; Hsiao, H. T.; Chen, C. M. Fabrication of Thermoplastic Polyurethane (TPU) / Thermoplastic Amide Elastomer (TPAE) Composite Foams with Supercritical Carbon Dioxide and Their Mechanical Properties. *J. Manuf. Process.* **2019**, *48* (September), 127–136. Doi: 10.1016/j.jmapro.2019.09.022.

23. Jabber, L. J. Y.; Grumo, J. C.; Alguno, A. C.; Lubguban, A. A.; Capangpangan, R. Y. Influence of Cellulose Fibers Extracted from Pineapple (Ananas Comosus) Leaf to the Mechanical Properties of Rigid Polyurethane Foam. *Mater. Today Proc.* **2020**. Doi: 10.1016/j.matpr.2020.07.566.

24. Członka, S.; Strąkowska, A.; Kairytė, A. Effect of Walnut Shells and Silanized Walnut Shells on the Mechanical and Thermal Properties of Rigid Polyurethane Foams. *Polym. Test.* **2020**, *87* (April). Doi: 10.1016/j.polymertesting.2020.106534.

25. Krundaeva, A.; Bruyne, G. De; Gagliardi, F. Dynamic Compressive Strength and Crushing Properties of Expanded Polystyrene Foam for Different Strain Rates and Different Temperatures. *Polym. Test.* **2018**, *55* (2016), 61–68. Doi: 10.1016/j. polymertesting.2016.08.005.

26. Tang, N.; Lei, D.; Huang, D.; Xiao, R. Mechanical Performance of Polystyrene Foam (EPS): Experimental and Numerical Analysis. *Polym. Test.* **2019**, *73* (November 2018), 359–365. Doi: 10.1016/j.polymertesting.2018.12.001.

27. Kwak, J. Il; An, Y. Iced Block Method : An Efficient Method for Preparation of Micro-Sized Expanded Polystyrene Foams. *Environ. Pollut.* **2020**, *263*, 114387. Doi: 10.1016/j.envpol.2020.114387.

28. Yassin, M. A.; Aziz, A.; Gad, M. Immobilized Enzyme on Modified Polystyrene Foam Waste : A Biocatalyst for Wastewater Decolorization. *J. Environ. Chem. Eng.* **2020**, *8* (5), 104435. Doi: 10.1016/j.jece.2020.104435.

29. Schulze, M.; Handge, U. A.; Rangou, S.; Lillepärg, J.; Abetz, V. Thermal Properties, Rheology and Foams of Polystyrene-Block-Poly(4-Vinylpyridine) Diblock Copolymers. *Polymer (Guildf).* **2015**, *70*, 88–99. Doi: 10.1016/j.polymer. 2015.06.005.

30. Wang, L.; Wang, C.; Liu, P.; Jing, Z.; Ge, X.; Jiang, Y. The Flame Resistance Properties of Expandable Polystyrene Foams Coated with a Cheap and Effective Barrier Layer. *Constr. Build. Mater.* **2018**, *176* (189), 403–414. Doi: 10.1016/j. conbuildmat.2018.05.023.

31. Turner, A. Polystyrene Foam as a Source and Sink of Chemicals in the Marine Environment : An XRF Study. *Chemosphere* **2021**, *263*, 128087. Doi: 10.1016/j. chemosphere.2020.128087.

32. Francesco, M.; Spadea, S.; Fabbrocino, F.; Lonetti, P. On the Elastic Properties of PVC Foam. *Procedia Struct. Integr.* **2020**, *28* (2019), 1503–1510. Doi: 10.1016/j. prostr.2020.10.123.

33. Zhang, Q.; Yang, Y.; Xiang, L. Post-Curing Effect on the Chemical Structure and Its Relationship with Heat Resistance and Thermal-Mechanical Properties of Crosslinked PVC Foams. *Polym. Test.* **2019**, *73* (November 2018), 418–424. Doi: 10.1016/j.polymertesting.2018.12.004.

34. Jiang, Z.; Yao, K.; Du, Z.; Xue, J.; Tang, T.; Liu, W. Rigid Cross-Linked PVC Foams with High Shear Properties : The Relationship between Mechanical Properties and Chemical Structure of the Matrix. *Compos. Sci. Technol.* **2014**, *97*, 74–80. Doi: 10.1016/j.compscitech.2014.04.005.

35. Assarar, M.; El, A.; Berthelot, J. Evaluation of the Dynamic Properties of PVC Foams under Flexural Vibrations. *Compos. Struct.* **2012**, *94* (6), 1919–1931. Doi: 10.1016/j.compstruct.2012.01.017.

36. Sadler, R. L.; Sharpe, M.; Panduranga, R.; Shivakumar, K. Water Immersion Effect on Swelling and Compression Properties of Eco-Core, PVC Foam and Balsa Wood. *Compos. Struct.* **2009**, *90* (3), 330–336. Doi: 10.1016/j.compstruct.2009. 03.016.

37. Li, Z.; Shahrajabian, H.; Amin, S. Effects of Nano-Clay Content, Foaming Temperature and Foaming Time on Density and Cell Size of PVC Matrix Foam by Presented Least Absolute Shrinkage and Selection Operator Statistical Regression via Suitable Experiments as a Function of MMT Content. *Physica A.* **2020**, *537*, 122637. Doi: 10.1016/j.physa.2019.122637.

38. Ren, P.; Tao, Q.; Yin, L.; Ma, Y.; Wu, J.; Zhao, W.; Mu, Z.; Guo, Z.; Zhao, Z. High-Velocity Impact Response of Metallic Sandwich Structures with PVC Foam Core. *Int. J. Impact Eng.* **2020**, *144*, 103657. Doi: 10.1016/j.ijimpeng.2020.103657.

39. Chen, Q.; Linghu, T.; Gao, Y.; Wang, Z.; Liu, Y.; Du, R.; Zhao, G. Mechanical Properties in Glass Fiber PVC-Foam Sandwich Structures from Different Chopped Fiber Interfacial Reinforcement through Vacuum-Assisted Resin Transfer Molding (VARTM) Processing. *Compos. Sci. Technol.* **2017**. Doi: 10.1016/j.compscitech.2017.03.033.

40. Kai, W.; Qiang, H. E. Progress on Study of Key Technologies for Polymethacrylimide Foam Core Sandwich Lifecycle. *Acta Mater. Compos. Sin.* **2020**, *37* (8), 1805–1822.

41. Yan, L.; Zhu, K.; Zhang, Y.; Zhang, C.; Zheng, X. Effect of Absorbent Foam Filling on Mechanical Behaviors of 3D-Printed Honeycombs. Polymers **2020**, 12(9), 2059.

42. Grace, I.; Pilipchuk, V.; Ibrahim, R.; Ayorinde, E. Temperature Effect on Non-Stationary Compressive Loading Response of Polymethacrylimide Solid Foam. *Compos. Struct.* **2012**, *94* (10), 3052–3063. Doi: 10.1016/j.compstruct.2012.04.022.

43. Zhou, H.; Liu, R. Quasi-Static Compressive Strength of Polymethacrylimide Foam-Filled Square Carbon Fiber Reinforced Composite Honeycombs. *J. Sandw. Struct. Mater.* **2020**, (2), 1–17. Doi: 10.1177/1099636220909819.

44. Zaid, N. Z. M.; Rejab, M. R. M.; Mohamed, N. A. N. Sandwich Structure Based On Corrugated-Core: A Review. In The 3rd International Conference on Mechanical Engineering Research (ICMER *2015*), Kuantan, Malaysia; **2015**; pp. 74–00029.

45. Dayyani, I.; Ziaei-rad, S.; Friswell, M. I. The Mechanical Behavior of Composite Corrugated Core Coated with Elastomer for Morphing Skins. *J. Compos. Mater.* **2014**, *48* (13), 1623–1636. Doi: 10.1177/0021998313488807.

46. Bartolozzi, G.; Pierini, M.; Orrenius, U. An Equivalent Material Formulation for Sinusoidal Corrugated Cores of Structural Sandwich Panels. *Compos. Struct.* **2013**, *100*, 173–185. Doi: 10.1016/j.compstruct.2012.12.042.

47. Chang, W.; Ventsel, E.; Krauthammer, T.; John, J. Bending Behavior of Corrugated-Core Sandwich Plates. *Compos. Struct.* **2005**, *70*, 81–89. Doi: 10.1016/j.compstruct.2004.08.014.

48. Rejab, M. R. M.; Cantwell, W. J. The Mechanical Behaviour of Corrugated-Core Sandwich Panels. *Compos. Part B.* **2013**, *47*, 267–277. Doi: 10.1016/j.compositesb.2012.10.031.

49. Hou, S.; Zhao, S.; Ren, L.; Han, X.; Li, Q. Crashworthiness Optimization of Corrugated Sandwich Panels. *J. Mater.* **2013**, *51*, 1071–1084. Doi: 10.1016/j.matdes.2013.04.086.

50. Cote, F.; Deshpande, V. S.; Fleck, N. A.; Evans, A. G. The Compressive and Shear Responses of Corrugated and Diamond Lattice Materials. *Int. J. Solids Struct.* **2006**, *43*, 6220–6242. Doi: 10.1016/j.ijsolstr.2005.07.045.

51. Rupani, S. V; Jani, S. S. Design, Modelling and Manufacturing Aspects of Honeycomb Sandwich Structures : A Review. *Int. J. Sci. Dev. Res.* **2017**, 2 (4), 526–532.

52. Bitzer, T. *Honeycomb Technology: Materials, Design, Manufacturing, Applications and Testing*; Springer Science and Business Media, 1997.

53. Burlayenko, V. N.; Sadowski, T. Analysis of Structural Performance of Sandwich Plates with Foam-Filled Aluminum Hexagonal Honeycomb Core. *Comput. Mater. Sci.* **2009**, *45* (3), 658–662. Doi: 10.1016/j.commatsci.2008.08.018.

54. Wadley, H. N. G.; Fleck, N. A.; Evans, A. G. Fabrication and Structural Performance of Periodic Cellular Metal Sandwich Structures. *Compos. Sci. Technol.* **2003**, *63*, 2331–2343. Doi: 10.1016/S0266-3538(03)00266-5.

55. Burman, M.; Zenkert, D. Fatigue of Foam Core Sandwich Beams — 1 : Undamaged Specimens. *Int. J. Fatigue.* **1997**, *19* (7), 551–561.

56. Burton, W. S.; Noor, A. K. Structural Analysis of the Adhesive Bond in a Honeycomb Core Sandwich Panel. *Finite Elem. Anal. Des.* **1997**, *26*, 213–227.

57. Chen, C.; Harte, A.; Fleck, N. A. The Plastic Collapse of Sandwich Beams with a Metallic Foam Core. *Int. J. Mech. Sci.* **2001**, *43*, 1483–1506.

58. Li, K.; Gao, X. Effects of Cell Shape and Cell Wall Thickness Variations on the Elastic Properties of Two-Dimensional Cellular Solids. *Int. J. Solids Struct.* **2005**, *42*, 1777–1795. Doi: 10.1016/j.ijsolstr.2004.08.005.

59. Shmorhunt, M.; Jamieson, A. M.; Simha, R. Free Volume Changes in Epoxy Adhesives during Physical Ageing : Fluorescence Spectroscopy and Mechanical Stress Relaxation *. *Polymer (Guildf).* **1990**, *31* (5), 812–817.

60. Wei, H.; Xia, J.; Zhou, W.; Zhou, L.; Hussain, G.; Li, Q.; Ken, K. Adhesion and Cohesion of Epoxy-Based Industrial Composite Coatings. *Compos. Part B.* **2020**, *193* (December 2019), 108035. Doi: 10.1016/j.compositesb.2020.108035.

61. Jin, F.; Li, X.; Park, S. Synthesis and Application of Epoxy Resins : A Review. *J. Ind. Eng. Chem.* **2015**, *29*, 1–11. Doi: 10.1016/j.jiec.2015.03.026.

62. Ahmadi, Z. Nanostructured Epoxy Adhesives : A Review. *Prog. Org. Coatings.* **2019**, *135* (March), 449–453. Doi: 10.1016/j.porgcoat.2019.06.028.

63. Hou, W.; Gao, Y.; Wang, J.; John, D.; Teo, S. Recent Advances and Future Perspectives for Graphene Oxide Reinforced Epoxy Resins. *Mater. Today Commun.* **2020**, *23* (December 2019), 100883. Doi: 10.1016/j.mtcomm.2019.100883.

64. Lutz, A.; Schneider, D. Toughened Epoxy Adhesive Composition, United States Patent - US 8.404,787 B2, 2013.

65. Ash, J. R.; Lambuth, A. L. Phenolic Adhesives and Method of Making Same, United States Patent – US 2,727,869, 1955.

66. HSe, C.-Y. Method of Bonding Particle Board and the Like Using Polyisocyanate/ Phenolic adhesive, United States Patent – US-4209433-A, 1980.

67. Higgins, A. Adhesive Bonding of Aircraft Structures. *Int. J. Adhes. Adhes.* **2000**, *20* (January), 367–376.

68. Kalami, S.; Arefmanesh, M.; Master, E.; Nejad, M. Replacing 100% of Phenol in Phenolic Adhesive Formulations with Lignin. *J. Appl. Polym. Sci.* **2017**, *45124*, 1–9. Doi: 10.1002/app.45124.

69. River, B. H. Relationship between Phenolic Adhesive Chemistry and Adhesive Joint Performance : Effect of Filler Type on Fraction Energy*. *J. Appl. Polym. Sci.* **1986**, *31*, 2275–2302.

70. Kalami, S.; Chen, N.; Borazjani, H.; Nejad, M. Comparative Analysis of Different Lignins as Phenol Replacement in Phenolic Adhesive Formulations. *Ind. Crop. Prod.* **2018**, *125* (August), 520–528. Doi: 10.1016/j.indcrop.2018.09.037.

71. Nordberg, A.; Antoni, P.; Montan, M. I.; Hult, A.; Holst, H. Von; Malkoch, M. Highly Adhesive Phenolic Compounds as Interfacial Primers for Bone Fracture Fixations. *Appl. Mater. Interf.* **2010**, *2* (3), 654–657. Doi: 10.1021/am100002s.

72. Das, A.; Mahanwar, P. A Brief Discussion on Advances in Polyurethane Applications. *Adv. Ind. Eng. Polym. Res.* **2020**, *3*, 93–101.

73. Eisen, A.; Bussa, M.; Röder, H. A Review of Environmental Assessments of Biobased against Petrochemical Adhesives. *J. Clean. Prod.* **2020**, *227*, 124277. Doi: 10.1016/j.jclepro.2020.124277.

74. Navarro-ban, V.; Martı, M. Addition of Nanosilicas with Different Silanol Content to Thermoplastic Polyurethane Adhesives. *Int. J. Adhes. Adhes.* **2006**, *26*, 378–387. Doi: 10.1016/j.ijadhadh.2005.04.004.

75. Çakır Çanakİ, T.; Serhatlı, İ. E. Synthesis of Fluorinated Urethane Acrylate Based UV-Curable Coatings. *Prog. Org. Coatings* **2013**, *76* (2–3), 388–399. Doi: 10.1016/j.porgcoat.2012.10.024.
76. Bluestein, C. Urethane Oligomers for Solventless Coatings, Adhesives, and Elastomers. *Polym. Plast. Technol. Eng.* **1981**, *17* (1), 83–93. Doi: 10.1080/03602558108067700.
77. Oprea, S.; Vlad, S.; Stanciu, A.; Macoveanu, M. Epoxy Urethane Acrylate. *Eur. Phys. J. Plus.* **2000**, *36* (41), 373–378.
78. Aziz, S. H.; Ansell, M. P.; Clarke, S. J.; Panteny, S. R. Science and Modified Polyester Resins for Natural Fibre Composites. *Compos. Sci. Technol.* **2005**, *65*, 525–535. Doi: 10.1016/j.compscitech.2004.08.005.
79. Gupta, A. K.; Mani, P.; Krishnamoorthy, S. Interfacial Adhesion in Polyester Resin Concrete. *Int. J. Adhes. Adhes.* **1983**, *3* (2), 149–154.
80. Grohens, Y.; Sire, O.; Baley, C. Influence of Chemical Treatments on Surface Properties and Adhesion of Flax Fibre – Polyester Resin. *Compos. Part A Appl. Sci. Manuf.* **2006**, *37*, 1626–1637. Doi: 10.1016/j.compositesa.2005.10.014.
81. Biswas, S.; Anurag, J. Fabrication of Composite Laminates. In *Reinforced Polymer Composites: Processing, Characterization and Post Life Cycle Assessment, Eds. Pramendra K. Bajpai, Inderdeep Singh*; 2020; pp 39–53.
82. Kelkar, A. D.; Tate, J. S.; Bolick, R. Introduction to Low Cost Manufacturing of Composite Laminates. In Proceedings of the *2003* American Society for Engineering Education Annual conference & Exposition; Nashville, Tennessee, 2003; pp 8.785.1–8.785.13.
83. Fong, T. C.; Saba, N.; Liew, C. K.; Silva, R. De; Hoque, M. E.; Goh, K. L. Yarn Flax Fibres for Polymer-Coated Sutures and Hand Layup Polymer Yarn Flax Fibres for Polymer-Coated Sutures and Hand Layup Polymer Composite. In *Manufacturing of Natural Fibre Reinforced Polymer Composites*; 2015; pp. 155–175. Doi: 10.1007/978-3-319-07944-8.
84. Ramasamy, M.; Ramesh, R. R.; Krishnan, P. Investigation on Static and Dynamic Mechanical Properties of Epoxy Based Woven Fabric Glass / Carbon Hybrid Composite Laminates ScienceDirect Investigation on Static and Dynamic Mechanical Properties of Epoxy Based Woven Fabric Glass / Carbon Hybrid Compo. In 12th Global Congress on Manufacturing and Management, GCMM 2014; 2014; pp. 459–468. Doi: 10.1016/j.proeng.2014.12.270.
85. Huber, T.; Pang, S.; Staiger, M. P. All-Cellulose Composite Laminates. *Compos. Part A.* **2012**, *43* (10), 1738–1745. Doi: 10.1016/j.compositesa.2012.04.017.
86. Kumaresan, M.; Sathish, S.; Karthi, N. Effect of Fiber Orientation on Mechanical Properties of Sisal Fiber Reinforced Epoxy Composites Effect of Fiber Orientation on Mechanical Properties of Sisal Fiber Reinforced Epoxy Composites. *J. Appl. Sci. Eng.* **2015**, *18* (3), 289–294. Doi: 10.6180/jase.2015.18.3.09.
87. Correlo, V. M.; Boesel, L. F.; Pinho, E.; Silva, M. L. A.; Bhattacharya, M.; Mano, J. F.; Neves, N. M.; Reis, R. L. Melt-Based Compression-Molded Scaffolds from Chitosan – Polyester Blends and Composites : Morphology and Mechanical Properties. *J. Biomed. Mater. Res. Part A.* **2008**, *91A* (2), 489–504. Doi: 10.1002/jbm.a.32221.
88. Aruchamy, K.; Pavayee, S.; Kumar, S. Study on Mechanical Characteristics of Woven Cotton / Bamboo Hybrid Reinforced Composite. *Integr. Med. Res.* **2019**, 1–9. Doi: 10.1016/j.jmrt.2019.11.013.
89. Lamontia, M. A.; Gruber, M. B.; Smoot, M. A.; Sloan, J.; Gillespie, J. W. Performance of a Filament Wound Graphite/Thermoplastic Composite Ring-Stiffened Pressure Hull Model. *J. Thermoplast. Compos. Mater.* **1995**, *8*, 15–36. Doi: 10.1177/089270579500800103.

90. Mertiny, P.; Ellyin, F.; Hothan, A. An Experimental Investigation on the Effect of Multi-Angle Filament Winding on the Strength of Tubular Composite Structures. *Compos. Sci. Technol.* **2004**, *64*, 1–9. Doi: 10.1016/S0266-3538(03)00198-2.

91. Aparna, M. L.; Chaitanya, G.; Srinivas, K.; Rao, J. A. Fabrication of Continuous GFRP Composites Using Vacuum Bag Moulding Process. *Int. J. Adv. Sci. Technol.* **2016**, *87*, 37–46.

92. Davim, J. P.; Reis, P. Drilling Carbon Fiber Reinforced Plastics Manufactured by Autoclave: Experimental and Statistical Study. *Mater. Des.* **2003**, *24*, 315–324. Doi: 10.1016/S0261-3069(03)00062-1.

93. Moschiar, S. M.; Reboredo, M. M.; Larrondo, H.; Vazquez, A. Pultrusion of Epoxy Matrix Composites : Pulling Force Model and Thermal Stress Analysis. *Polym. Compos.* **1996**, *17* (6), 850–858.

94. Miller, A. H.; Dodds, N.; Hale, J. M.; Gibson, A. G. High Speed Pultrusion of Thermoplastic Matrix Composites. *Compos. - Part A Appl. Sci. Manuf.* **1998**, *29A*, 773–782.

95. Fairuz, A. M.; Sapuan, S. M.; Zainudin, E. S.; Jaafar, C. N. A. Pultrusion Process of Natural Fibre- Reinforced Polymer Composites. In *Manufacturing of Natural Fibre Reinforced Polymer Composites*; 2015; pp 217–231. Doi: 10.1007/978-3-319-07944-8.

96. Velde Van De, K.; Kiekens, P. Thermoplastic Pultrusion of Natural Fibre Reinforced Composites. *Compos. Struct.* **2001**, *54*, 355–360.

97. Wong, K. V; Hernandez, A. A Review of Additive Manufacturing. *Int. Sch. Res. Netw. ISRN Mech. Eng.* **2012**, *2012*, 208760. Doi: 10.5402/2012/208760.

98. Guo, N.; Leu, M. C. Additive Manufacturing : Technology, Applications and Research Needs. *Front. Mater.* **2013**, *8* (3), 215–243. Doi: 10.1007/s11465-013-0248-8.

99. Najmon, J. C.; Raeisi, S.; Tovar, A. Review of Additive Manufacturing Technologies and Applications in the Aerospace Industry. In *Additive Manufacturing for the Aerospace Industry*; Elsevier Inc., 2019; pp 7–32. Doi: 10.1016/B978-0-12-814062-8.00002-9.

100. Frazier, W. E. Metal Additive Manufacturing : A Review. *J. Mater. Eng. Perform.* **2014**, *23* (6), 1917–1928. Doi: 10.1007/s11665-014-0958-z.

101. Busachi, A.; Erkoyuncu, J.; Colegrove, P.; Martina, F.; Watts, C. A Review of Additive Manufacturing Technology and Cost Estimation Techniques for the Defence Sector. *CIRP J. Manuf. Sci. Technol.* **2017**, *19* (November), 117–128. Doi: 10.1016/j.cirpj.2017.07.001.

2 Corrugated Core- and Fold Core-Based Sandwich Panels

A. Farrokhabadi
Tarbiat Modares University

M. Naghdi Nasab
Semnan University

CONTENTS

2.1   Introduction 27
2.2   Designs and Structures 29
     2.2.1   Corrugated Cores 29
     2.2.2   Fold Cores 30
2.3   Materials 33
     2.3.1   Corrugated Cores 33
     2.3.2   Fold Cores 34
2.4   Mechanical Properties 35
     2.4.1   Corrugated Cores 35
     2.4.2   Fold Cores 36
2.5   Manufacturing Methods 36
     2.5.1   Corrugated Cores 36
     2.5.2   Fold Cores 37
2.6   Applications 38
     2.6.1   Corrugated Cores 38
     2.6.2   Fold Cores 38
2.7   Conclusions 39
References 39

2.1   INTRODUCTION

Nowadays, one of the critical design factors for design of sophisticated engineering structures is the weight reduction. Among the numerous number of structures that are being used in modern technologies, sandwich panels have made their way into a range of applications from aerospace to marine and construction industries which require superior mechanical properties, high energy absorption capabilities

DOI: 10.1201/9781003143031-2

and incredible strength-to-weight ratios [1]. These kinds of structures mostly consist of two face sheets and a core, and based on the application of the sandwich panel, the face sheets and cores might have a range of different materials and design procedures. In the case of materials, different kinds of composites [2,3], metals [4,5], polymers [6] and even balsa wood and natural fibre/thermoset composites [7,8] are being used as the face sheet and core of sandwich panels. While face sheets are constructed from a fewer number of materials, cores are made of different materials and versatile geometries such as honeycomb [9,10], corrugated core-based [11,12], fold core-based [13,14], truss-shaped [15] and hybrid ones [16].

Some factors including the variety of attainable configurations, high ratio of specific strength to weight, improved mechanical properties and multifunctionality make sandwich-structured composites an extraordinary material of choice in civil, aerospace and automobile industries [17,18]. In many modern aircrafts, some critical sections such as control surfaces, overhead stowage compartments, fairings and many components inside the cabin are made of composite sandwich panels [19]. These structures which are subjected to tensile or compressive and bending loads are usually manufactured from steel, aluminium or composite materials. In sandwich panels, cores are mainly used to increase the bending stiffness of the panel by separating the face sheets from each other, carrying transverse shear loads and withstanding the compressive loads applied to the face sheets. As a result, the cores should be designed in a way to minimize the weight through either material selection, or spatial distribution of materials, or both [20,21]. This chapter is only focused on the corrugated core- and fold core-based sandwich panels which are among the most developing and commonly used sandwich panels in advanced industrial structures [22,23].

Corrugated core panels, due to their increasing strength-to-weight ratio, are an extremely interesting choice for the sandwich panels manufacturing [24,25]. In the corrugated sandwich panels, the stiffness of the employed material for the face sheets is mostly equal to or higher than the corrugated core's stiffness [26]. Such a structure represents higher ratios of bending, shear and tensile stiffnesses to weight than an equivalent panel that is only made of the face sheet or the corrugated core material [27]. Composites or metals can be used to make the corrugated cores and the face sheets of the sandwich panels. Overall, corrugated cores can be categorized based on their mechanical properties, the kind of core–face reinforcement that is used and the design formats that are used to manufacture them. Detailed categorization of these cores will be presented in the following sections.

In comparison with the traditional cores such as hexagonal honeycomb cores that can be found in almost every sandwich structure of aircrafts, recently, a newer cellular sandwich panel that can be employed in the structures of wings and fuselage of large civil aircrafts has attracted the most attention. Fold core or the so-called chevron core, origami core or folded core refers to a cellular core that was firstly investigated by the Japanese researcher Miura in 1972 [28,29]. However, the major interest in folded core sandwich panels including the engineering applications and extensive investigations on their manufacturing techniques, mechanical behaviours, possible geometries and modelling methods began later in the mid-1990s lasting until today. The geometry of diverse fold core patterns can be found in Refs. [30,31].

The main principle to make a fold core is that the original flat sheet should be folded in the form of a three-dimensional structure and without stretching the primary sheet. The geometry of the cell can be designed and optimized based on the specific load or the operational conditions that are defined for it. Cores with different heights and folding patterns can be generated; as a result, these kinds of cores do not have to be machined like the honeycomb ones. In addition to the low costs of manufacturing for the mentioned core structures in comparison with the honeycomb cores, a major advantage of the folded cores that makes them especially attractive for the aerospace industry is the fluid ventilation inside the core, which means the cells are not closed and fluid is not ensnared inside them [32].

Here in this chapter, a comprehensive overview of the corrugated core and fold core sandwich panels is presented. First, the possible design construction for these structures will be investigated, and then the materials used for manufacturing them and their mechanical properties will be examined. Finally, the manufacturing methods of these constructions and their applications in different industries will be explained.

2.2 DESIGNS AND STRUCTURES

2.2.1 CORRUGATED CORES

The term "corrugated structure" refers to any lattice with the shape of corrugation that is made by moulding, folding, 3D printing or any alternative manufacturing methods. In a general point of view, prevalent corrugated structures could be categorized into three branches including corrugated pipes, corrugated sheets and corrugated panels. Some of the main concepts of the corrugated cores are presented in Figure 2.1.

Almost all the corrugated structures have an extremely anisotropic behaviour with a high transverse stiffness to their corrugation direction [34]. Due to this significant feature, the corrugated structures have dramatically been used in the academic research and industrial applications.

By adding two face sheets to the upper and lower sides of the core, a novel geometry would be manufactured, which is known as the corrugated panel [35]. By choosing different geometrical shapes, materials and also dimensions for the core and face sheet skins for the corrugated sandwich structure, a wide diversity of in-plane/out-of-plane stiffness as well as strength at low weight ratios would be obtained. The structural characteristics of the manufactured sandwich panels exceedingly depend on the considered lightweight corrugated core which separates the face sheets from each other and contributes to the overall stiffness for the panel. It should be noted that different mechanical behaviours could be expected when the geometry of the panel is identical, but the face sheets and the corrugated cores have different stiffnesses. Table 2.1 briefly presents an overview of the common corrugated structures used in recent research activities.

(a) Corrugated core with elastomeric coating

(b) Twisted bi-stable corrugated core

(c) Curved corrugated sheet and some of its global deformations

(d) Bi-directional corrugated core

(e) Schematic of corrugated bi directional core

(f) Hierarchical corrugated core sandwich panel

(g) PMI foam filled hierarchical corrugated sheet

(h) Double wall corrugated concept

(i) Extruded pyramidal lattice truss sandwich structure

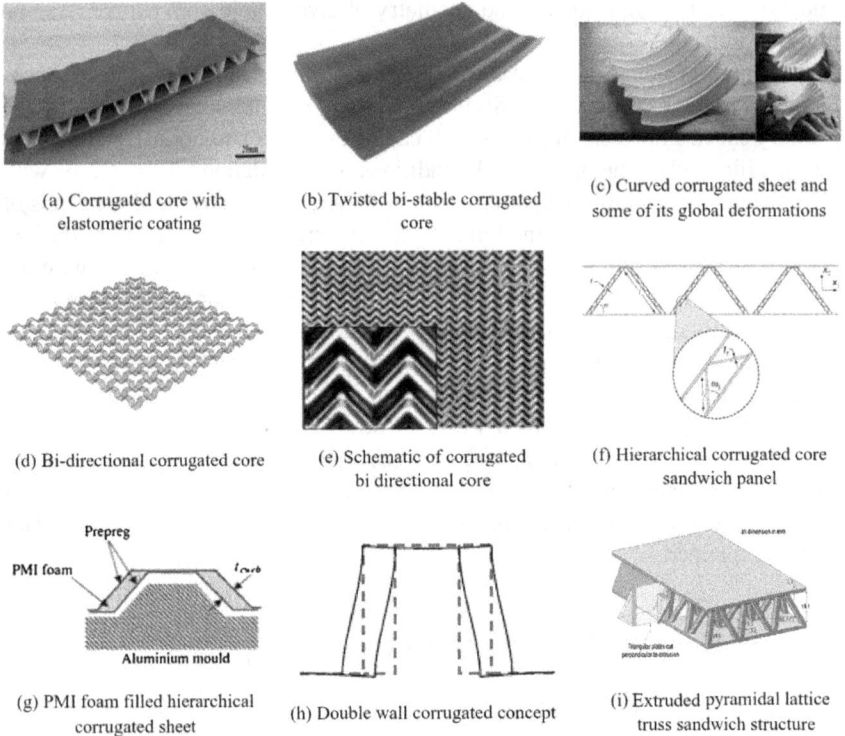

FIGURE 2.1 Schematic views of some of the main concepts of corrugated cores [33].

2.2.2 FOLD CORES

A folded core is actually an origami-like structure which is manufactured by using the origami technique to fold a planar base material into a three-dimensional structure; as a result, a numerous number of possible fold core cell designs with a wide variety of very complex geometries could be made [53]. It is worth noting that most of the folded core structures investigated in the works of recent decades are fabricated with relatively simple geometries with chevron or zigzag or with just simple improvement to them.

A possible set of fold core geometries is given in Refs. [54,55]. Cores with wedge-shaped structures are commonly used in aerodynamic surfaces of the aircrafts' flaps, spoilers and rudders.

Khaliulin et al. [56] developed simple practical solutions to such wedge-shaped cores by evaluating the structural and manufacturing capabilities of the folded structures used in sandwich structures. Then in another study, Hähnel et al. [57] enhanced the wedge-shaped folded sandwich cores for application in aircrafts. It was shown that to avoid a stepwise core height reduction and in order to allow for a continuous attachment to the skins, it is necessary to partly fold the top folding edges back into the core. Different patterns for the folding applicable in

TABLE 2.1
Common Corrugated Structures

Type	Characteristics	References
Corrugated structures with integrated woven corrugated cores	Making corrugated structures with enhanced resistance against face–core debonding, high flexural strength and energy absorption	[36,37]
Corrugated structures with hierarchical corrugated cores	Higher strength in compressive and shear loads in comparison with the structures of the same mass	[38,39]
Corrugated structures with a two-way corrugated core	Same mechanical response in the longitudinal and transverse directions of sandwich panel	[40,41]
Corrugated structures with bidirectional corrugated cores	Exhibition of quasi-isotropic bending behaviour	[42,43]
Corrugated structures with bidirectional corrugated strip core	High transverse shear stiffness in particular inclined bracing chord angles	[40]
Corrugated structures with egg-box core	Having a desirable energy absorption, exhibiting stable collapse behaviour like an ideal energy absorber	[44]
Corrugated structures with structural bi-cells	Providing considerable structural support	[45,46]
Corrugated structures with twisted bi-stable corrugated cores	Providing high strength-to-weight ratio	[47]
Corrugated structures with curved corrugated sheets	Structural integrity, capabilities of great shape change	[48]
Double-wall corrugated structures	Improving the in-plane shear stiffness as well as out-of-plane compressive properties	[49]
Pyramidal lattice corrugated structures	Making a lattice truss sandwich panel with improved mechanical properties	[50]
Corrugated structures used in morphing structures	Providing large shape change capabilities	[51,52]

advanced wedge-shaped cores and the manufacturing processes related to that are presented in another work [57].

Relatively large cell walls in the folded core geometries may result in the early stability failure of the structure under compression or shear loading modes. Thus, different methods were examined to enhance the stiffness of the cell walls. Mudra and Hachenberg investigated layered folded structures consisted of multiple core blocks [58]. Zakirov et al. [59] proposed geometrical modifications such as grooved cell walls and adding vertical cell walls to increase the strength and stability of the folded structures. The use of nanoparticles in the coating resin of the fold cores was investigated in another study to increase the stability and strength of the cell walls [60].

Another design option for folded core sandwich panels is using the dual-core or multi-core configurations. Dual-core panels consisting of one layer of carbon fibre-reinforced polymer (CFRP) and one layer of aramid fibre-reinforced polymer (ARFP) separated from each other using a third skin layer were investigated in some works [54]. In particular, dual-core and multi-core panels can be used in sound-absorbing and vibration-damping applications [61].

In an overall point of view, the main possible folding geometries can be categorized as follows:

- made of a unit cell,
- non-symmetric or non-uniform cores [55,59] or geometries made of curved folding lines [62],
- made of flat or curved panels or even made of doubly curved panels,
- with a wedge shape and a decreasing height,
- spiral-shaped materials [63] and
- dual-core or multi-core folded structures [54,60].

The schematic views of some folding geometries are depicted in Figure 2.2.

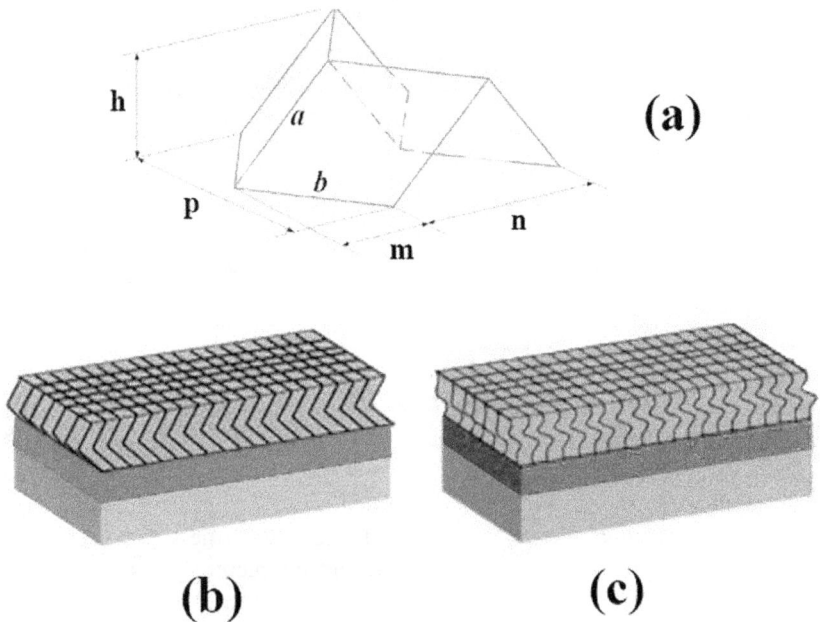

FIGURE 2.2   Schematic views of some folding geometries: (a) a unit cell, (b) a zigzag-shaped and (c) a spiral-shaped.

2.3 MATERIALS

2.3.1 CORRUGATED CORES

A range of different materials such as metals, composites, elastomers and foams can be used to make the cores, while using lightweight materials such as composites makes better opportunities for further development of corrugated cores. Composites are a very common material to fabricate the face sheets or even the core parts [64]. Having surprising features including the diversity of fibre orientations in different stacking sequences, fibre types and textile architecture of the woven plies made of the fibres, composite laminates can enhance the design space of the corrugated cores dramatically and consequently provide this chance to make more improvements through some affecting parameters of them. Kazemahvazi et al. [65] introduced a modern composite corrugation concept to exclude the buckling phenomenon in the core members.

Foams as a lightweight material have also been used in some studies, mainly as a filler in the core structures. As illustrated in Figure 2.3, Taghizadeh et al. [66] used PVC foams to characterize the compressive behaviour of corrugated cores under quasi-static loading. PMI foam-filled corrugated sheets were used in another study to examine different failure modes [67].

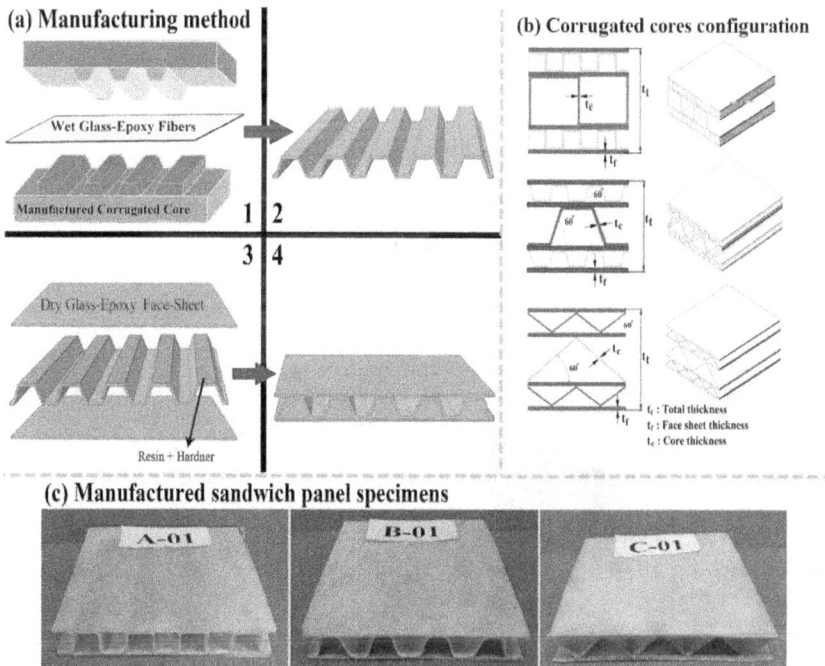

FIGURE 2.3 An overview of Taghizadeh et al.'s [66,77] work in which the compressive behaviour of corrugated cores was examined.

2.3.2 FOLD CORES

A wide variety of cell wall materials are also used for the folded cores. In a general point of view, all materials which are capable of being folded can be used. As a result, the ceramic materials which are too brittle are not suitable for this application. In contrast, the rubber materials with excellent elastic properties are an appropriate choice.

In academic and industrial applications, the materials mentioned below have been employed for the construction of fold core sandwich structures (Figure 2.4):

- metals such as aluminium, copper, steel, brass and titanium [39,73,92],
- plastics such as PEEK and PC [22,54],
- fibre-reinforced composites made of carbon, glass, Kevlar or even basalt fibres [41,54,56],
- aramid papers (Kevlar or Nomex) with different fibres and phenolic resins [40,54,56] and
- kraft papers [54,69].

A brief review of some materials used for fold core manufacturing is presented in [54]; Figure 2.4 also shows some folded core geometries and materials.

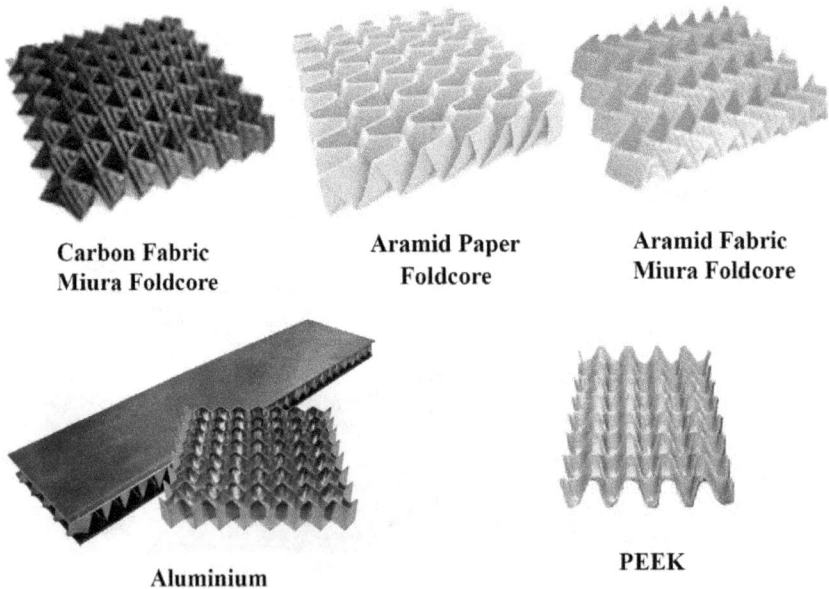

**Carbon Fabric
Miura Foldcore**

**Aramid Paper
Foldcore**

**Aramid Fabric
Miura Foldcore**

Aluminium

PEEK

FIGURE 2.4 An overview of different folded core geometries and the materials used in their manufacturing.

2.4 MECHANICAL PROPERTIES

2.4.1 CORRUGATED CORES

Mechanical tests are known as the major method to characterize the material properties of sandwich core structures. Since the design of corrugated structures is somehow an intrinsically multidisciplinary process, a prosperous design process must cover both the structural and practical necessities. For example, in aerospace structures the minimum weight of the structure is a priority, while vibration phenomenon, fatigue failure and damage tolerance are other major factors that should be considered as well. Actually, a comprehensive set of experimental, analytical as well as numerical investigations should be considered to examine the tensile/compressive, flexural, shear and out-of-plane strengths of corrugated panels. Recent studies have mainly examined the non-linear effects of the geometric parameters and material properties on diverse boundary/loading conditions [35], while analytical solutions also introduce a strong support for these investigations.

A brief categorization related to the concerned methods and tests is presented in Table 2.2.

TABLE 2.2
Categorization of Disciplines Concerned with Design of Corrugated Cores

Discipline	Findings	References
Bending tests	To investigate the flexural rigidities, web design and mechanical behaviour of corrugated structures	[42,70]
Tensile tests	Investigating the behaviour, local failure mechanisms and some properties such as transverse tensile elastic modulus	[70,71]
Shear and compression tests	Investigation of shear parameters such as shear stiffness, shear modulus and shear correction factors	[40,72]
Buckling	Using classical analytical methods, finite element analyses and homogenization techniques to investigate the buckling modes and complex deformations	[73,74]
Vibration	Investigating the effects of weight and damping on each other and homogenization techniques on vibration for different corrugated structures	[75,76]
Impact	Using low-velocity impact and quasi-static tests on different corrugated forms to investigate the effect of materials and layup on failure mechanisms and parameters such as energy absorption	[66,77]
Fatigue	Published works have investigated the girders with corrugated steel webs mainly used in highway bridges	[78,79]

2.4.2 Fold Cores

In this section, the necessary characterization tests for obtaining the mechanical properties of the folded cores are presented briefly. Traditional tests such as flat-wise compressive testing can be used to determine the out-of-plane compressive properties of sandwich core specimens. Transverse shear tests are being used to identify the properties of sandwich panels in longitudinal and lateral directions. Bending, edgewise compression and debonding tests are used to determine the properties and behaviours of the fold core structures [55,81]. Emergence of failure modes and their effects on the structural stability are other driving factors that should be investigated through mechanical tests [81].

Table 2.3 briefly categorizes the concerned methods and tests related to the fold core structures.

2.5 MANUFACTURING METHODS

2.5.1 Corrugated Cores

Reviewing the literature shows that appropriate design and manufacturing methods can optimize the tensional, compressional, torsional and flexural properties of composite panels greatly [87].

In general, manufacturing approaches of the corrugated core structures can be placed in two main categories, i.e. manufacturing approaches based on in situ bonding and manufacturing approaches based on ex situ bonding. As a part works related

---

TABLE 2.3
Categorization of Disciplines Concerned with Design of Fold Cores

Discipline	Findings	References
Quasi-static compression tests	Identifying the influence of geometrical parameters in the presence of quasi-static loading	[17,58,73]
High-rate dynamic compression testing	To examine the impact effects and energy-absorbing parameters efficiently	[69,83]
Shear testing	Investigation of shear parameters for different fold core geometries made of various materials	[58,68,73,92]
Vibration	Determining the shear stiffness in longitudinal and lateral directions	[82]
Bending tests	Three- and four-point bending tests have examined the effects of fold core samples geometry and orientation. Some studies have also measured the residual strength	[69,84]
Tensile tests	Mostly the bonding strength between face and core has been examined	[54,80,81,82]
Analytical approaches and homogenization techniques	Mostly to predict the mechanical properties such as stiffness, strength and transverse shear modulus	[85,86]

to in situ manufacturing, Banhart et al. [88] showed that the internal structure of the foam core can be resolved clearly and there is a strong dependence between the temperature and foaming kinetics and expansion behaviour [88]. Other works have investigated the effects of bonding strength, materials and other parameters in the manufacturing process [89]. However, in ex situ method face sheets should be attached to the core using adhesive bonding, brazing or diffusion bonding [90].

2.5.2 FOLD CORES

A range of approaches have been used for manufacturing fold core structures. The primary principle in folding is to fold/bend the ply in a way to prevent its elongation. As a result, very thin sheets of material should be used without any restriction towards the height of folding direction. An efficient manufacturing process should guarantee the mass production of folded core elements. Many research activities have investigated the fold core manufacturing processes, and a short review of them is presented by Schenk et al. [31].

Some of the main manufacturing processes for folded cores are explained in following:

Discontinuous manufacturing processes: These are the first techniques in fold core manufacturing research activities (Figure 2.5a). They are not as efficient as continuous ones and can produce a limited size of fold core

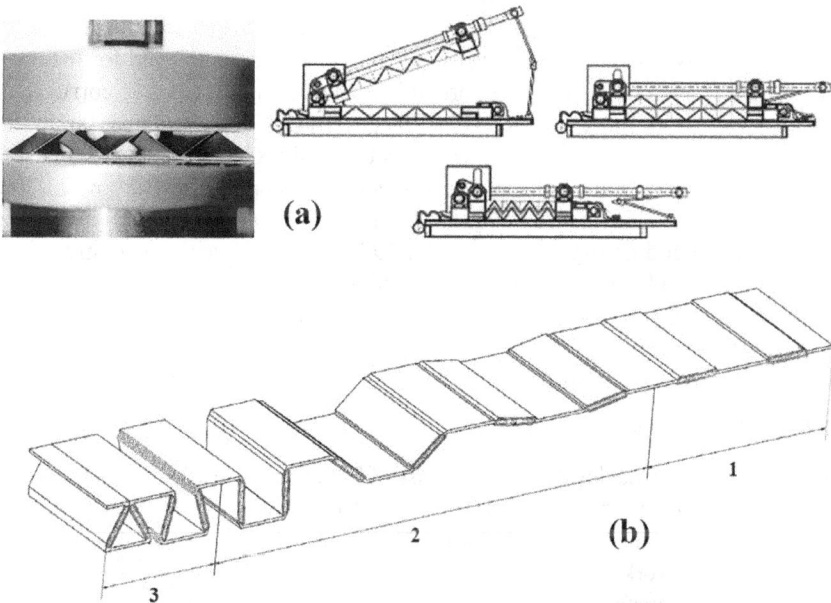

FIGURE 2.5 Schematic views of the (a) discontinuous and (b) continuous manufacturing methods of the fold cores.

structures. Explanations and details of these methods are presented in some research works [41,54,56].

Continuous manufacturing processes: As illustrated in Figure 2.5b, these methods have a higher efficiency than the previous ones and the manufacturing process is fulfilled continuously by automatic machines [54,55,82].

Synchronous manufacturing methods: These methods are more suitable to manufacture prototype fold core structures [91].

Gradual folding manufacturing methods: These are almost new manufacturing methods used for making lightweight sandwich panels [14,41]. As they are a continuous process, they have increased output rates in comparison with batch processes [31,92].

Pre-gathering methods: These methods mainly aim to overcome the coupled contraction in longitudinal and transverse directions. Different phases of these approaches, material deformation considerations and possible manufactured folded geometries are studied [69,93].

2.6 APPLICATIONS

2.6.1 Corrugated Cores

Corrugated core structures have applications ranging from traditional civil structures including corrugated steel beams through to the sophisticated morphing wing structures of aircrafts used in aerospace structures.

In packaging industry, lightweight corrugated sandwich panels with high stiffness and durability are being extensively used to produce rigid shipping containers [94]. Beams with corrugated webs, corrugated flanges employed for the roof of the houses and corrugated pipes are among the common usages of corrugated structures in civil and construction engineering [43,95]. Corrugated sandwich panels also have a range of design solutions to naval problems; they can be used in decks, helidecks, bulkheads and accommodation modules [96]. Above all the mentioned applications, corrugated sandwich panels are gaining a developing role in mechanical and aerospace engineering [97]. Advanced corrugated structures enhance the aerodynamic performance of flight surfaces, protect the spacecraft from extreme temperatures in the re-entry phase, and possess superior load-carrying capabilities [98].

2.6.2 Fold Cores

A major driver for the development and use of folded core structures is the aircraft industry. For a long time, fold core sandwich structures have been investigated as a potential candidate to be used in aircraft sandwich fuselage in numerous number of research works [54,99]. Fold cores can be used in other parts of aircrafts such as control surfaces and crew cabin components [54,57,61]. A large number of further potential applications of folded core structures in heat- and sound-insulating, shock-absorbing and packaging materials have been investigated in other research works [41,54,100].

2.7 CONCLUSIONS

Corrugated core and fold core structures have remarkable effects in engineering fields. High ratio of stiffness to weight, excessive anisotropic response and enhanced specific mechanical properties are among the superior characteristics that these structures provide. In this chapter, a detailed review of the studies investigating the design, materials, mechanical properties, manufacturing processes and applications of the corrugated and fold core structures was reported.

REFERENCES

1. Kujala P, Klanac A. Steel sandwich panels in marine applications. *Pentti Kujala Alan KLANAC Steel Sandwich Panels Marine Appl* 2005;56:305–14.
2. Taghizadeh SA, Liaghat G, Niknejad A, Pedram E. Experimental study on quasi-static penetration process of cylindrical indenters with different nose shapes into the hybrid composite panels. *J Compos Mater* 2019;53:107–23. Doi: 10.1177/0021998318780490.
3. Madadi H, Naghdinasab M, Farrokhabadi A. Numerical investigation of matrix cracking propagation in cross-ply laminated composites subjected to three-point bending load using concurrent multiscale model. *Fatigue Fract Eng Mater Struct* n.d.;n/a. Doi: 10.1111/ffe.13186.
4. Kozak J. Selected problems on application of steel sandwich panels to marine structures. *Polish Marit Res* 2009;16:9–15. Doi:10.2478/v10012-008-0050-4.
5. Kiliçaslan C, Güden M, Odaci IK, Taşdemirci A. Experimental and numerical studies on the quasi-static and dynamic crushing responses of multi-layer trapezoidal aluminum corrugated sandwiches. *Thin-Walled Struct* 2014;78:70–8. Doi:10.1016/j. tws.2014.01.017.
6. Bayat A, Liaghat G, Sabouri H, Ashkezari GD, Pedram E, Taghizadeh SA, et al. Experimental investigation on the quasi-static mechanical behavior of autoclaved aerated concrete insulated sandwich panels. *J Sandw Struct Mater* 2019:109963621985763. Doi: 10.1177/1099636219857633.
7. Atas C, Sevim C. On the impact response of sandwich composites with cores of balsa wood and PVC foam. *Compos Struct* 2010;93:40–8. Doi: 10.1016/j.compstruct. 2010.06.018.
8. Nagarajan R, Krishnasamy S, Siengchin S, Thiagamani SMK. *Mechanical and Dynamic Properties of Biocomposites*, 2021, Germany: Wiley.
9. Han B, Qin K, Yu B, Wang B, Zhang Q, Lu TJ. Honeycomb-corrugation hybrid as a novel sandwich core for significantly enhanced compressive performance. *Mater Des* 2016;93:271–82. Doi:10.1016/j.matdes.2015.12.158.
10. Tiwari G, Thomas T, Khandelwal RP. Influence of reinforcement in the honeycomb structures under axial compressive load. *Thin-Walled Struct* 2018;126:238–45. Doi:10.1016/j.tws.2017.06.010.
11. Zhang J, Supernak P, Mueller-Alander S, Wang CH. Improving the bending strength and energy absorption of corrugated sandwich composite structure. *Mater Des* 2013;52:767–73. Doi:10.1016/j.matdes.2013.05.018.
12. Nilsson P, Al-Emrani M, Atashipour SR. Transverse shear stiffness of corrugated core steel sandwich panels with dual weld lines. *Thin-Walled Struct* 2017;117:98–112. Doi:10.1016/J.TWS.2017.04.008.
13. Gattas JM, You Z. Quasi-static impact response of alternative origami-core sandwich panels. In: *Proceedings of the ASME International Design Engineering Technical Conferences*, vol. 6B, American Society of Mechanical Engineers; 2013. Doi:10.1115/DETC2013-12681.

14. Baranger E, Guidault PA, Cluzel C. Numerical modeling of the geometrical defects of an origami-like sandwich core. *Compos Struct* 2011;93:2504–10. Doi:10.1016/j. compstruct.2011.04.011.

15. Sypeck DJ, Wadley HNG. Cellular metal truss core sandwich structures. *Adv Eng Mater* 2002;4:759–64. Doi:10.1002/1527-2648(20021014)4:10<759::AID-ADEM759>3.0.CO;2-A.

16. Eyvazian A, Taghizadeh SA, Hamouda AM, Tarlochan F, Moeinifard M, Gobbi M. Buckling and crushing behavior of foam-core hybrid composite sandwich columns under quasi-static edgewise compression. *J Sandw Struct Mater* 2019:109963621989466. Doi:10.1177/1099636219894665.

17. Lehmhus D, Hünert D, Mosler U, Martin U, Weise J. Effects of eutectic modification and grain refinement on microstructure and properties of PM AlSi7 metallic foams. *Metals (Basel)* 2019;9:1241. Doi:10.3390/met9121241.

18. Schaedler TA, Carter WB. Architected cellular materials. *Annu Rev Mater Res* 2016;46:187–210. Doi:10.1146/annurev-matsci-070115-031624.

19. Heimbs S. IVW-Schriftenreihe Band 77 Sandwichstrukturen mit Wabenkern: Experimentelle und numerische Analyse des Schädigungsverhaltens unter statischer und kurzzeitdynamischer Belastung, Zugl.: Kaiserslautern, Techn. Univ., Diss., Kaiserslautern IVW, 2008.

20. Sun Y, Guo LC, Wang TS, Zhong SY, Pan HZ. Bending behavior of composite sandwich structures with graded corrugated truss cores. *Compos Struct* 2018;185:446–54. Doi:10.1016/j.compstruct.2017.11.043.

21. Birman V, Kardomateas GA. Review of current trends in research and applications of sandwich structures. *Compos Part B Eng* 2018;142:221–40. Doi:10.1016/j. compositesb.2018.01.027.

22. Fischer S, Drechsler K, Kilchert S, Johnson A.. Mechanical tests for foldcore base material properties. Composites A 2009; 40:1941–1952. Doi:10.1016/j. compositesa.2009.03.005.

23. Khoo ZX, Teoh JEM, Liu Y, Chua CK, Yang S, An J, et al. 3D printing of smart materials: A review on recent progresses in 4D printing. *Virtual Phys Prototyp* 2015;10:103–22. Doi:10.1080/17452759.2015.1097054.

24. Airoldi A, Fournier S, Borlandelli E, Bettini P, Sala G. Design and manufacturing of skins based on composite corrugated laminates for morphing aerodynamic surfaces. *Smart Mater Struct* 2017;26:045024. Doi:10.1088/1361-665X/aa6069.

25. Fischer FJC, Beyrle M, Thellmann A-H, Endrass M, Stefani T, Gerngross T, et al. Corrugated composites: production-integrated quality assurance in carbon fiber reinforced thermoplastic sine wave beam production. *Adv Manuf Polym Compos Sci* 2017;3:10–20. Doi:10.1080/20550340.2017.1283100.

26. Carlsson LA, Nordstrand T, Westerlind B. On the elastic stiffnesses of corrugated core sandwich. *J Sandw Struct Mater* 2001;3:253–67. Doi:10.1106/BKJF-N2TF-AQ97-H72R.

27. Mallick PK, Boorle R. Sandwich panels with corrugated core - A lightweighting concept with improved stiffness. *SAE Tech. Pap.* vol. 1, SAE International; 2014. Doi:10.4271/2014-01-0808.

28. Miura K. Zeta-core sandwich – its concept and realization. ISAS Report 480, Institute of Space and Aeronautical Science, University of Tokyo, 1972.

29. Miura K. New structural form of sandwich core. *J Aircr* 1975;12:437–41. Doi:10.2514/3.44468.

30. Lebée A, Sab K. Thick periodic plates homogenization, application to sandwich panels including chevron folded core, Dissertation, Université Paris-Est., 2011.

31. Schenk M, Allwood JM, Guest SD. Cold gas-pressure folding of miura-ori sheets, In: International conference on technology of plasticity (ICTP 2011), Aachen.

32. Basily B, Elsayed E. Design and development of lightweight sandwich structures with innovative sheet design and development of lightweight sandwich structures with innovative sheet folded cores, 2005

33. Dayyani I, Shaw AD, Saavedra Flores EI, Friswell MI. The mechanics of composite corrugated structures: a review with applications in morphing aircraft. *Compos Struct* 2015;133:358–80. Doi:10.1016/j.compstruct.2015.07.099.

34. Yokozeki T, Takeda SI, Ogasawara T, Ishikawa T. Mechanical properties of corrugated composites for candidate materials of flexible wing structures. *Compos Part A Appl Sci Manuf* 2006;37:1578–86. Doi:10.1016/j.compositesa.2005.10.015.

35. Gilchrist AC, Suhling JC, Urbanik TJ. Nonlinear finite element modeling of corrugated board, 1999; 231:101–6.

36. Kazemahvazi S, Khokar N, Technology SH. Confluent 3D-assembly of fibrous structures. *Elsevier* 2016;127:95–105.

37. Malcom AJ, Aronson MT, Wadley HN. Three-dimensionally woven glass fiber composite struts: characterization and mechanical response in tension and compression. *J Compos Mater* 2016;50:25–43. Doi:10.1177/0021998315569751.

38. Kooistra GW, Deshpande V, Wadley HNG. Hierarchical corrugated core sandwich panel concepts. *J Appl Mech Trans ASME* 2007;74:259–68. Doi:10.1115/1.2198243.

39. Fischer S. Aluminium foldcores for sandwich structure application: Mechanical properties and FE-simulation. *Thin-Walled Struct* 2015;90:31–41. Doi:10.1016/j.tws.2015.01.003.

40. Xiang XM, Lu G, You Z. Energy absorption of origami inspired structures and materials. *Thin-Walled Struct* 2020;157:107130. Doi:10.1016/j.tws.2020.107130.

41. Zang S, Zhou X, Wang H, You Z. Foldcores made of thermoplastic materials: Experimental study and finite element analysis. *Thin-Walled Struct* 2016;100:170–179. Doi:10.1016/j.tws.2015.12.017.

42. Seong DY, Jung CG, Yang DY, Moon KJ, Ahn DG. Quasi-isotropic bending responses of metallic sandwich plates with bi-directionally corrugated cores. *Mater Des* 2010;31(6):2804–12.

43. Dayyani I. Mechanical behavior of composite corrugated structures for skin of morphing aircraft. Thesis (Ph.D.)–Swansea University, Dissertation Abstracts International, Vol.76–06C, p. 169. 2015.

44. Yoo SH, Chang SH, Sutcliffe MPF. Compressive characteristics of foam-filled composite egg-box sandwich panels as energy absorbing structures. *Compos Part A Appl Sci Manuf* 2010;41:427–34. Doi:10.1016/j.compositesa.2009.11.010.

45. Schwingshackl CW, Aglietti GS, Cunningham PR. Experimental determination of the dynamic behavior of a multifunctional power structure. *AIAA J* 2007;45:491–6. Doi:10.2514/1.23894.

46. Schwingshackl CW, Aglietti GS, Cunningham PR. Parameter optimization of the dynamic behavior of inhomogeneous multifunctional power structures. *AIAA J* 2006;44:2286–94. Doi:10.2514/1.18599.

47. Norman A., Seffen K., Guest S. Multistable corrugated shells. *Proc R Soc A Math Phys Eng Sci* 2008;464:1653–72. Doi:10.1098/rspa.2007.0216.

48. Norman AD, Seffen KA, Guest SD. Morphing of curved corrugated shells. *Int J Solids Struct* 2009;46:1624–33. Doi:10.1016/j.ijsolstr.2008.12.009.

49. Previtali F, Delpero T, Bergamini A, Arrieta AF, Ermanni P. Extremely anisotropic multi-functional skin for morphing applications. *23rd AIAA/AHS Adaptive Structures Conference*, American Institute of Aeronautics and Astronautics Inc.; 2015. Doi:10.2514/6.2015-0787.

50. Queheillalt DT, Murty Y, Wadley HNG. Mechanical properties of an extruded pyramidal lattice truss sandwich structure. *Scr Mater* 2008;58:76–9. Doi:10.1016/j. scriptamat.2007.08.041.

51. Xiong J, Du Y, Mousanezhad D, Eydani Asl M, Norato J, Vaziri A. Sandwich structures with prismatic and foam cores. *A Review. Adv Eng Mater* 2019;21:1800036. Doi:10.1002/adem.201800036.

52. Dayyani I, Ziaei-Rad S, Friswell MI. The mechanical behavior of composite corrugated core coated with elastomer for morphing skins. *J Compos Mater* 2014;48:1623–36. Doi:10.1177/0021998313488807.

53. Nojima T, Saito K. Development of newly designed ultra-light core structures. *JSME Int J Ser A* 2006;49:38–42. Doi:10.1299/jsmea.49.38.

54. Heimbs S. Foldcore sandwich structures and their impact behaviour: An overview. *Solid Mech Its Appl* 2013;192:491–544. Doi:10.1007/978-94-007-5329-7_11.

55. Elsayed EA, Basily BB. A continuous folding process for sheet materials. *Int. J. Mater. Prod. Technol.* vol. 21, Inderscience Publishers; 2004:217–38. Doi:10.1504/ ijmpt.2004.004753.

56. Khaliulin VI, Batrakov VV, Menyashkin DG. On structural and manufacturing capabilities of folded structures for use in sandwich panels. In: SAMPE Europe International Conference, Paris, pp. 141–148, 2007..

57. Hähnel F, Wolf K, Hauffe A, Alekseev KA, Il'dus, MZ. Wedge-shaped folded sandwich cores for aircraft applications: from design and manufacturing process to experimental structure validation. *CEAS Aeronaut J* 2011;2:203–212 Springer n.d. Doi:10.1007/s13272-011-0014-8.

58. Mudra C, Hachenberg D. Alternative sandwich core structures – efficient investigation of application potential by using finite element modelling. In: SAMPE Europe International Conference, Paris, 444–448, 2003..

59. Zakirov IM, Nikitin A, Alexeev K, Mudra C; Folded structures: performance, technology and production. In: SAMPE Europe International Conference, Paris, pp. 234–239, 2006.

60. Zakirov IM, Alekseev KA, Kayumov RA, Gainutdinov IR. Some possible techniques for improving the strength characteristics of folded cores from sheet composite materials. *Russ Aeronaut* 2009;52:347–50. Doi:10.3103/S1068799809030143.

61. Wang Z, Engineering QX-J of V, 2006 undefined. Experimental research on soundproof characteristics for the sandwich plates with folded core. J Vib Eng 2006;19:65–69.

62. Drechsler K, Kehrle R. Manufacturing of folded core-structures for technical applications. In: SAMPE Europe International Conference, Paris, pp. 508–513, 2004.

63. Shi Y, Fluri A, Garbayo I, Schwiedrzik I.J, Michler J. et al. Zigzag or spiral-shaped nanostructures improve mechanical stability in yttria-stabilized zirconia membranes for micro-energy conversion devices. *Nano Energy* 2019;59:674–682. Doi:10.1016/j.nanoen.2019.03.017.

64. Banhart J, Seeliger HW. Aluminium foam sandwich panels: Manufacture, metallurgy and applications. *Adv Eng Mater* 2008;10:793–802. Doi:10.1002/adem.200800091.

65. Kazemahvazi S, Tanner D, Zenkert D. Corrugated all-composite sandwich structures. Part 2: Failure mechanisms and experimental programme. *Compos Sci Technol* 2009;69:920–5. Doi:10.1016/j.compscitech.2008.11.035.

66. Taghizadeh SA, Farrokhabadi A, Liaghat G, Pedram E, Malekinejad H, Mohammadi SF, et al. Characterization of compressive behavior of PVC foam infilled composite sandwich panels with different corrugated core shapes. *Thin-Walled Struct* 2019;135:160–72. Doi:10.1016/J.TWS.2018.11.019.

67. Kazemahvazi S, Tanner DZD. Corrugated all-composite sandwich structures. Part 2: failure mechanisms and experimental programme. *Compos Sci Technol* 2009:920–5.

68. Fischer S, Drechsler K. Aluminium foldcores for sandwich structure application. In: CELLMET2008, Cellular Metals for Structural and Functional Applications, International Symposium, Dresden, 2008.

69. Basily BB, Elsayed EA. Dynamic axial crushing of multilayer core structures of folded Chevron patterns. *Int. J. Mater. Prod. Technol.* Inderscience Publishers; 2004; 21:169–85. Doi:10.1504/ijmpt.2004.004750.

70. Farrokhabadi A, Ahmad Taghizadeh S, Madadi H, Norouzi H, Ataei A. Experimental and numerical analysis of novel multi-layer sandwich panels under three point bending load. *Compos Struct* 2020:112631. Doi:10.1016/j.compstruct.2020.112631.

71. hill C, Etches JA, Bond IP, Potter KD, Weaver PM. Corrugated composite structures for aircraft morphing skin applications. In: 18th International Conference of Adaptive Structures and Technologies, Ottawa, ON, 2007.

72. Rejab MRM, Cantwell WJ. The mechanical behaviour of corrugated-core sandwich panels. *Compos Part B Eng* 2013;47:267–77. Doi:10.1016/j.compositesb.2012.10.031.

73. Semenyuk NP, Zhukova NB, Ostapchuk V V. Stability of corrugated composite noncircular cylindrical shells under external pressure. *Int Appl Mech* 2007;43:1380–9. Doi:10.1007/s10778-008-0009-2.

74. Wang G, Sun H, Peng H, Uemori R. Buckling and ultimate strength of plates with openings. *Ships Offshore Struct* 2009;4:43–53. Doi:10.1080/17445300802479437.

75. Kumar N, Luding S, Magnanimo V. Macroscopic model with anisotropy based on micro-macro information. *Acta Mech* 2014;225:2319–43. Doi:10.1007/s00707-014-1155-8.

76. Yang J, Xiong J, Ma L, Feng L, - SW. Modal response of all-composite corrugated sandwich cylindrical shells. *Compos Sci Technol* 2015;115:9–20.

77. Taghizadeh S.A, Naghdinasab M, Madadi H.R, Farrokhabadi A. Investigation of novel multi-layer sandwich panels under quasi-static indentation loading using experimental and numerical analyses. *Thin-Walled Struct* 2021;160:107326. Doi:10.1016/j.tws.2020.107326.

78. Henderson D. Response of pierced fixed corrugated steel roofing systems subjected to wind loads. *Eng Struct* 2011;33:3290–8.

79. Kövesdi B, Dunai L. Fatigue life of girders with trapezoidally corrugated webs: An experimental study. *Int J Fatigue* 2014;64:22–32.

80. Cabo D, Autor A, Fernández L, Manuel C. Morphing Multiscale Textured Shells. In: Symposium of the International Association for Shell and Spatial Structures, Valencia, pp. 1767–79, 2009.

81. Heimbs S, Vogt D, Hartnack R, Schlattmann J, Maier M. Numerical simulation of aircraft interior components under crash loads. *Int J Crashworthiness* 2008;13:511–21. Doi:10.1080/13588260802221203.

82. Velea M.N, Schneider C, Lache S. Second order hierarchical sandwich structure made of self-reinforced polymers by means of a continuous folding process. *Mater Design* 2016;102:313–320. Doi:10.1016/j.matdes.2016.04.049.

83. Heimbs S, Schmeer S, Middendorf P, Maier M. Strain rate effects in phenolic composites and phenolic-impregnated honeycomb structures. Compos Sci Technol 2007;67:2827–37.

84. Klaus M, Reimerdes HG. Residual strength simulations of sandwich panels after impact. In: 17th International Conference on Composite Materials (ICCM-17), Edinburgh, 2009.

85. Lebee A, Sab K. Transverse shear stiffness of a chevron folded core used in sandwich construction. In: JNC16, 16`emes Journ´ees Nationales sur les Composites, Toulouse, 2009.

86. Lebée A, Sab K. Reissner–Mindlin Shear Moduli of a Sandwich Panel with Periodic Core Material, 2010:169–77. Doi:10.1007/978-1-4419-5695-8_18.

87. Saleh M, Luzin V, Toppler K, Kabir K. Response of thin-skinned sandwich panels to contact loading with flat-ended cylindrical punches: Experiments, numerical simulations and neutron diffraction. *Compos Part B Eng* 7 2015;78:415–30.

88. Banhart J, Stanzick H, Helfen L, Baumbach T, Nijhof K. Real-time x-ray investigation of aluminium foam sandwich production. 2001;3:407–11.

89. Böllinghaus T, Von Hagen H, Bleck W. Ermüdung von sandwichverbunden aus aluminiumschaum mit stahldeckblechen. *Materwiss Werksttech* 2000;31:488–92. Doi:10.1002/1521-4052(200006)31:6<488::AID-MAWE488>3.0.CO;2-X.

90. Styles M, Compston P, Kalyanasundaram S. Finite element modelling of core thickness effects in aluminium foam/composite sandwich structures under flexural loading. *Compos Struct* 2008;86:227–32. Doi:10.1016/j.compstruct.2008.03.024.

91. Akishev N, Zakirov I, AV. Nikitin - US Patent 7 487,658, 2009 undefined. Device for sheet material corrugation. Google Patents n.d. CA2546606C (Canada). https://patents.google.com/patent/CA2546606C

92. Kehrle R. - US Patent 6 913,570, 2005 undefined. Method and apparatus for producing a composite structural panel with a folded material core. Google Patents n.d.

93. Schenk M, Guest SD, McShane GJ. Novel stacked folded cores for blast-resistant sandwich beams. *Int J Solids Struct* 2014;51:4196–214. Doi:10.1016/j.ijsolstr.2014.07.027.

94. Singh SP, Chonhenchob V, Singh J. Life cycle inventory and analysis of re-usable plastic containers and display-ready corrugated containers used for packaging fresh fruits and vegetables. *Packag Technol Sci* 2006;19:279–93. Doi:10.1002/pts.731.

95. Dubina D, Ungureanu V, Gîlia L. Cold-formed steel beams with corrugated web and discrete web-to-flange fasteners. *Steel Constr* 2013;6:74–81. Doi:10.1002/stco.201310019.

96. Liang C, Yang M, Wu P. Optimum design of metallic corrugated core sandwich panels subjected to blast loads. *Ocean Eng* 2001;28:825–61.

97. Brown W. The suitability of various gasket types for heat exchanger service. *Am. Soc. Mech. Eng. Press. Vessel. Pip. Div. PVP*, American Society of Mechanical Engineers Digital Collection; 2002;433:45–51. Doi:10.1115/PVP2002-1081.

98. Martinez OA, Sankar BV, Haftka RT, Bapanapalli SK, Blosser ML. Micromechanical Analysis of Composite Corrugated-Core Sandwich Panels for Integral Thermal Protection Systems. *ArcAiaaOrg* 2007;45:2323–36. Doi:10.2514/1.26779.

99. Kolesnikov BY, Herbeck L. Carbon fiber composite airplane fuselage: concept and analysis. In: ILA International Conference 'Merging the efforts: Russia in European research programs on aeronautics', Berlin, 2004.

100. Zhang P, Li X, Wang Z, Zhao L, Yan X. Dynamic blast loading response of sandwich beam with origami-inspired core. *Results Phys* 2018;10:946–955. Doi:10.1016/j.rinp.2018.07.043.

3 Metallic Core- and Truss Core-Based Composite Sandwich Panels

Jin-Shui Yang and Shuang Li
Harbin Engineering University

CONTENTS

3.1 Introduction 45
3.2 Fabrication Approach and Performance Characterization of
Metallic Core-Based Composite Sandwich Panels................................... 46
3.3 Fabrication Approach and Performance Characterization of
Truss Core-Based Composite Sandwich Panels 52
References.. 58

3.1 INTRODUCTION

The rapid development of aerospace, automobile, ship, construction and other industries demands increasing load capacity, reducing energy consumption and fast transportation, which in turn has put forward the urgent requirement of a new generation of lightweight, high-strength, multifunctional structures. So far, foam sandwich structures [1] and honeycomb sandwich structures [2] and recently lattice sandwich structures [3], inspired by the porous structures in nature such as animal bones and plant rhizomes, have been gradually proposed and studied. The foam core sandwich structure is mostly made up by disordered unit cells, and then 2D periodic grid and honeycomb sandwich structures are proposed. In terms of the macroscopic strength of the foam sandwich structures, the bending deformation of cell wall is the dominant mode for the deformation of the structure. The macroscopic strength of 2D grid sandwich structures and honeycomb sandwich structures has been improved, but their closed unit cells limit the multifunctional application of the structures. In view of the above, 3D truss unit cells in the form of a two-force rod are designed and used as the core of the sandwich structure. The unit cells with trusses not only realize the macroscopic strength determined by the tensile and compressive strengths of the rod, but also realize the connection between the unit cells. The multifunctional integration space and potential inside the core are greatly improved. By filling the core with foam and pre-embedded materials, the lightweight sandwich structure has attracted much attention in the

DOI: 10.1201/9781003143031-3

fields of vibration and noise reduction, temperature control, impact [4] and explosion resistance [5]. To date, plenty of research on the metallic core- and truss core-based composite sandwich panels has mainly focused on the fabrication approach and performance characterization. The fabrication approaches of the structures are different according to different materials, and they are roughly divided into fabrication approaches for metal sandwich structures and composite sandwich structures. In order to investigate the lattice sandwich structure more comprehensively, scholars have carried out extensive research on the performance. It not only involves static characteristics such as compression [6], shear [7] and bending [8] and dynamic characteristics such as impact and vibration [9], but also includes multifunctional characteristics such as thermodynamics [10] and acoustics [11].

3.2 FABRICATION APPROACH AND PERFORMANCE CHARACTERIZATION OF METALLIC CORE-BASED COMPOSITE SANDWICH PANELS

The fabrication approach of the structure directly affects its performance characterization. Appropriate fabrication approach can enable the structure to achieve the designed performance characterization, and the simple, low-cost fabrication approach for mass production is the prerequisite for the industrial application of the structures. In recent years, a series of research on the fabrication approach of metal lattice sandwich structures has been carried out. Initially, Sypeck et al. [12] prepared a metal lattice sandwich structure using a wire weaving process. Subsequently, Kang [13] used the weaving process in the preparation of metal lattice sandwich structures. Two wire-woven bulk Kagome cores are shown in Figure 3.1. Due to the characteristics of the wire-woven process, the metal wire will be deformed during the preparation process, causing its strength and modulus to decrease, thus affecting the overall performance of the sandwich structure. Later, the investment casting method [14] was used to fabricate tetrahedral, pyramidal

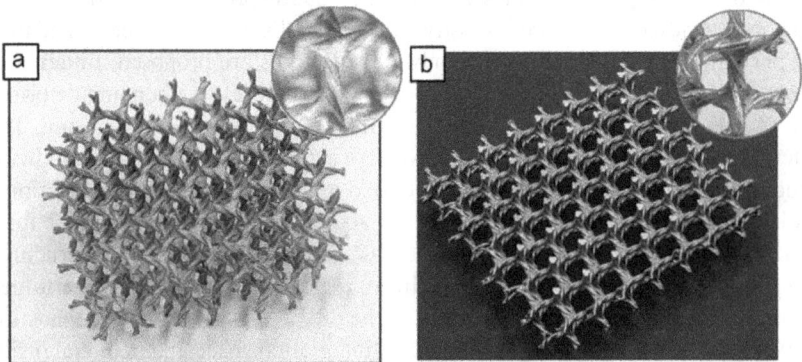

FIGURE 3.1 Two wire-woven bulk Kagome cores made of (a) spring steel and (b) SUS 304 stainless steel. (Kang [13].)

and Kagome lattice sandwich structures. However, this method requires the material to have high fluidity during casting; thus, the choice of material is limited. And the casting process will produce defects that affect the structural performance. Therefore, Sypeck and Wadley [15] proposed a method of bending at the nodes of suitably perforated metal sheets and welding. A stainless steel tetrahedral lattice sandwich structure was prepared by this method. It is suitable for various heat-treatable forged alloys to fabricate lattice sandwich structures. On this basis, Kooistra [16] further used the perforated sheet folding/brazing method to prepare an aluminum alloy tetrahedral lattice truss structure. The fabrication process of the aluminum alloy tetrahedral lattice truss cores by perforating and folding sheet is shown in Figure 3.2. The perforation and folding method has become a widely used metal lattice structure fabrication approach due to its simple preparation and mass production. However, the rod of trusses is bended during the folding process, so the strength of the rod is reduced. In view of this problem, Yang et al. [17] and Li et al. [18] used the cutting and slot-fitting method and fabricated an aluminum alloy pyramid lattice core. Figure 3.3 shows the preparation flowchart of structures. This method was firstly proposed by Finnegan et al. [19] to prepare the composite sandwich structure. In recent years, Feng et al. [20,21] have explored a preparation process of snap-fit and vacuum brazing approach and fabricated a novel Hourglass lattice sandwich structure with stainless steel sheets. Based on this preparation method, the lattice sandwich structures of aluminum alloy and stainless steel were prepared by Yang et al. [6] and Li et al. [9]. The fabrication process can be seen in Figure 3.4. First, the 304 stainless steel sheets were cut into truss strips with different configurations by wire electrical discharge machining.

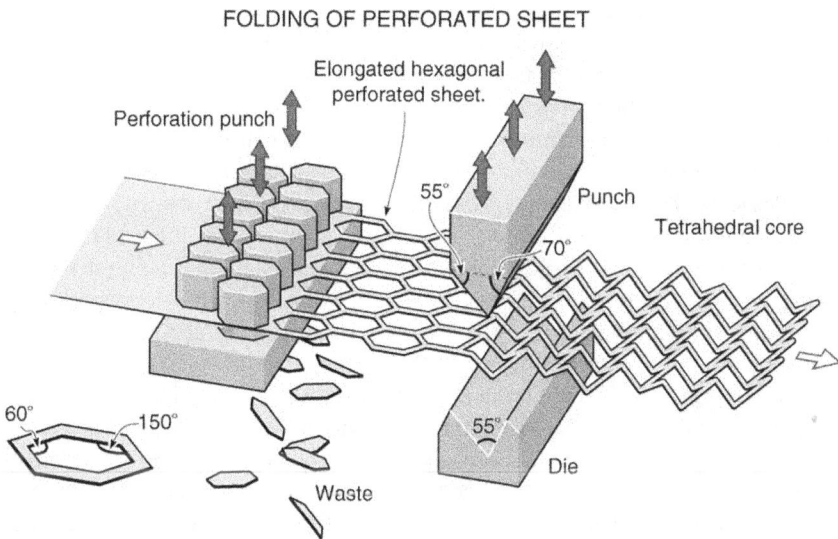

FOLDING OF PERFORATED SHEET

FIGURE 3.2 The fabrication process of the aluminum alloy tetrahedral lattice truss cores by perforating and folding sheet. (Kooistra [16].)

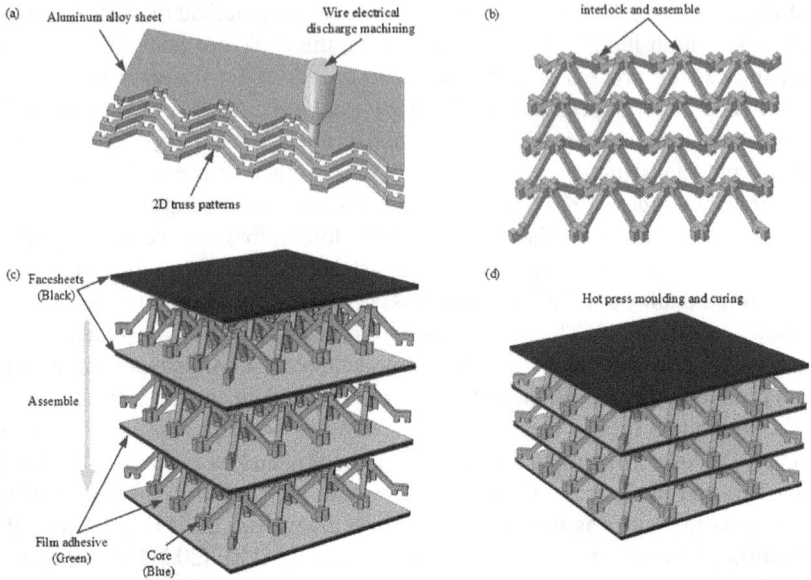

FIGURE 3.3   The preparation flowchart of multilayer gradient composite lattice sandwich panels. (Li et al. [18]).

As shown in Figure 3.4a and d, the truss strips were formed by the periodically arranged half unit cells. Then the two topological cores were formed by these assembled truss strips embedded into each other in perpendicular directions. Finally, by using the vacuum brazing approach, the face sheets and cores were welded into the sandwich lattice structure. To prepare the aluminum alloy lattice sandwich structure with thin face sheets, Yang et al. [22] improved the vacuum brazing approach. They proposed a method by combining the topology strengthening and material strengthening to fabricate the aluminum alloy lattice sandwich structure. The enhanced aluminum alloy Hourglass lattice sandwich structures were fabricated by this method. The structural strength of the enhanced Hourglass lattice sandwich structure decreases in a "cliff-like" manner as the load increases after reaching peak load. Hence, it is suitable for fields that require high structural strength. The node-interlocking assembly and welding process was applied by Hu et al. [23] to fabricate a 304 stainless steel hollow round tube pyramid lattice sandwich structure with millimeter-level diameter. The manufacturing route is shown in Figure 3.5a involving the node-interlocking assembly method. And the photograph of the pyramidal tube lattice structure is shown in Figure 3.5b.

In addition to the fabrication approaches, the mechanical properties of the lattice sandwich structure are also of great concern. The performance characterization of the lattice sandwich structure can be divided into static and dynamic properties. Compressive strength is the most basic static characteristic, and the compression characteristics of pyramidal, tetrahedral, 3D Kagome and other topologies have widely been studied by many scholars. In order to obtain a metal lattice sandwich

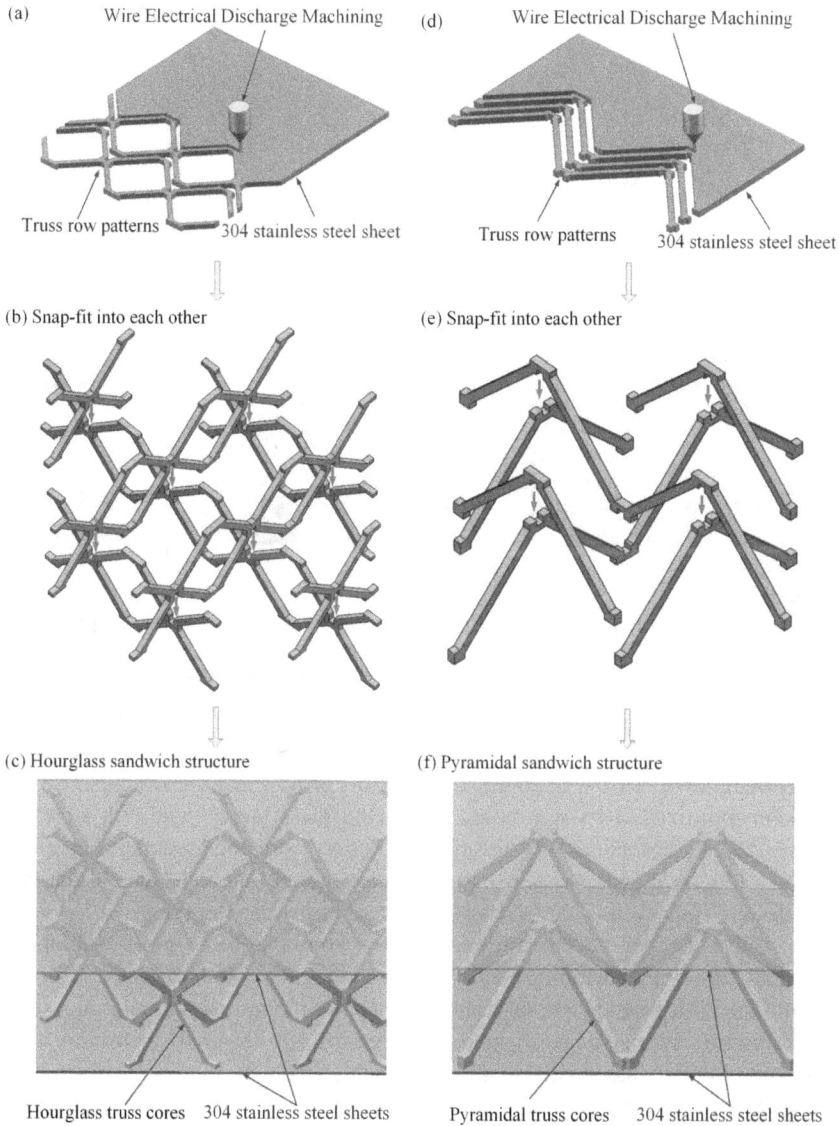

FIGURE 3.4 The wire EDM-interlocking assembly-vacuum brazing approach applied to fabricate (a–c) Hourglass sandwich structure and (d–f) pyramidal sandwich structure. (Li et al. [9].)

structure with higher specific compressive strength, Feng et al. [20] proposed an evolutionary structure of the pyramid lattice structure, and designed and prepared a 304 stainless steel Hourglass lattice sandwich structure. The evolutionary process is shown in Figure 3.6. The out-of-plane compression, in-plane compression [20], and in-plane shear and bending characteristics [7] of the metallic Hourglass

(a) Manufacturing route

(b)

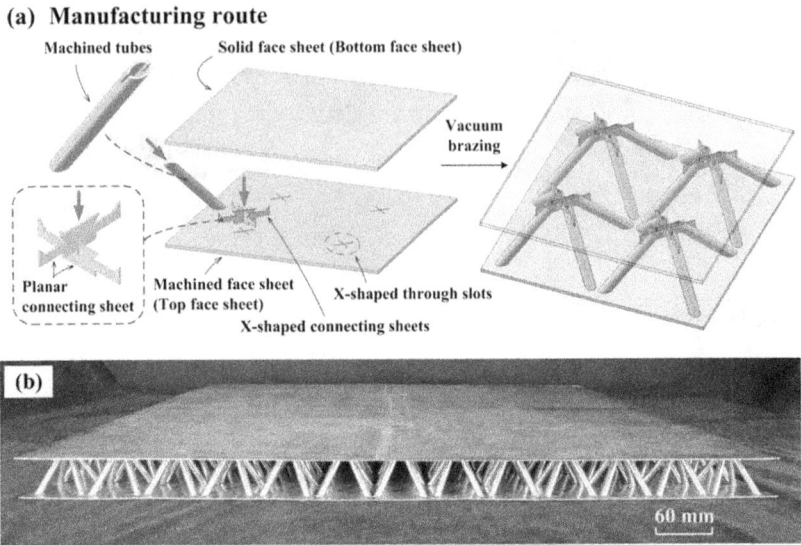

FIGURE 3.5 (a) The manufacturing route involving the node-interlocking assembly method and (b) the photograph of a pyramidal tube lattice structure. (Hu et al. [23].)

lattice sandwich structure were investigated. After analyzing the experimental results, it was found that the Hourglass lattice sandwich structure has higher compressive and shear strengths than the pyramidal lattice sandwich structure with the same relative density. In addition, the results proved that the structural design is more optimized. This was because the Hourglass lattice sandwich structure reduces the truss slenderness ratio and shortens the core node spacing. Therefore, the buckling resistance of the truss and the local buckling resistance of the face sheets were improved. That was why the strengths of the Hourglass lattice sandwich structure under out-of-plane compression, in-plane shear, in-plane compression and three-point bending loads were significantly better than those of the pyramidal lattice sandwich structure. The out-of-plane compression and shear performances were also investigated by Yang et al. [6]. The 6063 aluminum alloy enhanced lattice truss sandwich structures with different relative densities were designed and fabricated. The results showed that the enhanced lattice sandwich structure made of aluminum alloy was a potential material with low density and high strength. In addition, the out-of-plane compression and shear characteristics of the 304 stainless steel hollow tube pyramidal lattice sandwich structure were studied by Hu et al. [23]. In light of the results, it can be concluded that the specific strength and specific energy absorption of the hollow tube pyramidal lattice sandwich structure were better than those of solid tubes, other topological configurations made up by hollow tubes and honeycomb sandwich structures. This was because the hollow tube had a higher section moment of inertia, and the lattice sandwich structure composed of the hollow tube had more excellent elastic strength under the condition of buckling of the tube.

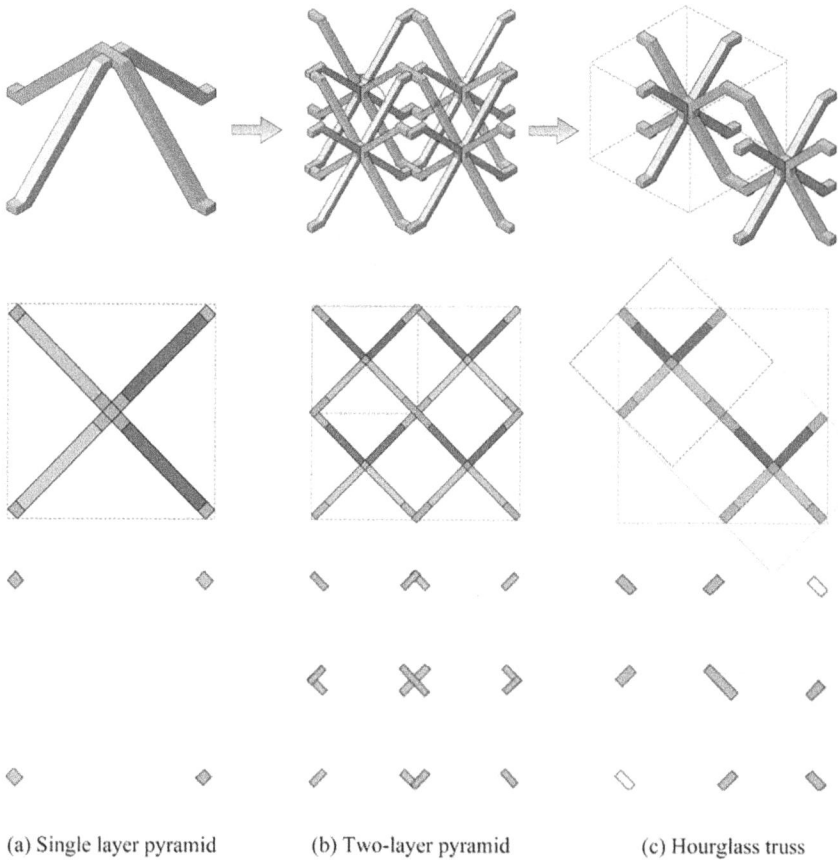

(a) Single layer pyramid   (b) Two-layer pyramid   (c) Hourglass truss

FIGURE 3.6 The evolutionary process from pyramidal truss core to Hourglass truss core: (a) single-layer pyramidal truss, (b) two-layer pyramidal truss and (c) Hourglass truss. (Feng et al. [20].)

Compared with the static performances, the dynamic characteristics such as modal characteristics, vibration isolation characteristics, anti-blast and other performances are relatively less investigated . The natural frequency of the structure is an important parameter in engineering applications as one of the most basic characteristics of the structure. Li et al. [9] studied the natural frequencies, vibration shapes, frequency response and damping characteristics of the new metallic sandwich panels with Hourglass truss cores by experiments and simulation. Furthermore, the influence of the inclination angle of the truss and the boundary conditions on the structural vibration characteristics was studied. The experimental and simulation results showed that the natural frequencies of the metallic sandwich panels with Hourglass truss cores were twice higher than that of the pyramidal lattice sandwich structure with the same relative densities under free boundary condition. The truss inclination angles and boundary conditions obviously affected the natural frequencies of the metallic Hourglass truss cores sandwich panels, and the influence law and frequency variation trend were presented. Feng et al. [21] investigated the underwater blast impact behaviors of enhanced metallic lattice truss sandwich panels by conducting an experimental study. It turned out that under the impact load of underwater explosion, the rear-face maximum deflection of the single-layer enhanced Hourglass lattice truss sandwich panels was distinctly smaller than that of the solid plate with same area and mass. That was because the Hourglass lattice truss core provides greater bending stiffness. More to the point, the front-face deflections of Hourglass lattice truss sandwich panels are less than those of the pyramidal lattice truss sandwich panels by about 22%–38%, and the stretching and tearing resistances of the face sheets are higher than those of the pyramidal lattice truss sandwich panels in identical mass per unit area. Yang et al. [24,25] investigated the blast resistance characteristics of multilayer gradient stainless steel Hourglass and pyramidal lattice sandwich structures. Subsequently, the performances of multilayer gradient stainless steel pyramidal lattice sandwich structure were studied through compression test and dynamic impact experiment by a large-diameter Hopkinson pressure bar device [26]. All of the results have proved that the multilayer lattice sandwich structures with suitable gradient were superior to non-gradient lattice sandwich structures in terms of blast resistance and anti-impact performance.

3.3   FABRICATION APPROACH AND PERFORMANCE CHARACTERIZATION OF TRUSS CORE-BASED COMPOSITE SANDWICH PANELS

Compared with the metallic truss sandwich structure, the fabrication of composite sandwich structure with truss core is relatively difficult, especially in the design of the mold, which requires comprehensive consideration of factors such as demolding, pressure and temperature field distribution. So far, many scholars have proposed a variety of fabrication approaches for truss core-based composite sandwich panels through continuous exploration and research. It is worth noting that since the face sheet–core interface of the truss core-based composite sandwich panels is generally bonded with film adhesive, and the bonding strength is generally low, optimizing the fabrication approach has become an important investigation content. To enhance the interface strength of the lattice sandwich structure between face sheet and core, technologies such as Z-pin and stitch have been gradually proposed. Wang et al. [27] proposed a new fabrication approach to improve the face sheet–core interface strength of the lattice sandwich structure. The foam sewing process was firstly used to prepare the fiber column-reinforced foam sandwich structure. The fiber column is thicker than the fiber bundle in the traditional Z-direction-reinforced structure, and the fiber column is generally less prone to buckling instability. In the compression and bending experiments, the thicker fiber column in the core mainly bears the external load and the load on the foam is very small. Furthermore, in order to fabricate a composite lattice core, a novel hot-press integrated molding process [28] was developed to prepare a carbon fiber composite straight-column lattice structure and the face sheets and cores were directly connected together by pre-embedding and curing. The preparation process mainly includes the following steps: (1) rolling the prepreg into a column with a certain diameter. (2) Placing the prepreg column into the mold and pre-embedding the upper and lower ends of the column in the face sheets. (3) Putting them in a hot press to heat and pressurize then, after curing and forming, demolding it to form a lattice structure. The hot-press integrated molding technology has many advantages: (1) no more mechanical processing. (2) Fiber reinforcement direction is consistent with the load direction. (3) The cores and face sheets are integrally formed, there is no node connection problem, and the connection weakness between the face sheets and cores is well overcome. It was found that this type of structure had similar limitations to the three-dimensional hollow sandwich structure; that is, the core topology was simple and the shear performance was poor. The slant-column lattice structure had a great improvement in shear performance compared to the straight-column lattice structure, but it only enhanced in one direction, while the shear performance in the other direction is still weak. To enhance the shear performance in both directions, the composite pyramidal lattice sandwich structures made of carbon fiber were prepared by the integrated process of hot pressing and molding. And this structure exhibited excellent performance. The fabrication process of the composite carbon fiber pyramidal lattice sandwich structure was very similar to the straight-column lattice sandwich structure, except that the rods need to be implanted at an angle when inserted into the mold. The integrated process of hot pressing and molding significantly improved the strength of the face sheet–core interface, but the partial embedding of the core into face sheets reduced the mechanical properties of the face sheets. Finnegan et al. [19] used the water cutting interlocking assembly process for the first time to prepare the pyramidal composite lattice structure, and the preparation process was gradually perfected. From the truss assembly process to the water cutting single truss splicing process, finally a mature water cutting interlocking assembly process was put forward. Xiong et al. [29] proposed a prepreg secondary molding hot-press method, which is evolved from the prepreg molding hot-press method. Since the prepreg integrated molding hot-press method would weaken the interlayer bonding strength of the face sheets, and the radial pressure of the fiber rod was difficult to ensure, the research team designed a special mold and used a secondary molding hot-press method to prepare a pyramidal composite lattice sandwich structure. Considering the characteristics of the mold assembly, it is often advantageous to form lattice cores with symmetrical forms, such as pyramids, eggshells and other even-numbered trusses. The core is integrally formed, and sufficient molding pressure can be applied through the upper and lower grids. The core of the pyramidal composite lattice structure prepared by the secondary molding hot-press process does not require mechanical processing and interlocking assembly, and the fiber is continuous in the core, which can significantly improve the compression and buckling resistances of the fiber trusses. On this basis, Yang et al. [30,31] using the secondary molding hot-press method prepared a pyramidal lattice sandwich structure with viscoelastic layer and carbon fiber composite sandwich cylindrical shell with pyramidal truss-like cores. The preparation flowchart is shown in Figure 3.7: (1) laying the pre-cut size composite prepregs in the mold cavity. (2) Assembling and tightening the rib shape mold to complete the pyramidal truss-like cores. (3) Laying the composite prepregs into the mold to form the outer face sheet. (4) Assembling and

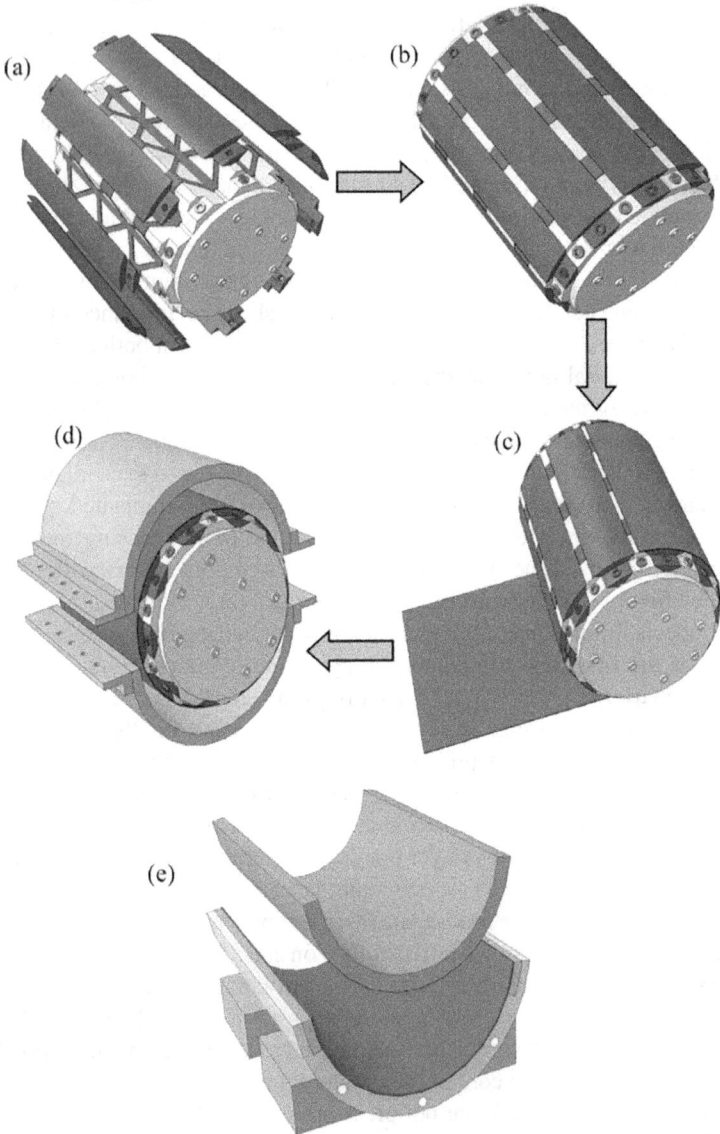

FIGURE 3.7 The flowchart of the secondary molding hot-press method used in the fabrication of composite sandwich cylindrical shells. (Yang et al. [31].)

tightening the outer cylindrical shape mold. (5) Laying the composite prepregs into the inner cylindrical shape and assembling to form the inner face sheet. The novel hot-press molding method was also applied to fabricate the all-composite corrugated sandwich cylindrical shells by Yang et al. [32]. With these previous works, Li et al. [33] cut and assembled the composite corrugated sandwich cylindrical shells and then used hot-press molding method to fabricate the sandwich cylindrical panels made of bidirectional corrugated strip cores. Figure 3.8 shows the fabrication process: (1) cutting the axial and (2) the circular corrugated cores into 2D corrugated partial strips, and (3) assembling, knitting and bonding the 2D corrugated partial strips with the two face sheets. In addition, Fan et al. [34] used the interleaved weaving process to prepare a carbon fiber composite lattice structure for the first time. This process is very similar to the previous wire weaving process. Yang et al. [17] and Li et al. [18] used a combination of cutting and embedding process and hot-press molding process to prepare a hybrid lattice sandwich structure with carbon fiber composite face sheets and aluminum alloy pyramidal truss lattice cores. Figure 3.3 shows the preparation process.

The sandwich structure of composite materials can reduce weight without losing strength compared with the lightweight sandwich structure of metal or aluminum alloy. Research on the performance characterization of composite lattice sandwich structures mainly focuses on the static properties, such as compression [18,20], shear [7] and bending [8]. There are still few studies on the damping, vibration isolation, impact and explosion resistance characteristics of composite sandwich structures. The damping characteristics of the composite pyramidal sandwich structures were improved by Yang et al. [30]. Viscoelastic layers

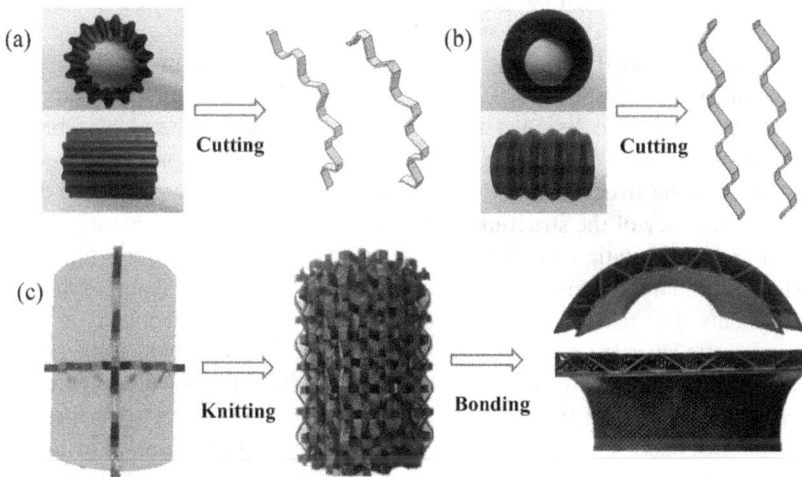

FIGURE 3.8 Preparation of bidirectional corrugated sandwich cylindrical panels. (a) Cutting the axial and (b) the circular corrugated cores into 2D corrugated partial strips, and (c) assembling, knitting and bonding the 2D corrugated partial strips with the two face sheets. (Li et al. [33].)

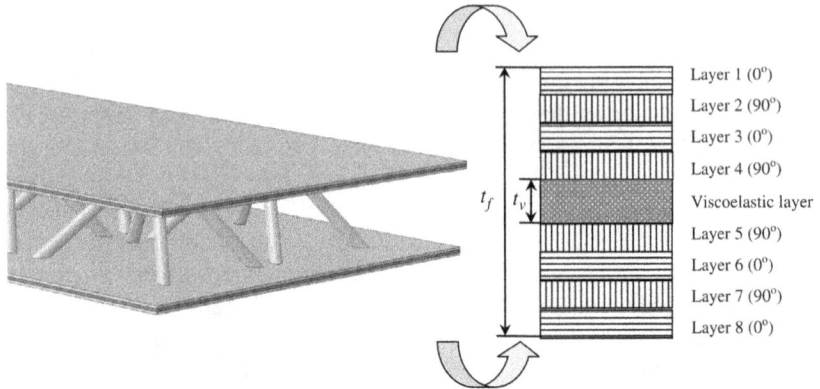

FIGURE 3.9 The fabrication process of sandwich panels whose face sheets are embedded with viscoelastic layers. (Yang et al. [30].)

were added to the face sheets of the composite pyramidal sandwich structures. This structure is shown in Figure 3.9. The natural frequencies, modal shapes and damping loss factors of the structure were measured through experiments. The damping characteristics of the hybrid sandwich structure were predicted based on the linear perturbation approach and the modal strain energy method. The results showed that the addition of the viscoelastic layers can significantly increase the damping loss factor, but the natural frequencies of the structure did not change significantly. As cylindrical shells are widely used in spacecraft, automobiles, ships and other fields, their working environment is relatively complicated, so it is necessary to study their basic vibration characteristics and damping characteristics. The natural frequencies and damping loss factors of the composite sandwich cylindrical shell with pyramidal truss-like cores were investigated under free boundary conditions [31]. There were three kinds of composite sandwich cylindrical shells were designed and prepared by carbon and glass fiber fabrics. It was proved that material mixing contributes to the improvement of the damping loss factor of the structure and the fiber ply angle had a greater influence on the natural frequency of the structure. Furthermore, Yang et al. [17] investigated the modal and quasi-static compressive performances of hybrid carbon fiber composite pyramidal truss sandwich panels with multiple damping configurations by experiments. The results showed that, when comparing the specimens in which a viscoelastic layer is inserted in the face sheets and those in which the cavities of the core are filled with silicone rubber, the structure filled with polyurethane foam in the cavities of the core can achieve the highest damping performance with the least mass. Taking advantage of material mixing and inspired by the idea of bionics, Li et al. [18] designed and prepared multilayer gradient hybrid carbon fiber composite pyramidal truss sandwich panels. The face sheets were made of orthogonal woven carbon fiber resin fabrics, and the truss cores were made of an aluminum alloy. The structural gradient design was realized by controlling the width of the core truss. The compression and hysteresis characteristics of the multilayer gradient hybrid structure were tested through experiments, and the compression failure mechanism of the structure was revealed with simulation. It was proved once again that adding foam to the core can improve the compressive strength of the structure and significantly improve the energy absorption performance of the structure. The hysteretic loops of the structures with and without foam are shown in Figure 3.10. The structural gradient caused the deformation processes of those structures were distinctly different under compression loads.

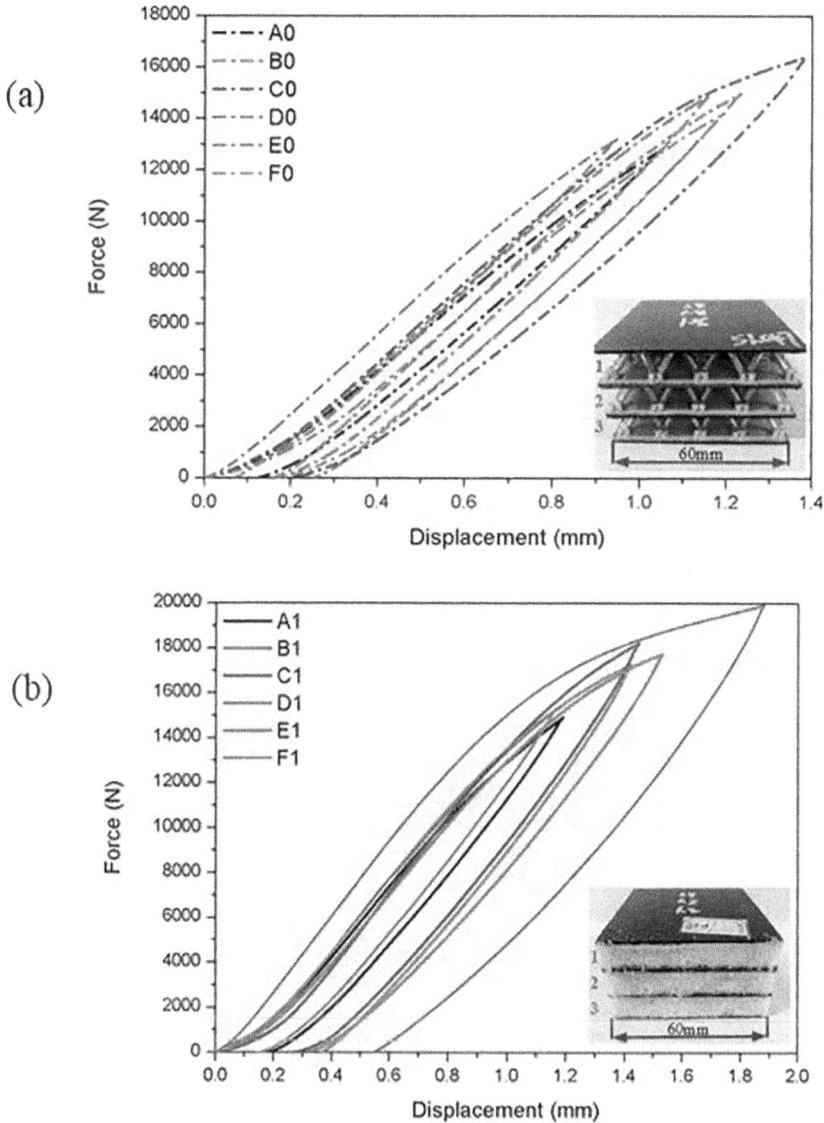

FIGURE 3.10 The hysteretic loops of the structures with and without foam. (Li et al. [18].)

By comparing the results of simulations and experiments, it was found that the face sheet which nearby the truss core with the thinnest truss was easier to fail, and the thinner the trusses, the more easier the core failure. The mechanism involved in these phenomena is that the large slenderness ratio causes the buckling of the truss. What's more, due to the small cross-sectional area of the truss, the stress of the face sheets where the node is located is great, so the face sheets fail first. It indicated that optimizing the design of the structural gradient and filling the cores with polyurethane foam were effective approaches to improving the hysteretic damping of the structure without sacrificing the mechanical properties. Moreover, Yang et al. [35] studied the impact properties of single-layer hybrid lightweight composite sandwich panels with aluminum alloy pyramidal truss cores. Due to the mixing of materials, the shock wave was effectively blocked. It is worth mentioning that the influence of several defects which easily appear during the preparation process on the structural modal characteristics and damping characteristics was investigated [36,37], and it was proved that the modified damage index method was more effective and accurate in locating the defects. This method was useful in vibration-based nondestructive evaluation of composite lattice sandwich structures. Based on the above description, it can be found that the key to restricting difficulties in the application and development of the lattice sandwich structure is the matching of the fabrication approach and the structural performance characteristics. In order to improve the structural performance, the fabrication approach needs to be continuously improved. The preparation cost of the metal lattice sandwich structure needs to be further reduced, and the preparation of the composite lattice structure needs to solve the problem of face sheet–core interface strength. In addition, the fabrication approach needs to be further simplified to achieve batch and standardized production. Moreover, it is anticipated to realize the preparation of structures with diverse topological configurations or large-size structural parts. The research on fatigue, vibration, damping, heat transfer, shock, defect sensitivity and other issues is still in the initial stage, and research work in this area needs to be strengthened in the future. In terms of structural topological configuration, it is necessary to develop new structural forms such as new lattice structures, especially for the large-size flat and curved lattice structures. The multifunctional integrated design of lightweight sandwich structures has become a trend, and a lattice sandwich structure with multiple functions is an urgent need.

REFERENCES

1. Hutchinson JW, He MY. Buckling of cylindrical sandwich shells with metal foam cores. *International Journal of Solids and Structures*, 2000, 37(46): 6777–6794.
2. Evans AG, Hutchinson JW, Fleck NA, et al. The topological design of multifunctional cellular metals. *Progress in Materials Science*, 2001, 46(3): 309–327.
3. Dong L, Deshpande V, Wadley H. Mechanical response of Ti–6Al–4V octet-truss lattice structures. *International Journal of Solids and Structures*, 2015, 60–61: 107–124.

4. Yang JS, Zhang WM, Yang F, et al. Low velocity impact behavior of carbon fibre composite curved corrugated sandwich shells. *Composite Structures*, 2020, 238: 112027.
5. Yang LH, Chen ZB, Dong YL, et al. Ballistic performance of composite armor with dual layer piecewise ceramic tiles under sequential impact of two projectiles. *Mechanics of Advanced Materials and Structures*, 2020: 1–14. Doi: 10.1080/15376494.2020.1749737
6. Yang W, Xiong J, Feng LJ, et al. Fabrication and mechanical properties of three-dimensional enhanced lattice truss sandwich structures. *Journal of Sandwich Structures & Materials*, 2018, 22(5): 1594–1611.
7. Feng LJ, Xiong J, Yang LH, et al. Shear and bending performance of new type enhanced lattice truss structures. *International Journal of Mechanical Sciences*, 2017, 134: 589–598.
8. Taghipoor H, Damghani NM. Axial crushing and transverse bending responses of sandwich structures with lattice core. *Journal of Sandwich Structures & Materials*, 2018, 22(3): 572–598.
9. Li S, Yang JS, Wu LZ, et al. Vibration behavior of metallic sandwich panels with Hourglass truss cores. *Marine Structures*, 2019, 63: 84–98.
10. Yang W, Xiong J, Wu LZ, et al. Methods for enhancing the thermal properties of epoxy matrix composites using 3D network structures. *Composites Communications*, 2019, 12: 14–20.
11. Wang DW, Ma L. Sound transmission through composite sandwich plate with pyramidal truss cores. *Composite Structures*, 2017, 164: 104–117.
12. Sypeck D J, Wadley HNG. Multifunctional microtruss laminates: Textile synthesis and properties. *Journal of Materials Research*, 2001, 16(3): 890–897.
13. Kang KJ. Wire-woven cellular metals: The present and future. *Progress in Materials Science*, 2015, 69. 213–307.
14. Chiras S, Mumm DR, Evans AG, et al. The structural performance of near-optimized truss core panels. *International Journal of Solids & Structures*, 2002, 39(15): 4093–4115.
15. Sypeck JD, Wadley HNG. Cellular metal truss core sandwich structures. *Advanced Engineering Materials*, 2002, 4: 10.
16. Kooistra G. Compressive behavior of age hardenable tetrahedral lattice truss structures made from aluminium. *Acta Materialia*, 2004, 52(14): 4229–4237.
17. Yang JS, Ma L, Schmidt R, et al. Hybrid lightweight composite pyramidal truss sandwich panels with high damping and stiffness efficiency. *Composite Structures*, 2016, 148: 85–96.
18. Li S, Yang JS, Schmidt R, et al. Compression and hysteresis responses of multilayer gradient composite lattice sandwich panels. *Marine Structures*, 2021, 75: 102845.
19. Finnegan K, Kooistra G, Wadley HNG, et al. The compressive response of carbon fiber composite pyramidal truss sandwich cores. *International Journal of Materials Research*, 2007, 98(12): 1264–1272.
20. Feng LJ, Wu LZ, Yu GC. An Hourglass truss lattice structure and its mechanical performances. *Materials & Design*, 2016, 99: 581–591.
21. Feng LJ, Wei GT, Yu GC, et al. Underwater blast behaviors of enhanced lattice truss sandwich panels. *International Journal of Mechanical Sciences*, 2019, 150: 238–246.
22. Yang W, Xiong J, Feng LJ, et al. Fabrication and mechanical properties of three-dimensional enhanced lattice truss sandwich structures. *Journal of Sandwich Structures & Materials*, 2020, 22(5): 1594–1611.

23. Hu B, Wu LZ, Xiong J, et al. Mechanical properties of a node-interlocking pyramidal welded tube lattice sandwich structure. *Mechanics of Materials*, 2019, 129: 290–305.

24. Yang LH, Han X, Feng LJ, et al. Numerical investigations on blast resistance of sandwich panels with multilayered graded hourglass lattice cores. *Journal of Sandwich Structures & Materials*, 2018, 22(7): 2139–2156.

25. Yang LH, Sui L, Li XY, et al. Sandwich plates with gradient lattice cores subjected to air blast loadings. *Mechanics of Advanced Materials and Structures*, 2021, 28(13):1355–1366.

26. Yang LH, Sui L, Dong YL, et al. Quasi-static and dynamic behavior of sandwich panels with multilayer gradient lattice cores. *Composite Structures*, 2021, 255: 112970.

27. Wang B, Wu LZ, Jin X, et al. Experimental investigation of 3D sandwich structure with core reinforced by composite columns. *Materials & Design*, 2010, 31(1): 158–165.

28. Wang B, Wu LZ, Ma L, et al. Fabrication and testing of carbon fiber reinforced truss core sandwich panels. *Journal of Materials Science and Technology -Shenyang*, 2009, 25(4): 547–550.

29. Xiong J, Ma L, Wu LZ, et al. Mechanical behavior and failure of composite pyramidal truss core sandwich columns. *Composites Part B: Engineering*, 2011, 42(4): 938–945.

30. Yang JS, Xiong J, Ma L, et al. Vibration and damping characteristics of hybrid carbon fiber composite pyramidal truss sandwich panels with viscoelastic layers. *Composite Structures*, 2013, 106(12): 570–580.

31. Yang JS, Xiong J, Ma L, et al. Study on vibration damping of composite sandwich cylindrical shell with pyramidal truss-like cores. *Composite Structures*, 2014, 117: 362–372.

32. Yang JS, Xiong J, Ma L, et al. Modal response of all-composite corrugated sandwich cylindrical shells. *Composites Science and Technology*, 2015, 115: 9–20.

33. Li S, Yang JS, Chen SY, et al. Modal response and vibration attenuation of carbon fiber composite sandwich cylindrical panels made of bi-directional corrugated strip cores. *Mechanics of Advanced Materials and Structures*, 2020, 1–15. Doi: 10.1080/15376494.2020.1758256

34. Fan HL, Meng FH, Yang W. Mechanical behaviors and bending effects of carbon fiber reinforced lattice materials. *Archive of Applied Mechanics*, 2006, 75(10–12): 635–647.

35. Yang JS, Chen SY, Li S, et al. Dynamic responses of hybrid lightweight composite sandwich panels with aluminium pyramidal truss cores. *Journal of Sandwich Structures & Materials*, 2021, 23(6): 2176–2195.

36. Yang JS, Ma L, Chaves-Vargas M, et al. Influence of manufacturing defects on modal properties of composite pyramidal truss-like core sandwich cylindrical panels. *Composites Science and Technology*, 2017, 147: 89–99.

37. Yang JS, Liu ZD, Schmidt R, et al. Vibration-based damage diagnosis of composite sandwich panels with bi-directional corrugated lattice cores. *Composites Part A: Applied Science and Manufacturing*, 2020, 131: 105781.

4 Composite Sandwich Panels with the Metallic Facesheets

Lin Feng Ng
Universiti Teknologi Malaysia

Kathiravan Subramaniam
Universiti Teknikal Malaysia Melaka

CONTENTS

4.1 Introduction 61
4.2 Classification of FMLs 63
4.3 Composite Sandwich Laminates 66
4.4 Adhesive Bonding of Metal–Composite Interface 68
4.5 The Manufacturing Process of FMLs 70
4.6 Conclusions 70
References 71

4.1 INTRODUCTION

Aluminium is one of the most popular metallic alloys which has been widely utilised in a broad range of applications primarily due to its high strength-to-weight ratio. Its applications mainly involve aircraft components, automotive structures, constructions, marine components, military equipment, electrical conductors, etc. (Yuvanarasimman and Malayalamurthi 2018). Lightweight is the driving factor that makes aluminium the successor to stainless steel. The high strength-to-weight ratio characteristic of aluminium improves energy efficiency, particularly in the transportation sectors. However, the stringent environmental rules and regulations have aroused the desire to further reduce the structural weight of aerospace and automotive components. One of the main drawbacks of these metallic alloys is their poor fatigue resistance. This drawback urges aerospace industries to search for advanced alternative materials to substitute metallic alloys.

The high demand for superior fatigue resistance materials has successfully ignited the development of a new range of sandwich materials comprising composite and thin metal layers, namely fibre metal laminates (FMLs). The aerospace industry triggers the attempt to develop FMLs to resolve the poor fatigue crack resistance of the metallic alloys. It was shown that the fatigue crack growth rate of FMLs is tremendously lower than that of metallic alloys due to the fibre bridging mechanism. The stiffness of the fibres tends to restrain the crack tip opening during the crack propagation, which reduces the stress intensity factor, and thus the fatigue crack growth rate of FMLs is lower than that of metallic alloys. It was reported that the fatigue crack growth rate is 10–100 times lower than aluminium (Vogelesang and Vlot 2000). Besides that, the previous research works have shown the excellent impact properties of these sandwich materials. The development of FMLs has successfully combined the high impact and fatigue properties inherited from the metals and composites, respectively. The sandwich concept in FMLs infers that the superiority of one constituent could offset the disadvantage of another constituent, developing an advanced material with enhanced performance. Apart from the excellent fatigue and impact properties, FMLs also offer additional advantages such as good corrosion resistance, enhanced fire resistance, lightweight and good residual and blunt notch strength (Alderliesten and Homan 2006; Abdullah, Prawoto, and Cantwell 2015). The lightweight characteristic of FMLs is another attractive feature that can improve the energy efficiency of an automobile. It had been shown that the weight of Airbus A380 has been successfully reduced up to 794 kg with the use of GLARE as the fuselage materials (Huang et al. 2015).

Various series of FMLs have been successfully developed, mainly depending on the different types of fibre and metal series. Aramid fibre-reinforced aluminium laminates (ARALL), carbon fibre-reinforced aluminium laminates (CARALL) and glass fibre-reinforced aluminium laminates (GLARE) are the three series of commercially available FMLs. ARALL is the first generation of FMLs which was introduced at Delft University in 1978. ARALL, as depicted in Figure 4.1, consists of aramid fibre prepreg bonded with aluminium layers. After that, carbon fibre was incorporated into FMLs to improve the mechanical performance of this sandwich material to a greater extent. However, the coalescence of carbon fiber and aluminium alloys might result in galvanic corrosion.

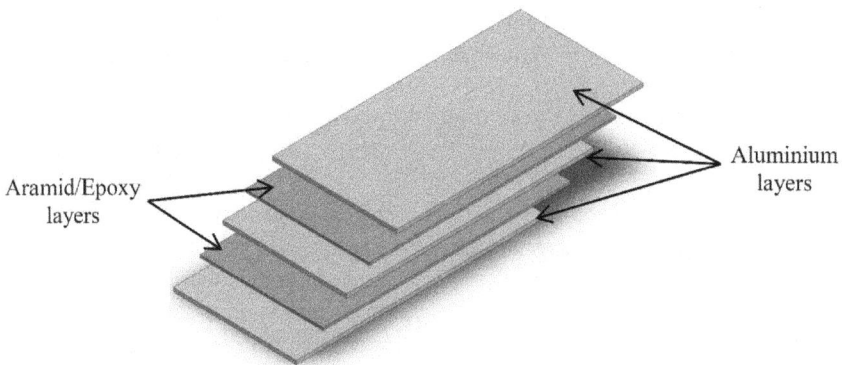

FIGURE 4.1   Schematic diagram of ARALL.

Finally, aramid fibre was replaced with glass fibre in the FMLs, leading to the third-generation FMLs, GLARE. Due to the excellent fatigue crack resistance of FMLs, these sandwich materials have been employed in the aerospace industries as the materials for upper fuselage skin and cargo interior. In fact, composite materials have been dominated by glass fibres as they are inexpensive and have high mechanical strength (Ng et al. 2020).

To date, FMLs have been dominated by both synthetic fibres and thermoset polymers. Although thermoset polymers have relatively high mechanical properties compared to thermoplastic polymers, the long processing time of thermoset polymers is considered their main disadvantage, which reduces the productivity of the materials and increases the production cost. Moreover, thermoset polymers are non-biodegradable and non-recyclable, leading to waste production and accumulation after their service life. Unlike thermoset polymers, thermoplastic polymers are gaining acceptance of a broad research community as they are recyclable, and their processing time is shorter since it only involves heating to processing temperature and cooling to ambient temperature. From the viewpoint of environmental-friendliness, the recyclability characteristic of thermoplastic polymers is particularly promising as it can avoid waste production and accumulation on the earth. Furthermore, a number of literature studies have shown that thermoplastic-based FMLs encompass superior interlaminar fracture toughness (Abdullah and Cantwell 2006; Cortes and Cantwell 2007). The achievement in the conventional FMLs has inspired the continuous exploration of the feasibility of thermoplastic-based FMLs in engineering applications. When further looking into the thermoplastic polymers, polypropylene is among the most popular semicrystalline polymers for composite materials due to its low density and low cost. Currently, polypropylene is widely employed in the transportation sectors for the interior part of a vehicle.

4.2 CLASSIFICATION OF FMLs

During the past decades, an effort has been made to search for high fatigue resistance and lightweight materials to replace metallic alloys in the aerospace industries. In this regard, composites have been found as an alternative candidate that meets the requirement for aerospace applications. However, the fracture toughness of composite materials is lower than that of metallic alloys (Sinmazçelik et al. 2011). In 1978, an achievement was reached in the field of sandwich materials by introducing aramid fibre into the metal–composite sandwich materials, resulting in an improvement in the fatigue resistance of the materials. As a result, ARALL was introduced at the Delft University of Technology as the first generation of FMLs. Later, different grades of ARALL, as shown in Table 4.1, were standardised. In an attempt to enhance the mechanical performance of FMLs, high-strength carbon fibre was incorporated into the sandwich materials, which was named CARALL. The idea of adding carbon fibre in the FMLs is due to the superior stiffness of the fibre, which could not be achieved by both glass and aramid fibres. Nevertheless, it was found that galvanic corrosion of aluminium alloys

TABLE 4.1

Several Grades of GLARE and ARALL

Grade	Fibre Orientation in Each Fibre Layer	Aluminium Thickness	Type of Aluminium
GLARE			
GLARE 1	0°/0°	0.3–0.4	7075-T6
GLARE 2A	0°/0°	0.2–0.5	2024-T3
GLARE 2B	90°/90°	0.2–0.5	2024-T3
GLARE 3	0°/90°	0.2–0.5	2024-T3
GLARE 4A	0°/90°/0°	0.2–0.5	2024-T3
GLARE 4B	90°/0°/90°	0.2–0.5	2024-T3
GLARE 5	0°/90°/90°/0°	0.2–0.5	2024-T3
GLARE 6A	+45/–45	0.2–0.5	2024-T3
GLARE 6B	–45/+45	0.2–0.5	2024-T3
ARALL			
ARALL 1	0°/0°	0.3	7075-T6
ARALL 2	0°/0°	0.3	2024-T3
ARALL 3	0°/0°	0.3	7475-T761
ARALL 4	0°/0°	0.3	2024-T81
CARALL			
CARALL 1	0°/0°	0.2–0.5	2024-T3
CARALL 2	90°/90°	0.2–0.5	2024-T3

has been the major drawback of CARALL. Indeed, carbon fibre is conductive and electrochemically noble, and hence the contact between metals and carbon fibre in an electrolyte leads to the occurrence of galvanic corrosion of the metals, which eventually deteriorates the overall mechanical strength of CARALL. Due to the high risk of galvanic corrosion between carbon fibre and aluminium alloys, increasing focus has been given to the development of GLARE. GLARE was developed in the year 1990 by adopting glass fibre to supersede aramid fibre in the FMLs. GLARE has been considered as the most successful sandwich material over ARALL and CARALL. GLARE exhibits excellent mechanical properties after impact compared to other materials (Meola, Boccardi, and Carlomagno 2017). The impact resistance of GLARE is even superior to both monolithic aluminium and carbon laminates (Abdullah, Prawoto, and Cantwell 2015). The excellent impact properties of GLARE are primarily attributed to the high failure strain of glass fibre in the laminates.

The early exploration of the lightweight and high damage tolerance materials had led to the development of three main types of FMLs: GLARE, ARALL and CARALL. GLARE, ARALL and CARALL can be subdivided into several respective grades. The grades of FMLs are defined according to the metal alloys, fibre orientations and the number of fibre layers. As shown in Table 4.1, the majority of GLARE are mainly based on aluminium 2024-T3 with a thickness of 0.2–0.5 mm. In comparison with other grades, GLARE 1 is incorporated with high-strength aluminium 7075 with a thickness of 0.3–0.4 mm. The number of fibre layers in GLARE differs according to the grades. GLARE 1, 2, 3 and 6 are composed of two layers of glass fibre prepreg, whereas GLARE 4 and 5 are incorporated with three and four layers of glass fibre prepreg, respectively. When looking into ARALL, the grades are defined based on the types of aluminium layers in the FMLs. Each of the ARALL has the same number of fibre layers, fibre orientations and aluminium thickness.

In addition to FMLs based on aluminium metallic layers, FMLs based on other metallic layers such as titanium and magnesium have also been developed (Cortés and Cantwell 2005; Ali et al. 2016; Zhang et al. 2017). The purpose of developing titanium-based FMLs is similar to conventional FMLs. Titanium-based FMLs were developed to improve the fatigue resistance of titanium alloys (Kolesnikov, Herbeck, and Fink 2008). It is envisaged since the fatigue crack growth resistance of adhesively bonded laminated structures is better than the single monolithic metal sheet. The adhesive in the laminated structure acts as a crack divider that impedes the crack from propagating to the neighbouring uncracked sheet, leading to the decline in the fatigue crack growth rate. Thus, the use of titanium-based FMLs in marine and offshore applications has been continuously augmenting because of their innumerable virtues such as high corrosion resistance, lightweight and high specific mechanical properties (Golaz et al. 2013; He, Chen, and Yang 2015). Titanium alloys are particularly favourable in marine sectors because of their high corrosion resistance. The oxide layer formed on the titanium surface is thicker and more durable compared to aluminium. It is worth mentioning that the combination of titanium and carbon fibre will not result in galvanic-related issues owing to their good electrochemical compatibility.

On the other hand, the potential of FMLs based on magnesium alloys and carbon fibre has been widely explored due to the various advantages offered by magnesium alloys. Low density, excellent corrosion resistance and superior electromagnetic shielding ability are the benefits provided by magnesium alloys (Cortés and Cantwell 2004). The density of magnesium alloys is 30% lower than that of aluminium alloys. In spite of the several advantages offered by magnesium alloys, possible applications of magnesium-based FMLs have still not yet been reported. In an attempt to employ magnesium alloys in FMLs, various limitations have arisen, which retards the use of magnesium-based FMLs in several possible applications. Similar to aluminium alloys, the combination of magnesium alloys and carbon fibres results in galvanic corrosion (Pan et al. 2015). Besides that, Pärnänen et al. (2012) showed that the energy required to fracture the magnesium/glass fibre-based FMLs is lower than that of GLARE, implying that the impact resistance of magnesium-based FMLs is lower than that of GLARE. The relatively low mechanical strength of magnesium alloys compared to aluminium alloys is also considered the shortcoming that limits the use of magnesium alloys in FMLs. In the case of galvanic corrosion, some techniques can be used to improve the galvanic corrosion resistance between carbon fibre and metal alloys. Pre-treatments such as anodising, laser or ion surface irradiation and plasma electrolytic oxidation are being used for the surface modification of metal alloys to improve the galvanic corrosion resistance (Rhee et al. 2003; Torres-Acosta and Sen 2005; Alfano et al. 2012; Peng and Nie 2013; Pan et al. 2015).

4.3   COMPOSITE SANDWICH LAMINATES

Composite sandwich laminates comprise thin facesheets and thick core constituent. The core materials are generally low-density and lightweight materials to improve the specific mechanical properties of the structure. The facesheet materials can be high-strength fibre-reinforced composites or metal layers. In addition to the standardised FMLs consisting of metal layers and the fibre-reinforced epoxy composites, several types of core materials such as honeycomb, balsa wood, and polymeric and metallic foam have been used to form the sandwich materials. Honeycomb is among the most widely explored core materials of the composite sandwich laminates because of its lightweight characteristic, high strength-to-weight ratio and excellent impact properties. The core materials maintain a certain distance between the facesheets and sustain the shear deformation when subjected to out-of-plane loadings. The combination of high-strength facesheets and lightweight core materials entails the sandwich materials with high bending stiffness and lightweight behaviour. Griškevičius et al. (2010) studied the impact properties of composite sandwich laminates with polypropylene honeycomb core and composite facesheets. They reported that the majority of the impact energy was absorbed by the honeycomb core compared to the composite facesheets. The results showed that the honeycomb core has a major impact on the impact properties of the sandwich materials. Apart from the types of facesheets and core materials, the performance of the sandwich materials can be further enhanced by changing the thickness of facesheet and core materials. The sandwich structures have an excellent moment of inertia due to the high strength and modulus of facesheets and excellent bending and shear rigidity, resulting in the superior structural integrity of the materials. The material selection to develop the composite sandwich materials depends on the requirements of particular applications. Fibre-reinforced composites are commonly used for structural applications owing to their high mechanical strength (Guo et al. 2015).

Besides the thickness of the facesheets and core materials, the mechanical properties of sandwich materials are governed by the lay-up sequences. 2/1, 3/2 and 4/3 are typical lay-up sequences of FMLs, which refer to the ratio of metal to composite layers. It had been shown that the mechanical properties of FMLs are highly affected by the lay-up sequences. When looking into GLARE, certain grades show multiple layers of composites and aluminium alloy. Ammar et al. (2019) investigated the effect of thickness of aluminium sheets (0.5, 0.7 and

1.0 mm) and the lay-up sequences (2/1, 3/2 and 4/3) on the tensile properties of FMLs based on chopped glass fibre-reinforced epoxy composites. As expected, the tensile properties of the FMLs increased with the increase in aluminium thickness because of the rise in the metal volume fraction. However, it is interesting to note that the tensile properties of FMLs dropped when increasing the number of aluminium and composite layers in FMLs. The decrease in tensile properties of FMLs with 3/2 lay-up compared to 2/1 lay-up is attributed to the decrease in metal volume fraction in 3/2 lay-up. The FMLs with 4/3 lay-up showed the lowest tensile properties, mainly due to the highest void content in the composites. Fan, Guan and Cantwell (2011) studied the low-velocity impact of glass fibre-reinforced epoxy FMLs with varying lay-up sequences (2/1, 3/2 and 4/3). They reported that the increase in metal and composite layers in FMLs increased the perforation energy. Nonetheless, it had been reported that the specific perforation energy of FMLs with 2/1 lay-up was superior to 3/2 and 4/3 lay-ups (Abdullah and Cantwell 2006).

Despite the outstanding mechanical properties of synthetic fibres, they possess limitations of non-degradability and difficulty in disposal, leading to immense pollution to the environment (Senthilkumar et al. 2018). Recently, natural fibres have started gaining wide acceptance as green materials to be embedded in composite materials (Senthilkumar et al. 2019). The utilisation of natural fibres is not limited to composite materials; they can also be used in metal–composite sandwich structures. Apart from the green characteristics of natural fibres, low fatigue sensitivity, lightweight and low energy consumption are the driving factors that make natural fibres gain their popularity for industrial applications. Feng et al. (2019) reported that incorporating natural fibres into FMLs could reduce the fatigue sensitivity coefficient, indicating the reduction in the fatigue strength degradation of FMLs. It is interesting to note that the density of natural fibres is approximately half of that of glass fibres, inferring that the energy consumption in the transportation sectors could be drastically reduced. Therefore, the energy efficiency of vehicles can be considerably enhanced. Besides that, it was also found that natural fibre-based FMLs exhibited higher absolute and specific mechanical strength than their composites (Feng et al. 2020). Today, kenaf, hemp, jute, flax, sisal and oil palm are among the most extensively explored natural fibres in the research communities. Figure 4.2 shows random short and plain weave kenaf fibre.

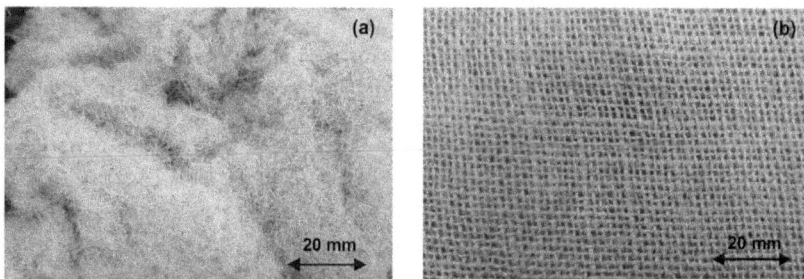

FIGURE 4.2   Kenaf fibre: (a) random short and (b) plain weave.

4.4   ADHESIVE BONDING OF METAL–COMPOSITE INTERFACE

Numerous factors such as fibre content, fibre properties, matrix properties, fibre orientation and types of metallic alloys have a decisive effect on the fatigue and impact properties of FMLs. However, FMLs exhibit excellent fatigue crack resistance and impact properties only when the adhesion of the metal–composite interface has reached an acceptable level. The structural integrity of the metal–composite interface also has a significant role in determining the performance of FMLs. Since the bridging mechanism in FMLs is associated with the stress transfer from the metal to the fibre, an excellent bonding is particularly vital to allow the efficient bridging effect. A poor adhesion level leads to the delamination between metallic and composite layers, deteriorating the mechanical performance of the materials. In general, different materials can be joined together using mechanical fastening or adhesive bonding methods. Although mechanical fastening has been shown to have great potential to assemble different materials, adhesive bonding is still the most commonly applied technique due to its several advantages. In comparison with mechanical fastening, the adhesive bonding method offers lower-weight structural joint, less stress concentration points and high corrosion resistance (Boutar et al. 2016). Moreover, additional benefits such as enhanced stiffness, improved damage tolerance and less vibration can be obtained through the adhesive bonding method (Barnes and Pashby 2000). The adhesive joining method has been extensively utilised in the aerospace industries over the past decades, primarily due to its low cost and lightweight characteristics. Although there are innumerable benefits that can be obtained through adhesive bonding, the failure of adhesive bonding is generally related to debonding and fracture, which could reduce its long-term performance.

The performance of adhesive bonding is governed by several factors such as material properties, adhesive thickness and surface roughness (Banea and Da Silva 2009). Hence, these parameters are critical in engineering applications to achieve optimum strength. Pascoe et al. (2020) revealed that an increase in adhesive thickness could increase the fatigue crack growth rate of epoxy adhesives. There is no doubt that the surface roughness of the adherend has a great influence on the bonding strength. A number of research studies revealed that the surface roughness of the adherend has a positive impact on the bonding strength (Budhe et al. 2015; Hwang et al. 2018; Ng et al. 2019). A rough surface can improve the bonding strength as microcolumn or mini-scarf joints increase the bonding area and allow the proper infiltration of the resin during the fabrication, creating a robust mechanical interlocking at the metal–composite interface. Thus, the surface treatment on the metal surface can improve the metal–composite adhesion and guarantee the structural integrity of FMLs. However, it should be emphasised that the increase in surface roughness beyond a critical limit may have a negative effect on the bonding strength. Once the surface roughness exceeds the critical limit, the resin does not infiltrate well to the mini-scarf joints, and there is a high possibility that the air can be entrapped in the asperities. Consequently, the inter-face is filled with not only the adhesive, but also the air which is entrapped at the interface due to the lack of wettability. This reduces the interaction area between the adhesive and the adherend, resulting in weak bonding strength. Boutar et al. (2016) concluded that the shear strength of the adhesively bonded aluminium laminates increased with the increase in surface roughness up to a critical value. The further increase in the surface roughness undermined the shear strength. Gonzalez-Canche et al. (2018) reported that surface pre-treatments improved the wettability and surface roughness of metal layers, thereby enhancing the tensile properties of FMLs.

The adhesion level of the metal–composite interface highly depends on the adhered surface condition. In order to have an optimum adhesion between the metal and the composites, pre-treatments are usually applied on the metal sur-face to improve the bonding site. Surface pre-treatment tends to eliminate the contaminants and activate the bonding sites, consequently improving bonding quality. Park et al. (2010) reported that surface treatment is the most criti-cal step which cannot be neglected. Several surface pre-treatment techniques can be applied on the metal surface of FMLs to maximise the effective bond-ing area. In the aerospace industry, mechanical, chemical, electrochemical, coupling agent and dry surface treatments are commonly employed for metal surface modification to improve the adhesion between metal and composite materials (Sinmazçelik et al. 2011; Feng and Dhar Malingam 2019). The clas-sification of surface pre-treatments of FMLs is demonstrated in Figure 4.3. Each surface pre-treatment aims at enhancing the metal–composite interfacial adhesion.

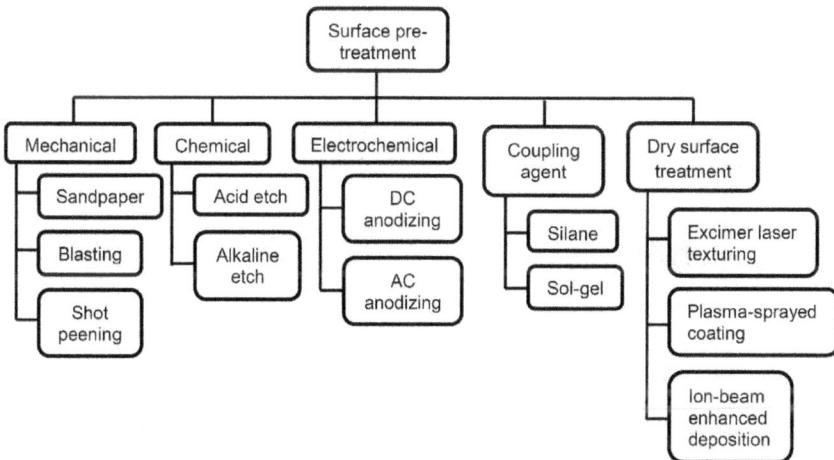

FIGURE 4.3 Classification of surface pre-treatment of FMLs.

4.5   THE MANUFACTURING PROCESS OF FMLs

Recently, thermoplastic-based FMLs have started gaining attraction of researchers and scientists due to their short processing duration and recyclability. The manufacturing process of thermoplastic-based FMLs involves only heating and cooling processes with applied pressure. However, the manufacturing process of thermoplastic- and thermoset-based FMLs is generally different. The standardised FMLs are prepared by stacking the metal and composite layers alternatively, followed by consolidation in the autoclave process. The autoclave pressure aids in removing the void content in the composites, thus improving the fibre volume fraction. The mechanical performance of the FMLs is greatly enhanced when the void content is minimised and fibre volume fraction is increased. The autoclave process has been the most common fabrication method for FMLs in aerospace industries, but this method often requires high processing cost. Due to the high cost and long processing time of this method, vacuum-assisted resin transfer moulding (VARTM) has been developed for FMLs production. The use of VARTM is considered more cost-effective due to the substitution of high-cost fibre/resin prepreg with the low-cost dry fabric (Ortiz de Mendibil et al. 2016).

Recently, there has been a growing interest in thermoplastic-based FMLs due to their relatively short processing time and the recyclability of thermoplastic matrices. However, the manufacturing process of thermoplastic-based FMLs is different from the conventional thermoset-based FMLs. Although the manufacturing process of thermoset- and thermoplastic-based FMLs is different, the surface pre-treatment of the metal layers prior to bonding is the critical step for both types of FMLs. In comparison with thermoset-based FMLs, the manufacturing process of thermoplastic-based FMLs is shorter and straightforward. Generally, the process involves heat compression moulding and rapid cooling. The metal layers are stacked alternatively with the composite layers in a mould. The adhesive films are incorporated at the metal–composite interfaces. The stack is then thermo-moulded for a certain duration, followed by cooling to ambient temperature. Inspection with the aids of ultrasound, X-rays, visual methods and mechanical tests is necessary for the thermoset- and thermoplastic-based FMLs to avoid any defects of the materials.

4.6   CONCLUSIONS

Composite sandwich materials with metallic facesheets have demonstrated their unique properties. The development of FMLs has resolved the weaknesses of poor impact properties in composite materials and high fatigue crack growth rate in metal alloys. FMLs offer advantages such as high damage tolerance and excellent fatigue crack resistance inherited from their individual constituents. In addition, the utilisation of FMLs to replace conventional alloys results in weight reduction, which directly improves the energy efficiency of vehicles. The major drawback of standardised FMLs is the long processing time due to the curing process for the thermoset matrix. In this context, there is a growing interest in thermoplastic-based FMLs due to their recyclability and ease of fabrication process. The virtues offered by thermoplastic polymer have largely incentivised the continuous exploration of the thermoplastic-based FMLs for engineering applications.

Surface pre-treatment of metal layers before bonding with composites is regarded as a single most critical step to ensure the structural integrity of FMLs. Surface pre-treatment guarantees excellent metal–composite adhesion in order to give an optimum mechanical performance to FMLs. Without proper surface pre-treatment, the weak interfacial adhesion results in delamination, which eventually deteriorates the mechanical performance of FMLs. The autoclave is the typical technique used to manufacture FMLs with a minimal amount of void or zero void. Nevertheless, the autoclave process in aerospace sectors to manufacture FMLs for the primary structure often requires high processing cost. Thus, the high processing cost of the autoclave process eventually hampers the mass production of FMLs. The introduction of the VARTM technique has provided an alternative avenue towards cost-effective FMLs. A key challenge of using VARTM is the ability to mimic the equivalent performance as the autoclave process. The development of an alternative manufacturing technique with low processing cost and high performance is the main challenge in the next millennium.

REFERENCES

Abdullah, M. R., and W. J. Cantwell. 2006. "The Impact Resistance of Fiber-Metal Laminates Based on Glass Fiber Reinforced Polypropylene." *Polymer Composites* 27: 700–708. doi:10.1002/pc.20240.

Abdullah, M. R., Y. Prawoto, and W. J. Cantwell. 2015. "Interfacial Fracture of the Fibre-Metal Laminates Based on Fibre Reinforced Thermoplastics." *Materials and Design* 66: 446–52. doi:10.1016/j.matdes.2014.03.058.

Alderliesten, R. C., and J. J. Homan. 2006. "Fatigue and Damage Tolerance Issues of Glare in Aircraft Structures." *International Journal of Fatigue* 28: 1116–23. doi:10.1016/j.ijfatigue.2006.02.015.

Alfano, M., G. Lubineau, F. Furgiuele, and G. H. Paulino. 2012. "Study on the Role of Laser Surface Irradiation on Damage and Decohesion of Al/epoxy Joints." *International Journal of Adhesion and Adhesives* 39: 33–41. doi:10.1016/j.ijadhadh.2012.03.002.

Ali, A., L. Pan, L. Duan, Z. Zheng, and B. Sapkota. 2016. "Characterisation of Seawater Hygrothermal Conditioning Effects on the Properties of Titanium-Based Fiber-Metal Laminates for Marine Applications." *Composite Structures* 158: 199–207. doi:10.1016/j.compstruct.2016.09.037.

Ammar, M. M., S. M. Osman, E. H. Hasan, and N. A. Azab. 2019. "A Tensile Characterisation of Random Glass Fiber/metal Laminates Composites." *Materials Research Express* 6: 085349. doi:10.1088/2053-1591/ab2994.

Banea, M. D., and L. F.M. Da Silva. 2009. "Adhesively Bonded Joints in Composite Materials: An Overview." *Journal of Materials: Design and Applications* 223: 1–18. doi:10.1243/14644207JMDA219.

Barnes, T. A., and I. R. Pashby. 2000. "Joining Techniques for Aluminum Spaceframes Used in Automobiles. Part II - Adhesive Bonding and Mechanical Fasteners." *Journal of Materials Processing Technology* 99: 72–79. doi:10.1016/S0924-0136(99)00361-1.

72 Sandwich Composites

Boutar, Y., S. Naïmi, S. Mezlini, and M. B. S. Ali. 2016. "Effect of Surface Treatment on the Shear Strength of Aluminium Adhesive Single-Lap Joints for Automotive Applications." *International Journal of Adhesion and Adhesives* 67: 38–43. doi:10.1016/j.ijadhadh.2015.12.023.

Budhe, S., A. Ghumatkar, N. Birajdar, and M. D. Banea. 2015. "Effect of Surface Roughness Using Different Adherend Materials on the Adhesive Bond Strength." *Applied Adhesion Science* 3. doi:10.1186/s40563-015-0050-4.

Cortés, P., and W. J. Cantwell. 2004. "Fracture Properties of a Fiber-Metal Laminates Based on Magnesium Alloy." *Journal of Materials Science* 39: 1081–83. doi:10.1023/B:JMSC.0000012949.94672.77.

Cortés, P., and W. J. Cantwell. 2005. "The Fracture Properties of a Fibre-Metal Laminate Based on Magnesium Alloy." *Composites Part B: Engineering* 37: 163–70. doi:10.1016/j.compositesb.2005.06.002.

Cortes, P., and W. J. Cantwell. 2007. "The Impact Properties of High-Temperature Fiber-Metal Laminates." *Journal of Composite Materials* 41: 613–32. doi:10.1177/0021998306065291.

Fan, J., Z. Guan, and W. J. Cantwell. 2011. "Structural Behaviour of Fibre Metal Laminates Subjected to a Low Velocity Impact." *Science China: Physics, Mechanics and Astronomy* 54: 1168–77. doi:10.1007/s11433-011-4261-9.

Feng, N. L., and S. D. Malingam. 2019. "Monotonic and Fatigue Responses of Fiber-Reinforced Metal Laminates." In *Mechanical and Physical Testing of Biocomposites, Fibre-Reinforced Composites and Hybrid Composites*, edited by Mohamad Jawaid, Mohamed Thariq, and Naheed Saba, 307–23. Woodhead Publishing. doi:10.1016/b978-0-08-102292-4.00016-3.

Feng, N. L., S. DharMalingam, K. A. Zakaria, and M. Z. Selamat. 2019. "Investigation on the Fatigue Life Characteristic of Kenaf/glass Woven-Ply Reinforced Metal Sandwich Materials." *Journal of Sandwich Structures and Materials* 21: 2440–55. doi:10.1177/1099636217729910.

Feng, N. L., S. D. Malingam, N. M. Ishak, and K. Subramaniam. 2020. "Novel Sandwich Structure of Composite-Metal Laminates Based on Cellulosic Woven Pineapple Leaf Fibre." *Journal of Sandwich Structures and Materials*. doi:10.1177/1099636220931479.

Golaz, B., V. Michaud, S. Lavanchy, and J. A. E. Månson. 2013. "Design and Durability of Titanium Adhesive Joints for Marine Applications." *International Journal of Adhesion and Adhesives* 45: 150–57. doi:10.1016/j.ijadhadh.2013.04.003.

Gonzalez-Canche, N. G., E. A. Flores-Johnson, P. Cortes, and J. G. Carrillo. 2018. "Evaluation of Surface Treatments on 5052-H32 Aluminum Alloy for Enhancing the Interfacial Adhesion of Thermoplastic-Based Fiber Metal Laminates." *International Journal of Adhesion and Adhesives* 82: 90–99. doi:10.1016/j.ijadhadh.2018.01.003.

Griškevičius, P., D. Zeleniakiene, V. Leišis, and M. Ostrowski. 2010. "Experimental and Numerical Study of Impact Energy Absorption of Safety Important Honeycomb Core Sandwich Structures." *Materials Science* 16: 119–23.

Guo, J., H. Lv, Z. Wang, X. Tang, W. Liang, and S. Zhang. 2015. "Thermo-Mechanical Property of Shape Memory Polymer/carbon Fibre Composites." *Materials Research Innovations* 19: 566–72. doi:10.1179/1432891715Z.0000000001750.

He, P., K. Chen, and J. Yang. 2015. "Surface Modifications of Ti Alloy with Tunable Hierarchical Structures and Chemistry for Improved Metal-Polymer Interface Used in Deepwater Composite Riser." *Applied Surface Science* 328: 614–22. doi:10.1016/j.apsusc.2014.12.081.

Huang, Y., J. Liu, X. Huang, J. Zhang, and G. Yue. 2015. "Delamination and Fatigue Crack Growth Behavior in Fiber Metal Laminates (Glare) under Single Overloads." *International Journal of Fatigue* 78: 53–60. doi:10.1016/j.ijfatigue.2015.04.002.

Hwang, J. H., C. K. Jin, M. S. Lee, S. W. Choi, and C. G. Kang. 2018. "Effect of Surface Roughness on the Bonding Strength and Spring-Back of a CFRP/CR980 Hybrid Composite." *Metals* 8: 716. doi:10.3390/met8090716.

Kolesnikov, B., L. Herbeck, and A. Fink. 2008. "CFRP/titanium Hybrid Material for Improving Composite Bolted Joints." *Composite Structures* 83: 368–80. doi:10.1016/j.compstruct.2007.05.010.

Meola, C., S. Boccardi, and G. M. Carlomagno. 2017. "Composite Materials in the Aeronautical Industry." *Infrared Thermography in the Evaluation of Aerospace Composite Materials*, 1–24. doi:10.1016/b978-1-78242-171-9.00001-2.

Ng, L. F., D. Sivakumar, X. J. Woo, S. Kathiravan, and I. Siva. 2019. "The Effects of Bonding Temperature and Surface Roughness on the Shear Strength of Bonded Aluminium Laminates Using Polypropylene Based Adhesive." *Journal of Advanced Manufacturing Technology* 13: 113–26.

Ng, L. F., S. D. Malingam, M. Z. Selamat, Z. Mustafa, and O. Bapokutty. 2020. "A Comparison Study on the Mechanical Properties of Composites Based on Kenaf and Pineapple Leaf Fibres." *Polymer Bulletin* 77: 1449–63. doi:10.1007/s00289-019-02812-0.

Ortiz de Mendibil, I., L. Aretxabaleta, M. Sarrionandia, M. Mateos, and J. Aurrekoetxea. 2016. "Impact Behaviour of Glass Fibre-Reinforced Epoxy/aluminium Fibre Metal Laminate Manufactured by Vacuum Assisted Resin Transfer Moulding." *Composite Structures* 140: 118–24. doi:10.1016/j.compstruct.2015.12.026.

Pan, Y., G. Wu, X. Cheng, Z. Zhang, M. Li, S. Ji, and Z. Huang. 2015. "Galvanic Corrosion Behaviour of Carbon Fibre Reinforced Polymer/magnesium Alloys Coupling." *Corrosion Science* 98: 672–77. doi:10.1016/j.corsci.2015.06.024.

Park, S. Y., W. J. Choi, H. S. Choi, H. Kwon, and S. H. Kim. 2010. "Recent Trends in Surface Treatment Technologies for Airframe Adhesive Bonding Processing: A Review (1995–2008)." *Journal of Adhesion* 86: 192–221. doi:10.1080/00218460903418345.

Pärnänen, T., R. Alderliesten, C. Rans, T. Brander, and O. Saarela. 2012. "Applicability of AZ31B-H24 Magnesium in Fibre Metal Laminates - An Experimental Impact Research." *Composites Part A: Applied Science and Manufacturing* 43: 1578–86. doi:10.1016/j.compositesa.2012.04.008.

Pascoe, J. A., N. Zavatta, E. Troiani, and R. C. Alderliesten. 2020. "The Effect of Bond-Line Thickness on Fatigue Crack Growth Rate in Adhesively Bonded Joints." *Engineering Fracture Mechanics* 229. doi:10.1016/j.engfracmech.2020.106959.

Peng, Z., and X. Nie. 2013. "Galvanic Corrosion Property of Contacts between Carbon Fiber Cloth Materials and Typical Metal Alloys in an Aggressive Environment." *Surface and Coatings Technology* 215: 85–89. doi:10.1016/j.surfcoat.2012.08.098.

Rhee, K. Y., S. G. Lee, N. S. Choi, and S. J. Park. 2003. "Treatment of CFRP by IAR Method and Its Effect on the Fracture Behavior of Adhesive Bonded CFRP/aluminum Composites." *Materials Science and Engineering A* 357: 270–76. doi:10.1016/S0921-5093(03)00207-7.

Senthilkumar, K., N. Saba, N. Rajini, M. Chandrasekar, M. Jawaid, S. Siengchin, and O. Y. Alotman. 2018. "Mechanical Properties Evaluation of Sisal Fibre Reinforced Polymer Composites: A Review." *Construction and Building Materials* 174: 713–29. doi:10.1016/j.conbuildmat.2018.04.143.

Senthilkumar, K., N. Rajini, N. Saba, M. Chandrasekar, M. Jawaid, and S. Siengchin. 2019. "Effect of Alkali Treatment on Mechanical and Morphological Properties of Pineapple Leaf Fibre/Polyester Composites." *Journal of Polymers and the Environment* 27: 1191–1201. doi:10.1007/s10924-019-01418-x.

Sinmazçelik, T., E. Avcu, M. Ö. Bora, and O. Çoban. 2011. "A Review: Fibre Metal Laminates, Background, Bonding Types and Applied Test Methods." *Materials and Design* 32: 3671–85. doi:10.1016/j.matdes.2011.03.011.

Torres-Acosta, A. A., and R. Sen. 2005. "Electrochemical Behavior of Coupled Carbon Fiber-Reinforced Polymer (CFRP) Rods in Concrete." *Journal of Applied Electrochemistry* 35: 529–37. doi:10.1007/s10800-005-1318-3.

Vogelesang, L. B., and A. Vlot. 2000. "Development of Fibre Metal Laminates for Advanced Aerospace Structures." *Journal of Materials Processing Technology* 103: 1–5. doi:10.1016/S0924-0136(00)00411-8.

Yuvanarasimman, P., and R. Malayalamurthi. 2018. "Studies on Fractures of Friction Stir Welded Al Matrix SiC-B4C Reinforced Metal Composites." *Silicon* 10: 1375–83. doi:10.1007/s12633-017-9614-1.

Zhang, X., Y. Zhang, Q. Y. Ma, Y. Dai, F. P. Hu, G. B. Wei, T. C. Xu, Q. W. Zeng, S. Z. Wang, and W. D. Xie. 2017. "Effect of Surface Treatment on the Corrosion Properties of Magnesium-Based Fibre Metal Laminate." *Applied Surface Science* 396: 1264–72. doi:10.1016/j.apsusc.2016.11.131.

5 Failure Behavior and Residual Strength of the Composite Sandwich Panels Subjected to Compression after Impact Testing

Dhaneshwar Mishra, Charanjeet Singh Tumrate, and Anoop Kumar Mukhopadhyay
Manipal University Jaipur

CONTENTS

5.1 Introduction 76
5.2 CAI Test Methods and Standards... 78
5.3 Experimental Investigations on CAI for Sandwich Composite Panels 79
    5.3.1 Non-Crimp Fabric (NCF) Sandwich Panels................................... 79
    5.3.2 CAI Behavior of Thin Laminates... 80
    5.3.3 CAI: Issue of Core Crushing Damage 81
    5.3.4 CAI by Column Compression... 81
    5.3.5 CAI of Honeycomb Core Structures... 82
    5.3.6 CAI: Issue of Low Temperature 82
    5.3.7 CAI Issue in Expendable Launch Vehicles.................................... 83
    5.3.8 CAI after Full-Scale Blast Experiments... 83
    5.3.9 Influence of Skin Thickness Increment on CAI 83
    5.3.10 Influence of Core Density Increment on CAI................................ 84
    5.3.11 Reinforcement Effect on CAI... 84
    5.3.12 Issue of Hybrid Face Sheet Effect on CAI.................................... 84
    5.3.13 Thickness and Ply-Stacking Sequence Effect on CAI 84
    5.3.14 Shape-Memory Alloy (SMA)–GFRP Hybrid Face Sheets and CAI Behavior... 85
5.4 Numerical Studies on Failure of Composite Sandwich Panels for Compression after Impact Loading 85
    5.4.1 Damage Zone Modeled as Equivalent Hole 86

DOI: 10.1201/9781003143031-5

        5.4.2   Modeling of Intralaminar and Interlaminar Damages 87
                5.4.2.1   Continuum Damage Model (CDM) 87
                5.4.2.2   Interlaminar Damage (Delamination) Modeling for
                          Composites... 89
                5.4.2.3   Modeling of Matrix Cracking and Interface
                          Delamination through Cohesive Elements 90
5.5   State-of-the-Art Numerical Models Available in the Literature to
        Study Intralaminar Failure and Delamination of Sandwich Composites..... 90
        5.5.1   Composite Panels and Laminates.. 90
        5.5.2   Foam Core Sandwich Panel under CAI..................................... 92
        5.5.3   CAI Test Model of Damaged Woven Fabric
                Sandwich Composite 92
        5.5.4   Stitched Composite under CAI 93
        5.5.5   Three-Dimensional Parametric Study for Impact
                Behavior of Honeycomb Sandwich Structure............................... 93
5.6   Single FE Model for Impact Damage and CAI 93
5.7   Conclusions.. 94
References.. 94

5.1   INTRODUCTION

The composite materials have become intrinsic part of engineering and have wide ranging applications in most of the industries. Their use is increasing rapidly due to their advantage of high strength-to-weight ratio. Sandwich composite structures made up of honeycomb core covered by thin face sheets are commonly used in aerospace, marine, automobile, and construction industries because of their high strength, stiffness, light weight, and ability to withstand extreme temperature conditions [1]. The increase in the use of composite materials as the primary structure material can be realized by the example of carbon fiber-reinforced plastic (CFRP) composite utilization of up to 50% in Boeing 787 Dreamliner [2,3]. The reduction in the overall weight of the transportation media including automotive, aerospace, and marine structures contributes to large amount of fuel saving, thereby reducing the pollution and saving the environment [4].

Sandwich panels are composites with outer skins made up of advanced composites such as metals (aluminum) or fiber-reinforced composites (FRPs) and honeycomb or foam-like lightweight core structure. The role of the facing is to withstand flexural load (bending) and compression, while the core withstands the shear loading and stabilizes the outer thin skin [5]. The mechanical properties of the sandwich composites depend on the core materials, its outer thin skin layer, and the manufacturing method employed to fabricate them [5]. The core material is lightweight; therefore, even it is thick, the sandwich composite structures are comparatively of less weight [5]. These composites also have good thermal insulating properties [1,5]. The application areas of these composites are ever increasing as there is further scope of reducing weight and saving cost with enhanced stiffness, strength, and thermal insulating properties.

Sandwich panels undergo various types of loadings such as bending, shear, and in-plane tensile or compressive loads during their operation. These loads acting in single mode or mixed mode (loading in real life is intrinsically mixed mode) give rise to various failure modes. Failure modes and their initiation and growth can be predicted by conducting various experiments and numerical studies. Possible failure modes include tensile or compressive failure of the facings, debonding at the core–facing interface, indentation failure under concentrated loads, core failure, wrinkling of the compression face, and global buckling [6]. Following initiation, a particular failure mode can lead to other modes and ultimately to the final failure path.

The impact loads whether with high velocity or low velocity cause damage to the sandwich panels. In case of high-velocity impact, the damage is critical and the failure can be seen by the naked eye. The effect of low-velocity impact (LVI) is important to understand as in many cases, the damage is highly localized in nature and barely visible (BV). These localized damages caused by LVIs can lead to significant loss of strength, especially compressive strength that can affect the overall performance [7].

Compression after impact (CAI) testing is widely carried out to understand the localized damage behavior leading to loss of residual strength in the composites subjected to compressive loading after impact. Research and investigations on CAI of different types of composites continuously highlight the need for developing a better understanding about this from experimental characterization standpoint [8], nondestructive evaluation standpoint [9], and the computational challenges [10]. The current research focuses on understanding the CAI behavior of different types of sandwich panels such as one with carbon/epoxy face sheet with Nomex honeycomb core subjected to LVIs at temperature in the range of −70°C to −151°C [11], and synthetic CFRP face sheet with/without lightweight honeycomb Nomex core material at room temperature [12]. These facts identify one basic disadvantage of composite sandwich materials. They inherently possess the low damage tolerance under impact load [13]. As discussed above, the impact caused by LVIs such as accidental drop of a tool during machining or due to a hit from spatial debris and/or environmental debris where the panel is in service can cause a lot of damage leading to failure. Some of the damages that can develop in the sandwich composite panels under impact load are face crushing, face sheet failure, in-plane failure of faces, flexural failure of faces, core crushing, core shear failure, and delamination between the core and face. As explained above, more often, such internal damages remain barely visible at the external surface of the sandwich composite panel and hence are usually identified as barely visible impact damage (BVID) [8–13]. The presence of BVID leads to reduction in residual strength. This reduction in residual strength ultimately compromises the structural integrity of the component. This fact defines the scope of the present work. It is evident therefore that global effort is directed toward developing a better understanding about CAI of sandwich composites in general [8–13,14–31].

This chapter starts with discussing various test standards for CAI test developed by various organizations worldwide, the experimental studies on this aspect of sandwich composites, and the numerical modeling and simulation schemes including the associated challenges. Finally, the concluding remarks are provided on the current state of the art on CAI, its strength and limitations, and the scope of future work.

5.2   CAI TEST METHODS AND STANDARDS

Conducting compression test of sandwich composites after being impacted either knowingly or unknowingly is very important to understand the nature of damage as well as to quantify the loss in residual strength. There are few standards developed by various standard organizations worldwide, such as standard developed by NASA on CAI [32], Airbus test standard (AITM 1–0010 [33]), European aerospace standard series (EN 6038 [34]), International Standard (ISO 18352 [35]), the standard developed by Suppliers of Advanced Composite Materials Association, the USA (SACMA SRM 2R-94 [36]), American standard for testing and materials (ASTM D7137/D7137M [37]), and the standard developed by Boeing (BSS 7260 [38]), for measuring the residual strength of composite laminates after impact. These standards provide test methods and supporting fixtures required to conduct the test. This test is conducted in three stages, namely developing impact, measuring the penetration by nondestructive techniques, and finally measuring the compressive residual strength [39]. The role of fixture design, especially the tightening of the test specimens to the CAI test device, and the realization of the free sliding edge required for compressing the specimen are important to correctly measure the CAI strength of the test specimen [39]. Among different test standards discussed above, the specifications of the standard developed by NASA and Boeing are generally used to measure the CAI strength [40]. The standard specifications of these tests are provided in Table 5.1 [40].

Various test standards are being developed by different world agencies to measure compressive or tensile strength of composites. There are reports and standard

TABLE 5.1
Summary of the Common CAI Test Methods and Details of the Test Standards [40]

Test Standards	NASA	Boeing
Material specimen thickness	6.35 mm	4–5 mm
Lay-up	(45,0,–45,90)	(–45,0,45,90)
Impact top diameter	12.7 mm	15.75 mm
Mass	4.5 kg	4.6–6.8 kg
Drop height	28	As needed
Area of specimen	127 mm^2	127 mm × 76 mm
Support	Clamped	Clamped at four points
Compression test specimen size	$h = 254$–317 mm	$h = 152$ mm
	$w = 178$ mm	$w = 102$ mm
Loading	End loading	End loading
Loading rate	1.27 mm/min	0.5 mm/min

TABLE 5.2

Toughened Resin Test and Hot-Wet Compression Test Geometries [41]

Test	Specimen Dimension (mm)	Ply Orientation	Required Data
Compression after impact	127×254	$[45°/0°/-45°/90°]_{6s}$	Strength Strain
Edge delamination tension test	38.1×254 (tabbed)	$[+35°/-35°/0°/90°]_s$	Interlaminar fracture toughness
Open hole tensile test	38.1×254 6.35 mm diameter hole	$[45°/0°/-45°/90°]_{6s}$	Strength Strain
Open hole compression test	127×254 25.4 mm diameter hole	$[45°/0°/-45°/90°]_{6s}$	Strength Strain
Double cantilever beam	38.1×228	$[0°]_{24T}$	Interlaminar fracture toughness

documents published by these standardization authorities in the literature [41–43]. To the reader's advantage, five different toughened resin test geometries and the parameters to be measured are provided in Table 5.2. It is advised to refer to the original literature [41–43] on test methods to find more details.

5.3 EXPERIMENTAL INVESTIGATIONS ON CAI FOR SANDWICH COMPOSITE PANELS

CAI test is important to correctly understand the behavior of composites, especially the sandwich composite panels undergoing sudden impact load. As discussed above in the Introduction section, the impact load, even if it is of low velocity, can contribute to the formation of localized damage that can affect structural integrity and lead to premature failure. The advantages and applications of using sandwich composites are enormous for different types of structures; therefore, the scientific communities worldwide are working on to better understand these phenomena and to accurately quantify the loss in the residual strength of these composite materials through various experimental investigations and numerical simulations. Generally, the experimental method requires a two-stage process to study the CAI behavior of sandwich composites (Figure 5.1 [44]). The subsections below explains different experimental investigations carried out by different researchers on sandwich composites to understand the failure mechanisms and CAI behaviors.

5.3.1 NON-CRIMP FABRIC (NCF) SANDWICH PANELS

These types of sandwich panels (e.g., 60 mm Divinycell H80 core and NCF-based face sheet made up of $[0/90/45/-45]_s$ lay-up sequence and 50 mm Divinycell H80 core and NCF-based face sheet with $[0/90/45/-45]_s/3$ lay-up sequence) are not the most common one [13]. Intrinsically, there is a certain amount of waviness

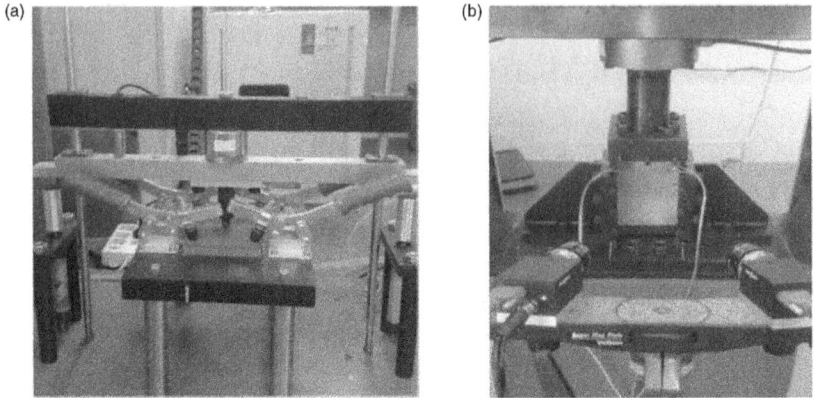

FIGURE 5.1 (a) Creating low-velocity impact on the sandwich composite test specimen; (b) test for loss of compressive strength due to LVI [44].

present in the structure of NCF sandwich panels. The experimentally measured uniaxial CAI test results reveal that for a given panel type there are insignificant differences in strain to failure between panels exposed to different impact energies. It is noted further that fiber micro-buckling in the impacted face sheet governs the CAI behavior of the NCF-based laminates.

5.3.2 CAI Behavior of Thin Laminates

There is not much work reported on the CAI behavior of thin laminates [14]. The reason is that there is yet to be any ASTM or other standards to have CAI test conduction for thin (e.g., 1.6–2.2 mm) CFRP laminates. Thus, two types of laminates are reported. For instance, one comprises 10 plies of woven AGP193-PW/8552 CFRP. The other type is reported [14] to comprise of AS4/3051-6 tape with two lay-ups. One lay-up sequence is, e.g., cross-ply $[0/90]_{3s}$. The other lay-up sequence is quasi-isotropic. Thus, it comprises a sequence of $[45/0/90]_s$. Finally, the woven laminate is 2.2 mm thick. The cross-ply laminate is also of the same thickness. But the quasi-isotropic laminate is even thinner at about 1.6 mm. All laminates of the same size, e.g., about 78 by 78 mm, are exposed to drop weight impact tests spanning the energy range of about 1–13 J. The CAI behavior exhibits that for all impact energies, the compressive strength decreases more drastically at relatively lower impact energy levels before saturating out at relatively higher impact energy levels. The absolute magnitude of the uniaxial compressive strength is the highest, e.g., about 360 MPa, for the 10-ply woven laminate. Similarly, the lowest absolute magnitude, e.g., about 190 MPa, of compressive strength is noted for the quasi-isotropic tape laminate. It is suggested that the 10-ply woven laminate performs better as the architecture of the reinforcement limits the spread of the damage from the impact much better as compared to other types of laminates. On the other hand, the quasi-isotropic laminates suffer the least reduction in residual compressive strength after impact. It happens because the surface plies shield the 0° plies from the usual damage caused by the drop weight impactor. It is reported further that the least damage-tolerant are the cross-ply laminates. It happens because the 90° plies have the least stiffness. These results imply that the 10-ply woven laminates possess the greatest compressive failure resistance, while the quasi-isotropic laminates possess the highest damage tolerance.

5.3.3 CAI: ISSUE OF CORE CRUSHING DAMAGE

The factor that governs the extent of damage in a sandwich panel is the amount of momentum that is transferred to the panel during the impact process. Therefore, the mass and the size of the impact emerge as critical factors. The other important factor is the velocity with which the impactor strikes the panel. Depending on the momentum of the impactor, material properties, and geometry of the sandwich panel, four types of damage mode outcomes can be anticipated. The first one is core crushing, especially when there is a relatively brittle core. The second and the most expected as well as observed damage mode is debonding that happens at the interface between the fiber and the core. The third damage mode is the well-researched process of delamination. The last of these is of course the most important. This is what comprises both fiber fracture and cracking of the matrix. In this section, the focus is kept on core crushing as it is an area of active research and development because often the face sheet masks the core crushing damage from being detected by visual inspections although it is important to note that the effect of core crushing damage on CAI behavior is significant [15–17].

For instance, research [18] on a sandwich panel consisting of 2.4 mm quasi-isotropic GFRP faces 50 mm Rohacell WF51 polymethacrylimide (PMI) foam core, and face sheets comprised of Devold DBLT600-E10 glass-fiber non-crimp fabric and vinyl ester Norpol Dion 9500 shows that for the impact energy range of 20–60 J, there is stable growth from the residual impact-induced inward dent. In other words, it happens toward the mid-plane of the panel until the final failure occurs in the CAI tests. The final failure happens when there is an abrupt separation between the face sheet and the panel. Postmortem analysis identifies traces of the foam core. This is reported to occur just at the outer boundary of the core crushing damage. Based on these observations, it is suggested that the out-of-plane tensile fracture of the foam core in the zones with outward displacements provides the basis for panel failure in CAI tests. As the impact damage size increases with an increase in impact energy, the CAI strength drops accordingly. Further, the face sheets do not fail from compressive failure in the CAI tests.

5.3.4 CAI BY COLUMN COMPRESSION

CAI experiments by column compression are also reported for 27.9 by 27.9 by 2.82 cm sandwich panels made from 25.4 mm thick dense closed-cell PVC foams and 1.27 mm thick face sheets made of 4-ply woven carbon fabric/epoxy laminates [19]. CAI test on panels impacted by drop weight impact tests conducted up to the maximum value of 108 J confirms that as high as 60% (e.g., 269 MPa for impacted and damaged panel and ca. 679 MPa for the undamaged panel) can occur due to delamination that initiates exactly from the position of the preexisting impact damage. The delamination progresses through the thickness of the laminate. As a result, instability occurs in the face sheet and final compressive failure occurs in that region through the formation of kink bands.

5.3.5 CAI of Honeycomb Core Structures

CAI behaviors of thick (e.g., 9.525 and 19.05 mm) honeycomb sandwich panels made of 4.76 mm cells (Plascore and Nomex) and plain weave carbon fabric/epoxy (PWCF) prepreg as well as fiberglass/epoxy (SWGF) prepreg systems are also reported [20]. These are made of three, e.g., [(0/45)n/core/(45/0)n], (n = 1, 2, 3), quasi-isotropic stacking sequences. Thus, there are 2-ply, 4-ply, and 6-ply face sheets. The impactors of 25.4 and 76.2 mm are used to ensure that impact energies span a wide range from about 6 to 31 J. Under in-plane compression tests, the CAI strengths of sandwich panels with 9.525 mm thick core are reported to be lower than those of the 19.05 mm core. It is expected because the thinner panel fails from global buckling failure. Further, in the cases of sandwich panels with 4-ply and 6-ply face sheets, both core thickness and impactor diameter affect CAI. For instance, if the impactor diameter is smaller, the damage is less severe and vice versa. However, these traits are not so well defined in the case of sandwich panels with only a 2-ply face sheet. Experimental observations confirm further that there can be three different failure mechanisms. The first one is of course net section fracture. It occurs across the width of the panel. In the second process, propagation of a dimple/dent due to earlier impact damage happens. The arresting process of such a propagation leads to a fracture of the net section of the sandwich panel. In the third process, it happens by the propagation of a dimple/dent that finally leads to buckling failure. Thus, a larger-diameter impactor causes a larger dent diameter and hence causes a more significant reduction in strength in the CAI tests. The reduction could be as high as even 40% of the strength measured for the undamaged sandwich panel.

5.3.6 CAI: Issue of Low Temperature

Since the temperature has a profound effect on the mechanical properties of polymers, the issue of strength of laminates in CAI at low temperature is a matter of paramount importance although it has seldom been investigated [21]. For instance, the CAI behaviors of three different carbon/epoxy laminates after LVI tests conducted at 1, 3, 4, and 6 J are reported at −60°C and −150°C. Thus, the CAI experiments are conducted after the aforementioned LVI tests on a 2.2 mm thick cross-ply laminate (0/90)$_{3s}$, a 1.6 mm quasi-isotropic (±45/0/90)$_s$ laminate, and a 2.2 mm thick 10-ply woven laminate. The results confirm that the influence of temperature on the CAI behavior of cross-ply laminates is far from significant, if at all. In the case of the 10-ply woven laminates, however, a slight decrease in strength is seen in CAI behavior as a function of temperature. Thus, in CAI behaviors the 10-ply woven laminates exhibit the highest residual strength, while the quasi-isotropic laminates exhibit the lowest. On the other hand, the cross-ply laminates exhibit the lowest tolerance against impact-induced damage.

5.3.7   CAI ISSUE IN EXPENDABLE LAUNCH VEHICLES

The sandwich composite panels used in the interstage of expendable launch vehicles are also examined for CAI issues [22]. For instance, the composite is made up of two equivalent face sheets. The face sheets are co-cured to a 3.175 mm cell size perforated Al honeycomb core of 2.858 cm thickness and 39.85 kg/m³ density. The bonding is done by an epoxy film adhesive. Two laminates are used. One is a 16-ply quasi-isotropic laminate. The other is an 18-ply directional laminate. Drop weight impact tests with different impactors are done to cover the impact energy range of about 2–20 J. The results confirm that the width rather than the depth of the damaged area is a much more important parameter as far as CAI strength is concerned. In contrast, other researchers [23] claim that the depth rather than the width of the damaged area is a much more important parameter as far as CAI strength is concerned. It is noted further that going by the width of the damage area criterion the 18-ply laminate possesses residual strength higher than the 16-ply laminate in CAI behavior [22]. It suggests that 18-ply laminates are more impact damage-tolerant than the 16-ply laminates. On the other hand, for a given impactor size and energy, the damaged area of the 16-ply laminates is always less than that of the 18-ply laminates. It implies that the 16-ply quasi-isotropic laminates are more impact damage-resistant than the directional 18-ply laminates.

5.3.8   CAI AFTER FULL-SCALE BLAST EXPERIMENTS

These studies are very rare [24]. The full-scale blast (100 TNT equivalent) studies reveal that CFRP sandwich panels perform better than equivalent GFRP panels. However, in terms of CAI behavior, it is notable that the percentage drop in the CFRP sandwich panels is greater than that in the GFRP sandwich panels. For instance, after the blast, the residual CAI strength is about 37% of the undamaged strength of CFRP, while the residual CAI strength is about 50% of the undamaged strength in CFRP sandwich panels [24].

5.3.9   INFLUENCE OF SKIN THICKNESS INCREMENT ON CAI

It is interesting to note that in situations where the panels are exposed to possible LVI events, an enhancement in skin thickness of Al skin from even 1 to 1.5 mm in honeycomb core sandwich panels [25] leads to reduction in CAI strength. Up to the impact energy of 50 J, the CAI strength reduction happens, but beyond that, it almost plateaus out.

5.3.10 Influence of Core Density Increment on CAI

As expected, it is reported [26] that the CAI strength of the sandwich panel with 0.5 mm thin face sheet with 6 lb/ft³ core density is higher than that of the sandwich panel with 0.5 mm thin face sheet with 3 lb/ft³ core density. The general trend is that even with slight impact energy damage, e.g., 1 lb/ft, there is a significant decrease in the given panel CAI strength. Further, as the impact energy increases more loss happens in the CAI strength of a given sandwich panel. However, one interesting observation is that the relative reduction in CAI strength value itself decreases as the impact energy is increased. In the case of a low-density core, local buckling of the face sheet happens as the core underneath collapses due to the impact damage. This leads to failure through the propagation of the indentation damage created due to the impact damage. In the case of a high-density core, the failure happens through the propagation of a transverse fiber crack. This crack propagates across the face sheet. It happens without any local buckling of the face sheet. It also happens even when there is no crushing of the core present underneath the face sheet.

5.3.11 Reinforcement Effect on CAI

Nanosilica and rubber particle reinforcement [27] in CFRP laminates leads to about 30% improvement in CAI strength after 20 J impact as compared to that of the control CFRP laminate. Even the delamination area reduces by about 50% in the case of the nanosilica- and rubber particle-reinforced CFRP laminates exposed to impact damage of 7 J energy.

5.3.12 Issue of Hybrid Face Sheet Effect on CAI

It is interesting to note that in terms of CAI behavior of sandwich panels with foam core and hybrid face sheets, e.g., [GFRP/CFRP]2/foam core/[CFRP/GFRP]2 perform much better than both [GFRP]2/foam core/[GFRP]2 and [CFRP]2/foam core/[CFRP]2 sandwich panels [28]. Further, it is important to appreciate that the sandwich panels with hybrid face sheets provide not only higher CAI strength, but also relatively higher compressive strength in the unimpacted condition. It is observed [23] that during failure in CAI tests there are two predominant modes of damage propagation. The first damage mode propagates through the formation of wrinkles in the face sheets. The second yet concurrent damage mode involves buckling of the underlying foam core, as expected.

5.3.13 Thickness and Ply-Stacking Sequence Effect on CAI

It has recently been reported [29] that for CFRP laminates, the CAI strength decreases due to enhancement in damage amount as impact energy increases, as is also observed by other researchers. Thicker laminates have higher CAI as they possess higher bending stiffness. Further, angle (±45) ply laminates have lower notch sensitivity and hence higher CAI strength and damage tolerance than cross-ply laminates. The presence of (±45) ply laminates delay the initiation of buckling because now in this case the central sub-laminate is more stable. The thicker laminates with ply cluster ensure delay in buckling initiation and hence ensure an improvement in CAI strength of CFRP laminates.

5.3.14 SHAPE-MEMORY ALLOY (SMA)–GFRP HYBRID FACE SHEETS AND CAI BEHAVIOR

A very recent work [30] attempted to understand the CAI behaviors of poly-amine foam core sandwich panels with hybrid face sheets, e.g., GFRP–SMA fiber (0° and 90° orientations) and GFRP–SS304 wire, with LVI damages caused by impact energy of up to about 22 J. It is reported that following impacts at energy values of 18 and 22 J, compared to CAI of Ply-I configuration (only polyamine core with GFRP face sheet) sandwich panels, the CAI behavior of Ply-VI config-uration sandwich panels (Ply-I reinforced with 0° and 90° SMA fibers as well as SS304 wire mesh) is higher by about 13 and 2 times, respectively. This happens because the super-elastic property of SMA ensures much better impact energy dissipation. It also transfers the impact energy of the impactor in a much more efficient manner to the outer face sheet. In terms of damage process, at first, the delamination initiates mainly at the impact site and there too, mainly at the interface between the foam core and the face sheet. This delamination propagates across the entire cross section of the sandwich panel to induce final failure in the CAI tests.

5.4 NUMERICAL STUDIES ON FAILURE OF COMPOSITE SANDWICH PANELS FOR COMPRESSION AFTER IMPACT LOADING

Numerical modeling of the CAI behaviors of sandwich composite panels is highly challenging because of the complex failure mechanism including a high degree of nonlinearity due to impact loading. Numerical modeling is important to extract more information on the complex failure mechanisms as it is difficult and very expensive to get all the intrinsic details through experimentations. There is the problem of repeatability of the experimental results and data scattering [3] as well in them. Numerical modeling of CAI problems necessitates employing an appropriate damage model that can take care of the severe damage caused by the impact load. There are mainly two types of failure in sandwich composite panels under impact, namely the intralaminar failure and the interlaminar failure. The intralaminar failure includes both fiber failure and matrix cracking. The interlam-inar failure is delamination between the face sheet and the core material. Thus, different modeling methodologies are utilized to model and simulate impact and CAI of sandwich composites. They can be classified into the following catego-ries: (1) impact zone modeled as equivalent elliptical inclusion/hole, (2) modeling of intralaminar and interlaminar damages in sandwich composites under impact and CAI, and (3) single finite element model to simulate both impact and CAI behavior.

5.4.1 DAMAGE ZONE MODELED AS EQUIVALENT HOLE

As mentioned earlier, impacted zone is modeled as equivalent hole/inclusion [45,47]. Puhui et al. [45] developed a method to study the CAI behaviors where the impact damage zone is modeled as an equivalent elliptical hole (Figure 5.2) based on a compressive failure mechanism. They developed a technique to find out the shape and size of the hole/inclusion. The complex potential method is employed to determine stress distribution around the hole. The failure criterion independent of lay-ups of the composite called load-bearing ply failure (LBPF) is utilized [45] to predict CAI performance. The out-of-plane deflection of the damaged zone increases with an increase in the applied load. The fiber near the damaged region gets heavily bent. This results in impact-induced fiber failure and bending in fiber. This phenomenon can lead to a sharp decrease in the load-carrying capacity. For this reason, the authors consider that the fibers within the damaged zone cannot sustain any load. The experiments are carried out on notched multi-directional laminates to determine the characteristic length, l_0 [46]. The failure criterion developed in [45] is verified by a large amount of test data. The damage width is found to be the major parameter that affects the CAI strength. The role of the damaged area, impact energy, impactor shape, and dimension in CAI strength can be characterized by their effects on the damage width. There is negligible effect of the material toughness on the compressive strength of the damaged laminates [45]. The dent diameter has also very little effect in comparison with the width of the damage region. The residual strength of the laminate is not affected by the notch shape.

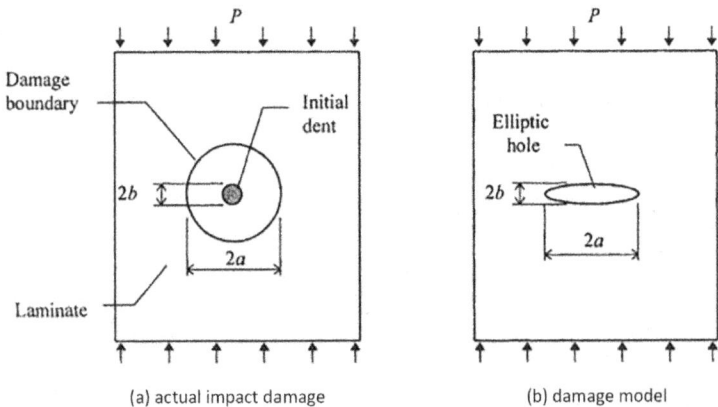

(a) actual impact damage   (b) damage model

FIGURE 5.2 Impact zone modeling through equivalent elliptical hole [45].

Romanoa et al. [47] also used the equivalent hole method to study the CAI behavior of composite panels. In their model, Romanoa et al. [47] considered BVID as the equivalent hole of ¼ inch diameter. The dimension of the equivalent hole is obtained by experiment based on the progressive failure criteria [47]. The residual strength data are evaluated and verified by experimental results for different sizes of holes up to one inch exposed to 50 J impact energy. It is reported further [47] that 1 inch (25.4 mm) hole size correlates with the 50 J impact energy in the experimental result.

5.4.2 Modeling of Intralaminar and Interlaminar Damages

Continuum damage model (CDM) with progressive damage criterion [48–51] is normally utilized to model the intralaminar failure, while cohesive zone models [50–54] are utilized to model the delamination (interlaminar failure). Recently, some of the works have used cohesive elements with mixed-mode failure criteria to study matrix cracking [52–54]. We would like to discuss the basic theory of the CDM development and different updating being done to account for the intralaminar failure in sandwich composite, and the cohesive zone formulations utilized to model delamination in the sandwich composites discussed in the preceding subsections. Recent developments in numerical modeling show that a single model can be used for both impact study and CAI behavior study.

5.4.2.1 Continuum Damage Model (CDM)

The CDM has been developed [55] to account for quasi-brittle fracture of laminated composite structure that is capable of handling the initiation and evolution of intralaminar failure mechanism in composite noncritical failure mechanism including the final collapse. This model also accounts for crack arrest in case of reversal of the load. The details of the constitutive model [55] are discussed here for better understandability.

The constitutive relation for the laminated composites is based on thermodynamic irreversibility [55]. This model uses fiber tensile/compressive failure criteria for damage activation. It defines a scalar function for the complementary free energy in the material. This scalar function must be positive definite, and zero at the origin with respect to the stresses. The ply complementary free energy density can be expressed as

$$G = \frac{\sigma_{11}^2}{2(1-d_1)E_1} + \frac{\sigma_{22}^2}{2(1-d_2)E_2} - \frac{v_{12}}{E_1}\sigma_{11}\sigma_{22} + \frac{\sigma_{12}^2}{2(1-d_6)G_{12}}$$

$$+ (\alpha_{11}\sigma_{11} + \alpha_{22}\sigma_{22})\Delta T + (\beta_{11}\sigma_{11} + \beta_{22}\sigma_{22})\Delta M \qquad (5.1)$$

where E_1, E_2, v_{12}, and G_{12} are the in-plane elastic orthotropic properties of a unidirectional lamina. The subscripts 1 and 2 denote the longitudinal (fiber) and the transverse (matrix) directions, respectively. Further, the quantities α_{11} and α_{22} are the thermal expansion coefficients in the longitudinal and transverse directions, respectively, while β_{11} and β_{22} are the coefficients of hygroscopic expansion in the longitudinal and transverse directions. Furthermore, the quantities ΔT and ΔM are the difference in temperature and moisture content with the reference. The stress tensor σ corresponds to average stress tensor over a representative volume assumed to be much larger than the diameter of a fiber. The damage variable d_1 is associated with longitudinal (fiber) failure, whereas d_2 is the damage variable associated with transverse matrix cracking; d_6 is a damage variable influenced by longitudinal and transverse cracks. The strain tensor can be determined as

$$\varepsilon = \frac{\partial G}{\partial \sigma} = H : \sigma + \alpha \Delta T + \beta \Delta M. \tag{5.2}$$

The lamina compliance tensor in terms of Voigt notation can be expressed as

$$H = \frac{\partial^2 G}{\partial \sigma^2} = \begin{bmatrix} \dfrac{1}{(1-d_1)E_1} & -\dfrac{v_{21}}{E_1} & 0 \\[3mm] -\dfrac{v_{12}}{E_1} & \dfrac{1}{(1-d_2)E_2} & 0 \\[3mm] 0 & 0 & \dfrac{1}{(1-d_6)G_{12}} \end{bmatrix} \tag{5.3}$$

The unilateral effect caused due to closure of transverse cracks under load reversal has been considered by defining the damage variables associated with longitudinal and transverse damage. The longitudinal and transverse damage variables can be expressed as

$$d_1 = d_{1+} \frac{\langle \sigma_{11} \rangle}{|\sigma_{11}|} + d_{1-} \frac{\langle -\sigma_{11} \rangle}{|\sigma_{11}|}$$

$$d_2 = d_{2+} \frac{\langle \sigma_{22} \rangle}{|\sigma_{22}|} + d_{2-} \frac{\langle -\sigma_{22} \rangle}{|\sigma_{22}|} \tag{5.4}$$

where $\langle\ \rangle$ is the Macaulay operator defined as $\langle x \rangle = (x + |x|)/2$. This model takes care of the damages caused by tensile and compressive loads, separately. Depending upon the sign of the normal load, the corresponding part of the damage model becomes active or passive. The model makes sure that the closure does not affect the shear damage variable d_6 as transverse normal loading can cause closure, not shear. Friction effect is neglected in the model.

The damage evolution criteria are important to predict the failure in composites. The CDM [52] thus considers the four surfaces of elastic domain, each accounting for one failure mechanism, namely the longitudinal and transverse fracture under tension and compression. The damage activation and evolution function based on LaRC03 and LaRC04 failure criteria represents the physical process accurately. The longitudinal tensile failure (Figure 5.3a) of fiber is

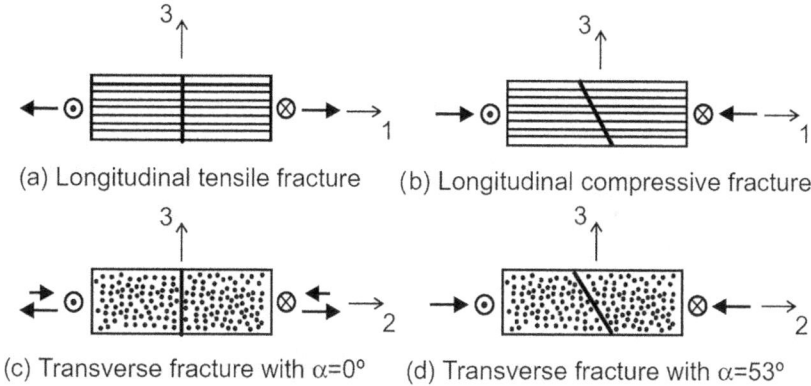

(a) Longitudinal tensile fracture   (b) Longitudinal compressive fracture

(c) Transverse fracture with $\alpha=0°$ (d) Transverse fracture with $\alpha=53°$

FIGURE 5.3 Ply fracture planes considered by Maimi et al. [55].

governed by the fracture criterion LaCR04, while the longitudinal compressive fracture (Figure 5.3b) is governed by the failure criterion LaCR03. The transverse fracture caused by in-plane shear stress and transverse tensile stress (Figure 5.3c), and the transverse fracture for high transverse compressive stress (Figure 5.3d) are governed by LaCR04 [55]. LaRC03 criterion can be employed to predict the failure of FRP laminated panels with in-plane stress states [55]. The LaRC04 failure criterion is for three-dimensional FRP composites and considers all the possible failure modes and nonlinearity in matrix shear behavior [55].

5.4.2.2 Interlaminar Damage (Delamination) Modeling for Composites

Interlaminar damage has been predicted by modeling the interfaces using interface elements in FEM formulation, such as cohesive elements [3,50–54]. The cohesive element follows traction–separation constitutive relation. The damage initiation is induced in the model by considering different criteria such as maximum stress, quadratic stress [3,50–54], maximum principal stress, maximum strain, quadratic strain, and maximum principal strain. The interfaces can be modeled either as explicitly cohesive zone (Figure 5.4, [3]) or as cohesive surfaces as used by W. Tan et al. [51]. The cohesive elements formulation is based on the cohesive constitutive law that relates the traction and separation at the

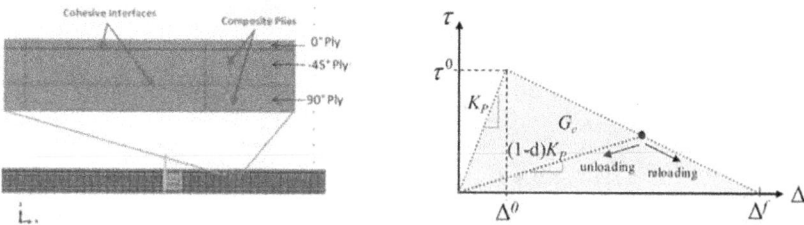

FIGURE 5.4 (a) Composite laminates modeling interface by cohesive zone element [3]; (b) traction–separation law for cohesive elements [3].

interface. The behavior of interface prior to damage is described as liner elastic and degrades under tensile and shear loading, while it is intact under compressive loading (Figure 5.3b). The cohesive zone can still transfer loads after damage initiation. For pure mode I, mode II, or mode III loading, once the interfacial normal or shear strength is attained, the stiffness gradually reduces to zero [3]. The area under the traction–separation curve represents the fracture energy in different modes of loadings. The delamination is simulated by reducing the ply stiffness and degrading the material properties of the cohesive elements to very low value [3]. The damage evolution is removed in the damaged cohesive element as these elements are completely failed [3].

5.4.2.3 Modeling of Matrix Cracking and Interface Delamination through Cohesive Elements

Some finite element analysis models use cohesive elements to model the matrix cracking behavior [52–54]. Thus, the fiber failure is modeled as continuum damage in [52] and as volume element degradation in [53]. Further, the permanent indentation is modeled by considering the introduction of localized plasticity in the matrix cracking interface element [53]. In another related work [54], the fiber failure is considered based on mode I energy criteria in the volume element (Figure 5.5). The buckling of the impact-damaged area, and crack propagation from the vicinity of the impacted region are also studied. Finally, the residual strength is also estimated [54].

5.5 STATE-OF-THE-ART NUMERICAL MODELS AVAILABLE IN THE LITERATURE TO STUDY INTRALAMINAR FAILURE AND DELAMINATION OF SANDWICH COMPOSITES

5.5.1 Composite Panels and Laminates

There are several studies available in the literature on numerical modeling techniques that employ CDM with progressive failure criteria for the intralaminar damage and the cohesive elements failure to model the interface failure through

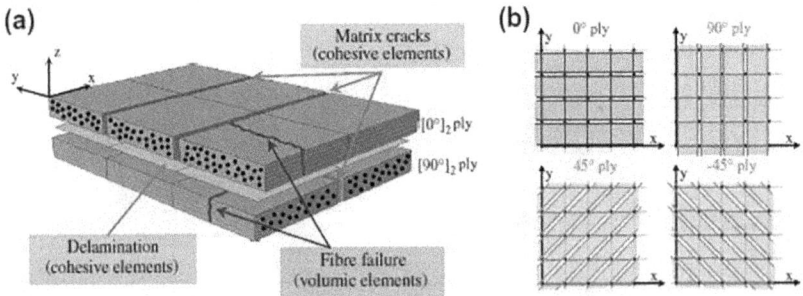

FIGURE 5.5 Modeling of matrix cracking through interface elements [54].

delamination. Thus, sandwich panels with carbon/epoxy skin and aluminum honeycomb core are modeled for impact and CAI tests using finite element method [49]. It uses CDM and progressive fiber fracture in the skin and elasto-plastic deformation in compression and shear in the honeycomb core. The combinations of thick core and thin face sheets, and thick face sheet with thin core are both considered. The impact damage is captured in dynamic simulation, and the damaged material is subjected to compression loading. It predicts how the failure propagates from the damage zone. The thick-skin, thin-core panel withstands higher impact energy (120 J) and suffers 34% loss in compressive strength. The thin-skin, thick-core panel suffers failure (e.g., front face perforated and back face deboned) at lower impact energy (80 J) and hence causes reduction in the CAI residual strength to about 32% of the undamaged value.

A numerical model to analyze CAI of thin face sheet honeycomb core sandwich panel [26] includes the progressive failure analysis based on continuum damage model [55], and a homogenized, nonlinear material model. It considers ply as orthotropic material, and corresponding damage modes are employed to study the CAI behavior. Reduced error is observed due to improvement in damage modeling via progressive damage model in this case. The effect of core density on the CAI strength is studied. It affects the nature of failure from indentation damage to crack propagation. It is found that the strength increases with an increase in core density. Enhancement in strength can be realized by small increment in mass when the composite undergoes large CAI.

The study in [49] also presents a numerical model to study the CAI behavior of the composite sandwich panel. It includes the intralaminar damage, core crushing, and delamination. The FE model is developed and analyzed in commercial software ABAQUS/Explicit solver for the delamination initiation and evolution in the impacted outer face sheet of the sandwich panel subjected to quasi-static indentation and edge compression. The initial indentation and CAI load are applied through rigid surface through self-contact module available in ABAQUS. The interface damage is modeled by using cohesive surface elements as a surface contact between the top and bottom layers. Intralaminar damage in face sheet is included by Hashin damage criteria.

Damage and failure mechanism of thin composite laminates under LVI and CAI has been studied by Tuao et al. [51] through the 3D damage model by considering both interlaminar and intralaminar damages. Energy-based CDM is utilized to predict intralaminar damage. For fiber damage, maximum strain criterion is used, while for matrix damage, the improved Puck criterion is utilized. According to the results obtained through this analysis, it is found that the damage always starts from the impact point and extends instantaneously across the laminate to the two outer edges. The area of fiber damage is found to be narrow, and the matrix damage area is observed to be large enough to contain the section with fiber damage. The area of delamination damage is obtained as the largest, and it exhibits an irregular shape. A good correlation was found between the thermal field and the damage initiation and evolution of composite laminates during the impact and CAI tests.

A user material subroutine in ABAQUS was developed by W. Tan and his research group [51] to model a 3D composite damage model for predicting the LVI and CAI behavior of composite laminates. It employs tensile and compressive failure mechanism to predict matrix and fiber failure based on continuum damage modeling, while cohesive elements are utilized to model interface delamination. In this case, three stages of analysis are carried out, namely capturing impact damage, using new boundary conditions to stabilize the geometry after impact load, and finally predicting the CAI strength. The model developed is found to be capable of capturing both intralaminar damage and the delamination. The permanent indentation is also evaluated. It is claimed that the model allows us to evaluate the residual strength with high accuracy. The nonlinear shear formulation is used for modeling intralaminar damage which contributes in capturing the permanent indentation.

5.5.2 Foam Core Sandwich Panel under CAI

Wang et al. [56] modeled foam core sandwich panels to investigate the CAI behavior by implementing the progressive damage material model. They utilized Schapery theory-based constitutive relations that take care of intra-ply shear inelasticity and fiber kinking that leads to fracture ultimately. The plasticity is also considered to represent elastic–plastic in-plane shear behavior of woven fabric plies. They utilized the impact damage solution developed separately for CAI analysis. It is concluded that micro-buckling can lead to kink bending which can be the reason for the failure of face sheets. The foam core sandwich panel under LVI and CAI is also studied by Bae et al. [44] both experimentally and by a developing numerical model. Their study considers the structural damages due to different impact shapes and impact depths on a foam core sandwich panel. The numerical model developed using commercial FEM software ABAQUS in this work provides energy absorption mechanism of the sandwich structure and damage within the structure. The boundary condition for LVI is considered as quadrilateral clamping, while for CAI, it is simply supported at both sides, clamped at the bottom and compressive load at the top. As discussed above, the interface is modeled using cohesive elements. The outer skin layer of the sandwich composite is modeled by continuum shell element. The facility to restart the analysis available in ABAQUS is used, and impact damage is introduced as the initial condition for CAI study. The foam core can be treated as an isotropic material, which considerably simplifies the material modeling and yet gives results that are comparable to the experimental findings. The major limitation of this type of work is that one has to repeat all the processes once the material of the core or skin as well as geometry changes.

5.5.3 CAI Test Model of Damaged Woven Fabric Sandwich Composite

The semi-empirical numerical models to estimate the residual strength of impact-damaged sandwich composites with woven fabric carbon/epoxy face sheets and Nomex honeycomb cores under CAI loading are also reported [57,58].

Nondestructive inspection and destructive sectioning of the damaged sandwich panel are utilized for establishing initial condition of the damage. Honeycomb core crush results are used to formulate nonlinear constitutive formulation. Nonlinear finite element model includes the geometry of the sandwich structure and proper damage specification obtained from the experiment. The effect of face sheet property degradation on the strain redistribution in the damaged panel is evaluated.

5.5.4 STITCHED COMPOSITE UNDER CAI

K. T. Tan et al. [3] developed a FE model to study the CAI behavior of stitched composites. They used continuum shell elements for meshing the stitched composite under CAI. Hashin failure criteria were employed for damage modeling. The interfaces were modeled as specified cohesive zone, which is different from the cohesive surface used by W. Tan [51]. The reduced material properties of the damaged cohesive zone are utilized to model interface damage due to delamination. Spring elements are utilized to represent through-thickness stitch thread. It was found that the compressive strength increases with a decrease in impact-induced delamination area. CAI strength was found to increase with the incorporation of the stitching of the composite, same as in case of experiments.

5.5.5 THREE-DIMENSIONAL PARAMETRIC STUDY FOR IMPACT BEHAVIOR OF HONEYCOMB SANDWICH STRUCTURE

The CAI behavior of titanium honeycomb sandwich structures was studied by developing numerical models using finite element method and by utilizing 3D parametric analysis [59]. This work confirms that the LVID could cause the change in failure modes and 9%–15% decrease in the compressive failure strength. This prediction matches with the results obtained from the experimental investigation.

5.6 SINGLE FE MODEL FOR IMPACT DAMAGE AND CAI

Recently, a single FE model [60] has been developed to study both the indentation damage and CAI, without losing the complexity of the damage parameters. CDM with stiffness degradation is utilized for intralaminar damage (fiber and matrix failure). The interface is modeled by cohesive elements. Quadratic traction–separation criteria are utilized. This aspect requires consideration of normal traction, shear traction, and interface normal and shear strength. The results identify that the reason for damage growth can be attributed to buckling induced by the impact. Reduction in damage size can increase buckling load and vice versa. Residual strength is found to be sensitive to fracture toughness and compressive strength. An increase in mode II interlaminar fracture toughness can lead to a decrease in delamination size during impact. As a result, it leads to an increase in damage tolerance. It is noted further that the material with higher compressive strength and compressive fracture toughness provides reduced damage size during impact. Hence, it possesses increased residual strength in the CAI behavior.

5.7 CONCLUSIONS

The state-of-the-art literature on the sandwich composite under LVI and loss in compressive strength due to these impacts is reviewed. The test standards for CAI, experimental studies to investigate different types of damage and their impact on compressive strength, and the numerical studies in this regard are considered. The strengths, limitations, and future scope on this very important topic are also discussed in detail. This chapter will help the readers better understand the sandwich composites in terms of their failure and loss of compressive residual strength under LVIs. This knowledge will help in taking appropriate measures while designing structures made up of sandwich composites.

REFERENCES

1. B. Vijaya Ramnath, K. Alagarraja, C. Elanchezhian, "Review on sandwich composite and their applications", *Materials Today: Proceedings* Vol. 16, 2019, pp. 859–864.
2. M.J. Li, S.L. Soo, D.K. Aspinwall, D. Pearson, W. Leahy, "Influence of lay-up configuration and feed rate on surface integrity when drilling carbon fibre reinforced plastic (CFRP) composites", *Procedia CIRP*, Vol. 13, 2014, pp. 399–404.
3. K.T. Tan, N. Watanabe, Y. Iwahori, "Finite element model for compression after impact behavior of stitched composites", *Composites: Part B*, Vol. 79, 2015, pp. 53–60.
4. S.N.A. Safri, M. Thariq, H. Sultan, M. Jawaid, K. Jayakrishna, "Impact behaviour of hybrid composites for structural applications: A review", *Composites: Part B*, Vol. 133, 2018, pp. 112–121.
5. A. Krzyhak, M.B. Mazur, M. Gajewski, K. Droz, A. Komorek, P. Przybylek, "Sandwich structured composites for aeronautics: Methods of manufacturing affecting some mechanical properties", *International Journal of Aerospace Engineering*, Vol. 2016, 2016, pp. 1–10.
6. R. Olsion, Methodology for predicting the residual strength of impacted sandwich panels, Technical Report, FFA TN 1997-09.
7. C. Bouvet, S. Rivallant, J.J. Barrau, "Low velocity impact modeling in composite laminates capturing permanent indentation", *Composites Science and Technology*, Vol. 72, 2012, pp. 1977–1988.
8. N. S. Fatima, G. S. Dhaliwal, G. Newaz, "Influence of interfacial adhesive on impact and post-impact behaviors of CFRP/end-grain balsawood sandwich composites", *Composites - Part B*, Vol. 212, 2021, pp. 108718 (1–17).
9. W. Nsengiyumva, S. Zhong, J. Lin, Q. Zhang, J. Zhong, Y. Huang, "Advances, limitations and prospects of nondestructive testing and evaluation of thick composites and sandwich structures: A state-of-the-art review", *Composite Structures*, Vol. 256, 2021, pp.112951 (1–52).
10. S. Lin, A. M. Waas, "Accelerating computational analyses of low velocity impact and compression after impact of laminated composite materials", *Composite Structures*, Vol. 260, 2021, pp. 113456 (1–23).
11. M. J. St-Laurent, M.-L. Dano, M.-J. Potvin, "Compression after impact behavior of carbon/epoxy composite sandwich panels with Nomex honeycomb core subjected to low velocity impacts at extreme cold temperatures", *Composite Structures*, Vol. 261, 2021, pp. 113516 (1–11).

12. C. Zhang, K.T. Tan, "Low-velocity impact response and compression after impact behavior of tubular composite sandwich structures", *Composites Part B*, Vol. 193, 2020, pp. 108026 (1–9).

13. F. Edgren, L. E. Asp, P. H. Bull, "Compressive failure of impacted NCF composite sandwich panels – characterisation of the failure process", *Journal of Composite Materials*, Vol. 38-6, 2004, pp. 495–513.

14. S. Saez, E. Barbero, R. Zaera, C. Navarro, "Compression after impact of thin composite laminates", *Composites Science and Technology*, Vol. 65-13, 2005, pp. 1911–1919.

15. D.P.W. Horrigan, R.R. Aitken, G. Moltschaniwskyj, "Modelling of crushing due to impact in honeycomb sandwiches", *Journal of Sandwich Structures and Materials* Vol. 2-2, 2000, pp. 131–151.

16. T.E. Lacy, Y. Hwang, "Numerical modelling of impact-damaged sandwich composites subjected to compression-after-impact loading", *Composite Structures*, Vol. 61, 2003, pp. 115–128.

17. A. Shipsha, S. Hallstr̈om, D. Zenkert, "Failure mechanisms and modelling of impact damage in sandwich beams—A 2D approach: Part I—experimental investigation", *Journal of Sandwich Structures and Materials*, Vol. 5-1, 2003, pp. 7–32.

18. A. Shipsha, D. Zenkert, "Compression-After-Impact Strength of Sandwich Panels with Core Crushing Damage", *Applied Composite Materials*, Vol. 12–3, 2005, pp. 149–164.

19. P. M. Schubel, J.-J. Luo, I. M. Daniel, "Impact and post impact behavior of composite sandwich panels", *Composites: Part A*, Vol. 38, 2007, pp.1051–1057.

20. K.S. Raju, B.L. Smith, J.S. Tomblin, K.H. Liew, J.C. Guarddon, "Impact damage resistance and tolerance of honeycomb core sandwich panels", *Journal of Composite Materials*, Vol. 42-4, 2008, pp. 385–412.

21. S. Sa´nchez-Sa´ez, E. Barbero, "Compressive residual strength at low temperatures of composite laminates subjected to low-velocity impacts", *Composite Structures*, Vol. 85, 2008, pp. 226–232.

22. A. Nettles, J. Jackson, "Compression After Impact Testing of Sandwich Composites for Usage on Expendable Launch Vehicles", *Journal of Composite Materials*, Vol. 44-6, 2010, pp. 707–738.

23. S.R. Finn, G.S. Springer, "Compressive Strength of Damaged and Repaired Composite Plates", In: *Proceedings of the Ninth DOD/NASA/FAA Conference on Fibrous Composites in Structural Design*, Lake Tahoe, NV, 1991, Vol. 2, pp. 1083–1095.

24. H. Arora, M. Kelly, A. Worley, P. Del Linz, A. Fergusson, P A Hooper, J. P. Dear, "Compressive Strength after Blast of Sandwich Composite Materials", *Philosophical Transaction of Royal Society*, Vol. A 372, 2014, pp. 20130212.

25. A. Gilioli, C. Sbarufatti, A. Manes, M. Giglio, "Compression after impact test (CAI) on Nomex™ Honeycomb Sandwich Panels with Thin Aluminum Skins", *Composites: Part B*, Vol. 43, 2014, pp. 313–325.

26. T. D. McQuigg, R. K. Kapania, S. J. Scotti, S. P. Walker, "Compression After Impact Experiments on Thin Face Sheet Honeycomb Core Sandwich Panels", *Journal of Spacecrafts and Rockets*, Vol. 51-1, 2014, pp. 253–266.

27. B. Nikfar, J. Njuguna, "Compression-after-impact (CAI) performance of epoxy-carbon fibre-reinforced nanocomposites using nanosilica and rubber particle enhancement", *2nd International Conference on Structural Nano Composites (NANOSTRUC 2014) IOP Publishing, IOP Conf. Series: Materials Science and Engineering*, Madrid, Spain, Vol. 64, 2014, pp. 012009.

28. B. Yang, Z. Wang, L. Zhou, J. Zhang, L. Tong, W. Liang, "Study on the low-velocity impact response and CAI behavior of foam-filled sandwich panels with hybrid face sheet", *Composite Structures*, Vol. 132, 2015, pp. 1129–1140.

29. M.A. Caminero, I. García-Moreno, G.P. Rodríguez, "Experimental study of the influence of thickness and ply-stacking sequence on the compression after impact strength of carbon fibre reinforced epoxy laminates", *Polymer Testing*, Vol. 66, 2018, pp. 360–370.

30. Y. Wu, Y. Wan, "The low-velocity impact and compression after impact (CAI) behavior of foam core sandwich panels with shape memory alloy hybrid face-sheets", *Science and Engineering of Composite Materials*, Vol. 26-5, 2019, pp. 517–530.

31. F. Flügge, Konstruktion und Validierung einer modifizierten, Compression-After-Impact Prüfvorrichtung für Dünnwandige Compositplatten. Master's Thesis, Hamburg University of Applied Sciences, Hamburg, Germany, 2019.

32. A.T. Nettles, "A lowcost method for testing compression after impact of composite laminates, N 92-22679", NASA report by NASA Marshall Space Flight Center Polymers and Composites Branch.

33. Airbus Test Method: Fibre Reinforced Plastics—Determination of Compression Strength after Impact; 0,10, 2005. Reference Number AITM1–0010; Airbus S.A.S.: Blagnac, France, 2005.

34. European Standard: Aerospace Series—Fibre Reinforced Plastics—Test Method—Determination of the Compression Strength after Impact; German and English Version EN 6038: 0,2,2016; Beuth Verlag GmbH: Berlin, Germany, 2016.

35. International Standard ISO 18352 Carbon-Fibre-Reinforced Plastics—Determination of Compression after Impact Properties at a Specified Impact-Energy Level; Reference Number ISO 18352: 30,09,2009 (E); Beuth Verlag GmbH: Berlin, Germany, 2009.

36. SACMA Recommended Test Method for Compression after Impact Properties of Oriented Fiber-Resin Composites SRM 2R-94; Suppliers of Advanced Composite Materials Association: Arlington, VA, 1994.

37. ASTM Test Method: Standard Test Method for Compressive Residual Strength Properties of Damaged Polymer Matrix Composite Plates; 0,0,2017. ASTM D7137/D7137M; American Society for Testing and Materials (ASTM): West Conshohocken, PA, 2017.

38. Boeing Advanced Composite Compression Tests. *Boeing Specification Support Standard BSS 7260*; The Boeing Company, Seattle, WA, 1988.

39. M. Linke, F. Flügge, A. J. Olivares-Ferrer, "Design and Validation of a Modified Compression-After-Impact Testing Device for Thin-Walled Composite Plates", *Journal of Composite Science*, Vol. 126-4, 2020, pp. 1–16.

40. S. Abrate, *Impact on Composite Structures*; Cambridge University Press, New York, 1998.

41. S. Lee, R.F. Scott, P. C. Gaurdert, W.H. Ubbink, C. Poon, "Mechanical testing of toughened resin composite materials", *Composites*, Vol. 19-4, 1988, pp. 300–310.

42. Standard Tests for Toughened Resin Composites, Compiled by ACEE Composite Project Office; Langley Research Center, Hampton, Virginia, NASA Reference Publication 1092, 1983.

43. Test Methods for Composites a Status Report, Vol. 2: Compression Test Methods; US Department of Transportation, Federal Aviation Administration, New Jersey, 1993.

44. R. Bai, J. Guo, Z. Lie, D. Liu, Y. Ma, C. Yan, "Compression after impact behavior of composite foam-core sandwich panels", *Composite Structures*, Vol. 225, 2019, pp. 111181 (1–10).

45. C. Puhui, S. Zhen, W. Junyang, "A New Method for Compression After Impact Strength Prediction of Composite Laminates", *Journal of Composite Materials*, Vol. 36-5, 2002, pp. 589–610.

46. C. Puhui, S. Zhen, W. Junyang, "Prediction of the strength of notched fiber-dominated composite laminates", *Composite Science and Technology*, Vol. 61, 2001, pp. 1311–1321.

47. F. Romanoa, F. D. Capriob, U. Mercurio, "Compression After Impact Analysis of Composite Panels and Equivalent Hole Method", *Procedia Engineering*, Vol. 167, 2016, PP. 182–189.

48. G.A.O. Davies, D. Hitchings, T. Besant, A. Clarke, C. Morgan, "Compression after impact strength of composite sandwich panels", *Composite Structures*, Vol. 63, 2004, pp. 1–9.

49. C. T. James, A. Watson, P. R. Cunningham, "Numerical modelling of the compression after-impact performance of a composite sandwich panel", *Journal of Sandwich Structures and Materials*, Vol. 17-4, 2015, pp. 376–398.

50. H. Tuo, Z. Lua, X. Maa, J. Xing, C. Zhang, "Damage and failure mechanism of thin composite laminates under low-velocity impact and compression-after-impact loading conditions", *Composites Part B*, Vol. 163, 2019, pp. 642–654.

51. W. Tan, B. G. Falzon, L. N. S. Chiu, M. Price, "Predicting low velocity impact damage and Compression-After-Impact (CAI) behavior of composite laminates", *Composites Part A: Applied Science and Manufacturing*, Vol. 71, 2015, pp. 212–226.

52. C. Bouvet, B. Castanié, M. Bizeul, J.-J. Barrau, "Low velocity impact modelling in laminate composite panels with discrete interface elements", *International Journal of Solids and Structures*, Vol. 46, 2009, pp. 2809–2821.

53. C. Bouvet, S. Rivallant, J.J. Barrau, "Low velocity impact modeling in composite laminates capturing permanent indentation", *Composites Science and Technology*, Vol. 72, 2012, pp. 1977–1988.

54. S. Rivallant, C. Bouvet, N. Hongkarnjanakul, "Failure analysis of CFRP laminates subjected to Compression after Impact: FE simulation using discrete interface elements", *Composites Part A: Applied Science and Manufacturing*, Vol. 55, 2013, pp. 83–93.

55. P. Maimi, P. P. Camanho, J. A. Mayugo, C. G. Da´vila, "A continuum damage model for composite laminates: Part I – Constitutive model", *Mechanics of Materials*, Vol. 39, 2007, pp. 897–908.

56. J. Wang, H. Wang, B. Chen, H. Huang, S. Liu, "A failure mechanism-based model for material modeling the compression after impact of foam-core sandwich panels", *Composite Science and Technology*, Vol. 151, 2017, 258–267.

57. T. E. Lacy, Y. Hwang, "Numerical modeling of impact-damaged sandwich composites subjected to compression-after-impact loading", *Composite Structures*, Vol. 61, 2003, pp. 115–128.

58. Y. Hwang, T. E. Lacy, "Numerical Estimates of the Compressive Strength of Impact-damaged Sandwich Composites", *Journal of Composite Materials*, Vol. 41-3, 2007, pp. 367–388.

59. W. Zhao, Z. Xie, X. Li, X. Yue, J. Sun, "Compression after impact behavior of titanium honeycomb sandwich structures", *Journal of Sandwich Structures and Material*, Vol. 20-5, 2017, pp. 1–9.

60. M.R. Abir, T.E. Tay, M. Ridha, H.P. Lee, "Modelling damage growth in composites subjected to impact and compression after impact", *Composite Structures*, Vol. 168-15, 2017, pp. 13–25.

6 Low-Velocity Impact Response of the Composite Sandwich Panels

Vishwas Mahesh
Siddaganga Institute of Technology
Indian Institute of Science

Vinyas Mahesh
Indian Institute of Science
National Institute of Technology Silchar
Nitte Meenakshi Institute of Technology

Dineshkumar Harursampath
Indian Institute of Science

CONTENTS

6.1   Introduction 99
6.2   Materials and Methods 103
6.3   Results and Discussion 104
        6.3.1   Effect of Density of Core on Structural Response 104
        6.3.2   Force and Energy Response... 105
        6.3.3   Damage Analysis 107
        6.3.4   Taguchi Analysis 108
6.4   Conclusions..110
Acknowledgment 111
References..112

6.1   INTRODUCTION

Sandwich composites belong to a special class of composite materials having a lightweight core in between two stiff composite layers (skins). The transfer of load between the components is facilitated by joining the skin to the core. Sandwich structures are designed based on the principle of I-beam. In case of I-beam, the

DOI: 10.1201/9781003143031-6

flanges are located at a distance from the neutral axis. Most of the material is provided on the web to make the flanges work together and resist shear and buckling
loads. Comparing the sandwich structure with the I-beam, the skin of the sandwich structure acts as the flanges of the I-beam and the core acts the web. The
only difference is in the materials used. In case of sandwich structure, the materials used for skin and core are different, whereas in case of I-beam, the material
remains the same for the flange and web. The core used in sandwich structure
resists the shearing load and avoids buckling of skins. To withstand the shear and
tensile stresses between the skin and the core, the bonding between them should
be sufficiently strong and thus the adhesive that bonds the core and skin plays
an important role. The analogy between the I-beam and sandwich structure is
presented in Figure 6.1.

The recent trend witnesses the usage of sandwich composites in almost all
fields of engineering ranging from aerospace to sports [1,2] since they exhibit better mechanical properties such as enhanced specific strength, stiffness and resistance to corrosion [3]. For structural applications where high plane strength and
flexural rigidity are needed, sandwich composites are found to be suitable [4], but
such sandwich composites are prone to impact loading [5–9] due to poor resistance
to localized impact, which is a major concern in the usage of such composites.

Composites comprising carbon fibers reinforced in epoxy matrix have emerged
as the potential materials for various structural applications [10]. In case of sandwich structures, the core which is lighter in weight is accountable for dissemi-
nating the shear forces between the face sheets [11]. Bonding between the face
sheets and core material plays a prominent role in the construction of the sandwich structure. The bond between them should be stiff, strong and tough so that
higher loads are sustained by sandwich structures for long time [12].

Naturally available fibers (NFs) are notably considered as potential replace-
ments for man-made fibers in composites as reinforcement because they exhibit
acceptable mechanical properties such as better specific stiffness, powerful acoustic insulation, vibration damping and environmental-friendliness. Thus, NFs are

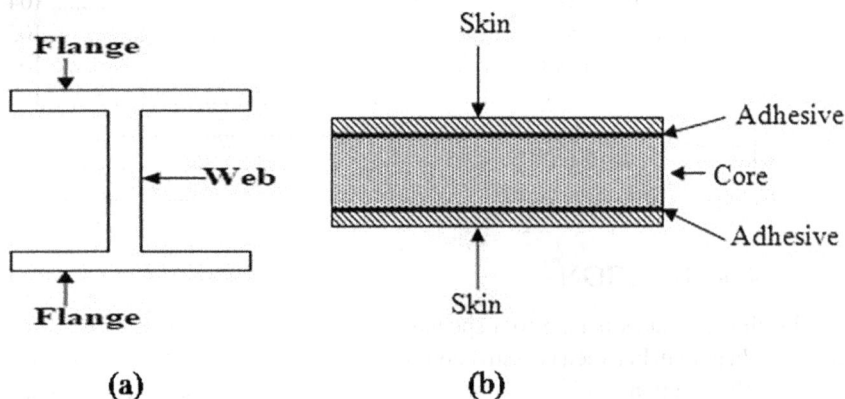

FIGURE 6.1   Analogy between (a) I-beam and (b) sandwich structure.

being extensively used in automobiles, aircraft, transportation and sports goods industries [13–21]. Out of all the existing NFs, jute turns out to be the most encouraging fiber to be used as reinforcement in composite due to its better properties such as tensile strength, elastic modulus and density [22].

Composites with particles as reinforcements are extensively used in sandwich structures as cores next to honeycomb cores due to the advantage of tailored mechanical, electrical and magnetic properties that can be achieved. Hence, they find a wide range of applications in weight-sensitive structural applications such as aircrafts, automobiles and sports goods. Sandwich structures making use of particulate composites as core materials possess enhanced specific strength of compression and bending stiffness [23]. Depending on the desired performance of the sandwich composites, a broad range of materials for core are available [24]. Few among them are the following:

1. Woods and foams having low density.
2. Honeycomb materials which exhibit high density.
3. Corrugated forms such as truss and corrugated sheets.

The contact between skin and core is affected by the core structure used in the composite. Low-density solid materials usually provide larger and continuous contact area of the core with the skin as opposed to expanded high-density core materials where the area of contact of the core with the skin is minimal. The selection of appropriate core material also depends on various other design parameters as per the requirement. Making use of cores with closed cell structures foams provides some advantage over the foams which are closed structures in terms of specific compressive strength since it is more in closed cell compared to open cell. Also closed cell absorbs less moisture compared to open cell [25].

The various studies available in the literature [26–28] concentrate on the impact response of monolithic laminated composites in low-velocity regime. However, it is found that there is relatively less work available on assessing the behavior of the sandwich composites subjected to impact and comprising polymeric core materials which exhibit much more complicated damage mechanism [29–33]. Out of the various types of cores available, less attention has been paid to assessing the effect of density of foam core on the impact response of the sandwich composites [6,7,32,34–36]. The study carried out by Caprino and Teti [34] on the sandwich composites comprising glass epoxy face sheets and polyvinyl chloride (PVC) foam core concluded that the density of the core played a prominent role in deciding the extent of damage in the sandwich composite since the contact force was found to increase with the density of core. The work reported by Anderson and Madenci [6] on carbon/PMMI foam core sandwich composite showed that the sandwich composite with a high-density core exhibited higher resistance to damage compared to sandwich composite having a core with low density. The experimental study carried out on the sandwich composites fabricated with PVC, poly urethane (PU) and polyethylene terephthalate (PET) as core materials with different densities showed that the character and degree of damage in face sheets are found to be dependent on the material properties of core [7,37]. The energy required for the initiation of damage [7] and the energy required for perforation [37] were observed to increase with an increase in density. Sandwich with PVC core possessing a density of 200 kg/m³ exhibited resistance to perforation eight times higher than the sandwich with PVC foam core having a density of 60 kg/m³. Comparative work executed by Vishwas et al. [38] exhibited the fact that the sandwich composites help in achieving better impact response compared to composite laminates.

A comparative study on the impact response of sandwich composite with glass as face sheet and PVC foam (density of 62 kg/m³) or balsa wood (density of 157 kg/m³) as core was carried out by Atas and Sevim [36]. The results revealed that the sandwich comprising PVC foam as core exhibited larger deflections and higher areas of delamination compared to its counterpart having balsa wood core. The literature [32,39,40] suggests that the impact resistance of sandwich composites comprising foam core is found to increase with increased strength and hardness of core material. The impact of core density on the impact response of any sandwich composite is therefore very important to evaluate.

Energy balance model is of primary importance to assess the behavior of sandwich composites subjected to impact loading with an assumption that the behavior of the target will be quasi-static and in bending, shear and contact deformations, the kinetic energy of the target is absorbed as in Eq. (6.1).

$$\frac{1}{2}mv^2 = E_{bs} + E_c \tag{6.1}$$

where the suffix b represents the bending stiffness, s represents the shear stiffness, and c represents the contact stiffness. The relation between force and displacement in case of a sandwich is represented by Eq. (6.2).

$$\delta = P\left[\frac{L^3}{48D} + \frac{L}{4AG}\right] \tag{6.2}$$

where G represents the modulus of shear related to foam core, L is the length, D represents the flexural rigidity of the face sheets, and A represents the geometrical parameter depending on the core thickness, material of face sheet and width of beam. In Eq. (11.2), the first term within the bracket $\frac{L^3}{48D}$ represents the bending of face sheets and the second term $\frac{L}{4AG}S$ represents the shear effect in the core. The energy absorption in bending and shear effects at maximum displacement is given by Eq. (6.3).

$$E_{bs} = \frac{P_{max}^2}{2}\left(\frac{L^3}{48D} + \frac{L}{4AG}\right) \tag{6.3}$$

Meyer's contact law is used to model the contact effects given by Eq. (6.4).

$$P = C\alpha^n \tag{6.4}$$

where C and n are experimentally determined constants.

The energy absorbed in contact stiffness is given by Eq. (6.5), and the energy balance for sandwich composite is given by Eq. (6.6).

$$E_c = \int_0^{\alpha_{max}} P\,d\alpha = \frac{C(P_{max}/C)^{(n+1)/n}}{n+1} \tag{6.5}$$

$$\frac{1}{2}mv^2 = \frac{P_{max}^2}{2}\left(\frac{L^3}{48D} + \frac{L}{4AG}\right) + \frac{C(P_{max}/C)^{(n+1)/n}}{n+1} \tag{6.6}$$

The above equation can be used to determine the maximum contact force once the face sheet and core properties are determined.

The present work deals with low-velocity impact behavior of sandwich structures consisting of jute/epoxy laminate as the face sheet and PVC foam core with core densities of 50, 80 and 120 kg/m³, and it also analyzes the peak force and energy absorption capabilities. Further, the statistical approach Taguchi's design of experiments (DOE) is made use of for analyzing the impact of select factors (core density and impact energy) on the energy absorption and peak force of the proposed sandwich composites. Signal-to-noise (SN) ratio, analysis of variance (ANOVA) and regression approach is adopted to select the factor that has a remarkable outcome on the results of the proposed composites.

6.2 MATERIALS AND METHODS

The composites consisting of jute/epoxy face sheets and PVC core with density varying from 50 to 120 kg/m³ are manufactured using a hand layup process. The schematic of the prepared composites is shown in Figure 6.2. The jute fabrics used to prepare the face sheets are in the form of bidirectional woven fabric having a density of 1.45 g/cm³.

After curing, the samples are cut to a dimension of 150×150 mm and subjected to impact testing under drop weight impact testing machine having a hemispherical impactor of mass 3.5 kg with an indenter diameter of 12 mm. Three drop heights of 400, 600 and 800 mm were selected so as to obtain impact velocities of 2.8, 3.43 and 3.96 m/s to provide impact energies of 13.72, 20.58 and 27.44 J,

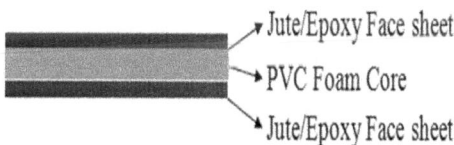

FIGURE 6.2   Schematic of the proposed sandwich composite.

TABLE 6.1

Factors and Their Levels Used in Taguchi's DOE

Factors	Level 1	Level 2	Level 3
Core density (kg/m³)	50	80	120
Impact energy (J)	13.72	20.58	27.44

respectively. The peak force and the energy data were obtained from the data acquisition system.

Further, in order to analyze the effect of core density and impact energy on the low-velocity impact behavior of the proposed sandwich composites, Taguchi's DOE method is adopted with L9 orthogonal array. The details about the factors and their levels are presented in Table 6.1.

6.3   RESULTS AND DISCUSSION

6.3.1   EFFECT OF DENSITY OF CORE ON STRUCTURAL RESPONSE

Figure 6.3 shows the stress–strain behavior of the PVC core with different core densities, which can be further used to assess the effect of density of core on the structural response of the PVC core.

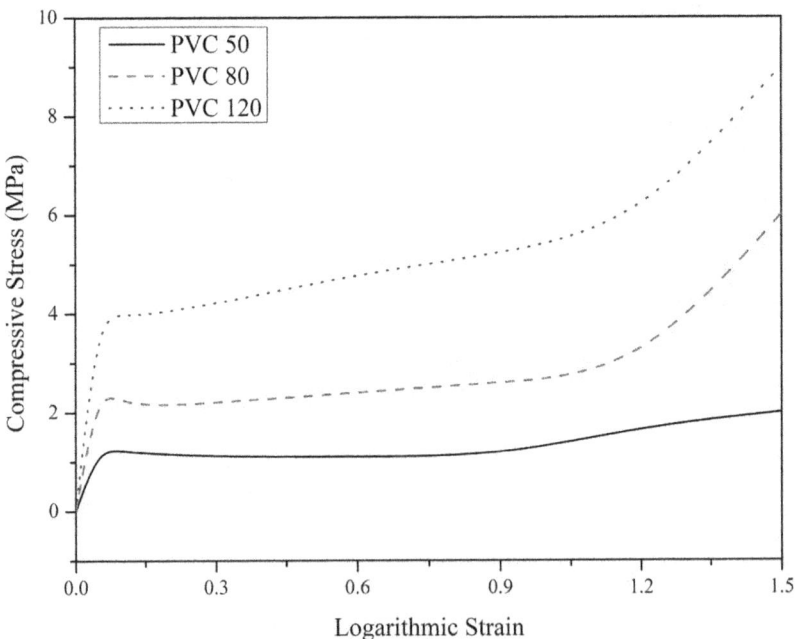

FIGURE 6.3   Stress–strain behavior of PVC core with different densities.

In order to assess the compressive properties of the PVC foam material with different densities, a nominal strain rate of 0.1/s is used. From Figure 6.3, it can be seen that the material exhibits initial elastic behavior followed by long stress plateau associated with buckling of walls of cell, also known as crushing. The final stage represents a steep increase in stiffness, indicating a densification stage, since the compaction of the material takes place. The stress–strain curve represents that the mechanical response of the PVC foam is extensively dependent on the density of the material. An increase in the density of the material results in a significant increase in the elastic modulus and plateau stress.

6.3.2 Force and Energy Response

The energy absorption and peak force obtained from the experimental results are made use of to analyze the impact behavior of the sandwich composites. The three different composites used along with their designations are presented in Table 6.2.

The trend of force vs time graph obtained for the proposed sandwich composites at an impact energy of 13.72 J is shown in Figure 6.4.

It can be visualized from the force curves that the proposed sandwich composite with higher core density exhibits higher peak force and as the core density of the proposed sandwich composite reduces, the peak force reduces. The peak forces of CVPC120, CVPC80 and CVPC50 are 4,605.98, 3,900 and 3,274.14 N, respectively. This indicates that as the density of the core increases, the resistance to damage of the proposed composite is enhanced. Thus, the density of the core plays a vital role in deciding the structural behavior of the sandwich composites. The contact duration of the proposed sandwiches also reduces with enhanced core density. The trend of variation of peak force at other impact energies of 20.58 and 27.44 J remains same. The comparison of the peak forces at different impact energies is shown in Figure 6.5. It can be seen that with an increase in impact energy, the peak force of all the proposed sandwiches is increased. However, the increases in case of CVPC80 and CVPC120 are more significant compared to CVPC50. This indicates that the density of the core plays a significant role in mitigating the damage.

The energy proposed by the proposed sandwich composites at different impact energies is presented in Figure 6.6. The impact velocities and impact energies are calculated according to Eqs. (6.7) and (6.8), respectively.

TABLE 6.2
Sandwich Composites and Their Designations

Sandwich Composite	Designation
Jute/epoxy face sheet with PVC core of 50 kg/m³	CPVC50
Jute/epoxy face sheet with PVC core of 80 kg/m³	CPVC80
Jute/epoxy face sheet with PVC core of 120 kg/m³	CPVC120

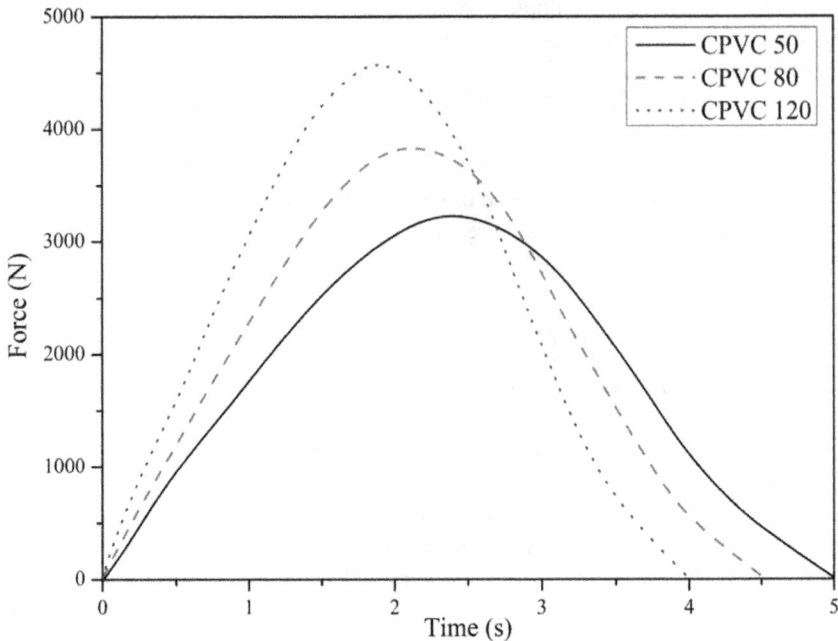

FIGURE 6.4 Force trend for various proposed sandwich composites.

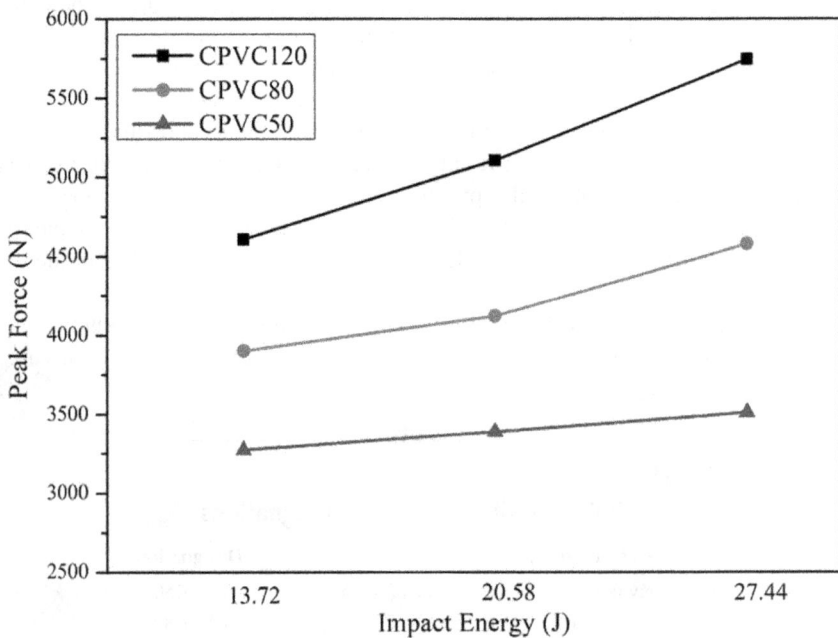

FIGURE 6.5 Variation of peak force with impact energy.

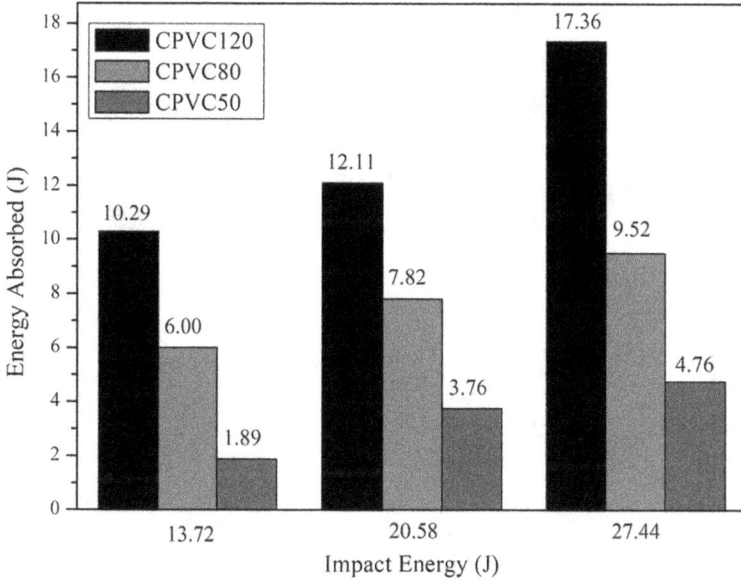

FIGURE 6.6   Energy absorbed by the proposed sandwich composites at different impact energies.

$$v = \sqrt{2gh} \tag{6.7}$$

$$E_i = \frac{1}{2}mv^2 \tag{6.8}$$

where v is the impact velocity in m/s, g is the acceleration due to gravity (9.81 m/s²), h is the drop height in m, m is the mass of impactor in kg (3.5 kg), and E_i is the impact energy in joule.

It can be seen that as the impact energy increases, the energy absorbed by the proposed sandwich composites is enhanced, with CVPC120 absorbing more energy compared to its counterparts. The denser core helps in absorbing more amount of energy. It can be seen that the CVPC120 absorbs 1.8 times and nearly 3.6 times more energy compared to CVPC80 and CVPC50 at an impact energy of 27.44 J.

6.3.3   DAMAGE ANALYSIS

The damage analysis of the proposed sandwich structure is presented in and analyzed through Figure 6.7.

It is evident from the damage morphology that the top face sheets of the proposed composites undergo damage/failure due to matrix cracking and breakage. However, due to the presence of the core material, the mitigation of damage is minimized to the bottom face sheet. It is clearly visible that the degree of damage

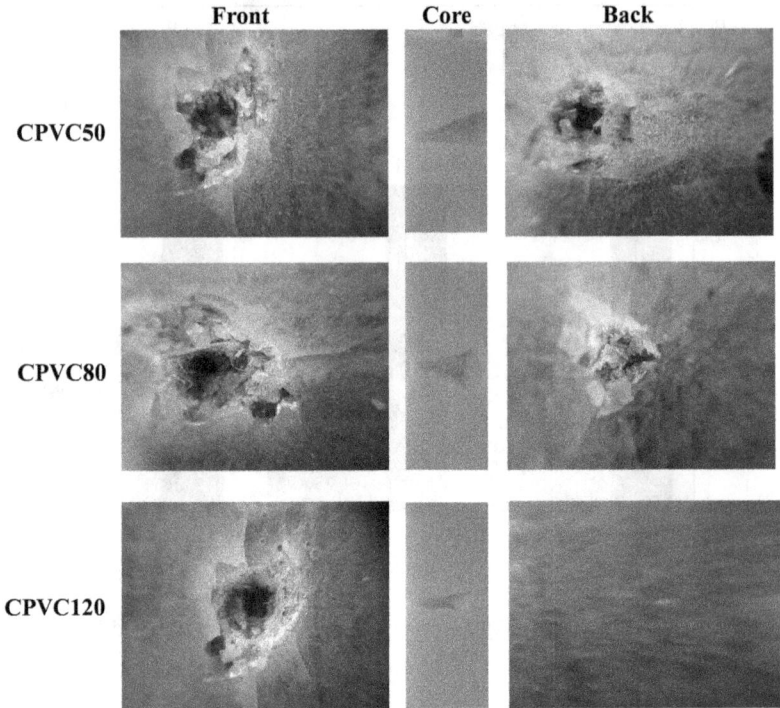

FIGURE 6.7 Damage morphology of the proposed sandwich composites.

in the bottom face sheet of CPVC50 is maximum compared to its counterparts. The extent of damage in the bottom face sheet of CPVC80 is minimized compared to CPVC50. Further, it is seen that there is no damage to the bottom face sheet of CPVC120. This indicates that the core density plays a prominent role in mitigating the damage in case of sandwich composites. The sign of delamination is also found on the back face sheets of CPVC80 and CPVC50, which is an indication of core crushing. This damage morphology is supported by the peak force curve which indicates that the proposed sandwich composite CPVC120 exhibits better resistance to damage compared to its counterparts.

6.3.4 TAGUCHI ANALYSIS

The SN ratio is made use of to determine the most influential factor on the response (absorbed energy and peak force) of the proposed composites. Since maximum energy absorption and higher peak force are desired, the SN ratio with higher the better criteria is calculated using Eq. (6.3)

$$S/N = -10\log_{10}\left(\sum \left(1/y\right)^2 / n\right) \tag{6.3}$$

The response table for SN ratio is provided in Table 6.3.

TABLE 6.3

SN Ratio Response for Absorbed Energy

	Core Density	Impact Energy	Core Density	Impact Energy
Level	Energy Absorption		Peak Force	
1	10.20	13.78	70.60	71.80
2	17.67	17.01	72.44	72.35
3	22.23	19.31	74.20	73.10
Delta	12.04	5.52	3.60	1.30
Rank	1	2	1	2

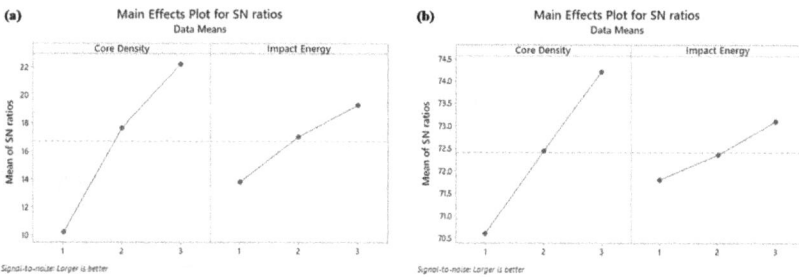

FIGURE 6.8 Main effect plot for SN ratio of (a) energy absorbed and (b) peak force.

Looking at Table 6.3, it can be concluded that the core density is the most influential factor that affects the energy absorption and peak force compared to impact energy. The main effect plot for SN ratio for both energy absorption and peak force is shown in Figure 6.8.

The main effect plots indicate that for both energy absorption and peak force, the composite having a higher core density provides better response at a higher impact energy. Thus, the proposed sandwich composite CVPC120 proves to be the better one among the counterparts. The conclusions put forth by SN ratio analysis are supported by ANOVA, which is presented in Table 6.4.

The percentage contribution of each of the factors for the absorbed energy and peak force is determined using Eq. (6.4).

$$\% \text{ Contribution} = \left(\frac{SS_f}{SS_t} \right) \times 100 \qquad (6.4)$$

where SS_f is the summation of squares for a factor and SS_t is the total sum of squares.

It can be seen from the percentage contribution that the core density is the factor that affects both the absorbed energy and peak force compared to impact energy. Coefficient of determination of the corresponding equation is measured through the R^2 value. An R^2 value more than 95% is an indication of good model

TABLE 6.4
ANOVA for Energy Absorption and Peak Force

Source	Response	DF	SS	MS	F value	P Value	% Contribution
Core density	Absorbed energy	2	144.220	72.110	41.95	0.002	**79.41**
	Peak force	2	4664461	2332231	44.81	0.002	**83.6**
Impact energy	Absorbed energy	2	30.513	15.256	8.88	0.034	16.80
	Peak force	2	706830	353415	6.79	0.052	12.66
Error	Absorbed energy	4	6.875	1.719			3.78
	Peak force	4	208175	52044			3.73
Total	Absorbed energy	8	181.608				
	Peak force	8	5579467				

Absorbed energy: $S = 1.31105$, $R^2 = 96.21\%$, R^2 (adj) $= 92.43\%$
Peak force: $S = 228.131$, $R^2 = 96.27\%$, R^2 (adj) $= 92.54\%$

The bold values indicate that their percentage contribution is high

being developed and can be confidently used to predict the response within the experiment conditions.

The models pertaining to regression are further arrived at for predicting the responses (absorbed energy and peak force). Corresponding regression equations for absorbed energy and peak force are given through Eqs. (6.5) and (6.6), respectively.

$$\text{Absorbed energy} = -6.10 + 4.891 \text{ core density} + 2.243 \text{ impact energy} \qquad (6.5)$$

$$\text{Peak force} = 1,803 + 880.7 \text{ core density} + 341.3 \text{ impact energy} \qquad (6.6)$$

It can be analyzed from the above equations that the core density has the highest coefficient, indicating that it is the factor that affects corresponding response compared to absorbed energy. The experimental and the predicted values obtained through regression equations are tabulated in Table 6.5. It can be seen that the experimental and predicted values agree well with each other, indicating that the developed regression equations are valid.

6.4 CONCLUSIONS

The present study concentrates on the low-velocity impact behavior of sandwich composites with three different core densities and three different energies of impact. Based on the experimental studies carried out, it is concluded that:

- The proposed sandwich composite with higher core density (CPVC120) exhibits higher peak force. This indicates that as the density of the core increases, the resistance to damage of the proposed composite is enhanced.

TABLE 6.5
Experimental and Predicted Values of Absorbed Energy and Peak Force

Core Density (Levels)	Impact Energy (Levels)	Exptl. Energy Absorbed (J)	Predicted Energy Absorbed (J)	Exptl. Peak Force (N)	Predicted Peak Force (N)
1	1	1.89	1.03	3,274.14	3,025
1	2	3.76	3.28	3,387.67	3,366.3
1	3	4.76	5.52	3,508.31	3,707.6
2	1	6.00	5.93	3,900.00	3,905.7
2	2	7.82	8.17	4,119.62	4,247
2	3	9.52	10.41	4,576.51	4,588.3
3	1	10.29	10.82	4,605.98	4,786.4
3	2	12.11	13.06	5,105.67	5,127.7
3	3	17.36	15.30	5,742.83	5,469

- The increase in peak force is more drastic in case of CVPC80 and CVPC120 compared to CVPC50. This indicates that the density of the core plays a significant role in mitigating the damage.
- As the impact energy increases, the energy absorbed by the proposed sandwich composites is enhanced, with CVPC120 absorbing more energy compared to its counterparts. It can be seen that the CVPC120 absorbs 1.8 times and nearly 3.6 times more energy compared to CVPC80 and CVPC50 at an impact energy of 27.44 J.
- The top face sheets of the proposed composites undergo damage/failure due to matrix cracking and breakage. However, due to the presence of the core material, the mitigation of damage is minimized to the bottom face sheet with damage in bottom face sheet of CPVC50 being maximum compared to its counterparts. The extent of damage in bottom face sheet of CPVC80 is minimized compared to CPVC50. Further, it is seen that there is no damage to the bottom face sheet of CPVC120. This indicates that the core density plays a prominent role in mitigating the damage in case of sandwich composites. The sign of delamination is also found on the back face sheets of CPVC80 and CPVC50, which is an indication of core crushing.
- The statistical approach adopted shows that the core density is the factor that affects the energy absorption and damage mitigation significantly compared to impact energy in case of the proposed sandwich composites.

ACKNOWLEDGMENT

The author Vinyas Mahesh acknowledges the support of Indian Institute of Science, Bangalore, through C.V. Raman Post-doctoral Fellowship R(IA)/CVR-PDF/2019/1630, under Institution of Eminence scheme.

REFERENCES

1. Ferri R, Sankar B. Static indentation and low velocity impact test on sandwich plates. *Proc. ASME Aerosp. Div.* 1997;55:485–490.
2. Carlsson LA, Sendlein LS, Merry SL. Characterization of face sheet/core shear fracture of composite sandwich beams. *J. Compos. Mater.* 1991;25:101–116.
3. Fatt H, Park K. Dynamic models for low velocity impact damage of composite sandwich panels – Part A: deformation. *Compos. Struct.* 2001;52:335–351.
4. Daniel IM, Abot JL. Fabrication, testing and analysis of composite sandwich beams. *Compos. Sci. Technol.* 2000;60:2455–2463.
5. Zee RH, Hsieh CY. Energy absorption processes in fibrous composites. *Mater. Sci. Eng. A.* 1998;246:161–168.
6. Anderson T, Madenci E. Experimental investigation of low-velocity impact characteristics of sandwich composites. *Compos. Struct.* 2000;50:239–247.
7. Akil Hazizan M, Cantwell WJ. The low velocity impact response of foam-based sandwich structures. *Compos. Part B Eng.* 2002;33:193–204.
8. Wardle MW, Tokarsky EW. Drop weight impact testing of laminates reinforced with Kevlar® aramid fibers, e-glass, and graphite. *J. Compos. Technol. Res.* 1983;5:4–10.
9. Mahinfalah M, Skordahl RA. The effects of hail damage on the fatigue strength of a graphite/epoxy composite laminate. *Compos. Struct.* 1998;42:101–106.
10. Adams DF, Miller AK. An analysis of the impact behavior of hybrid composite materials. *Mater. Sci. Eng.* 1975;19:245–260.
11. Bhuiyan MA, Hosur M V., Jeelani S. Low-velocity impact response of sandwich composites with nanophased foam core and biaxial (± 45°) braided face sheets. Compos. *Part B Eng.* [Internet]. 2009;40:561–571. Available from: Doi: 10.1016/j.compositesb.2009.03.010.
12. Burton WS, Noor AK. Structural analysis of the adhesive bond in a honeycomb core sandwich panel. *Finite Elem. Anal. Des.* 1997;26:213–227.
13. Shah DU. Natural fibre composites: Comprehensive Ashby-type materials selection charts. *Mater. Des.* [Internet]. 2014;62:21–31. Available from: Doi: 10.1016/j.matdes.2014.05.002.
14. Dicker MPM, Duckworth PF, Baker AB, et al. Green composites: A review of material attributes and complementary applications. *Compos. Part A Appl. Sci. Manuf.* [Internet]. 2014;56:280–289. Available from: Doi: 10.1016/j.compositesa.2013.10.014.
15. Faruk O, Bledzki AK, Fink HP, et al. Biocomposites reinforced with natural fibers: 2000–2010. *Prog. Polym. Sci.* [Internet]. 2012;37:1552–1596. Available from: Doi: 10.1016/j.progpolymsci.2012.04.003.
16. Ahmad F, Choi HS, Park MK. A review: Natural fiber composites selection in view of mechanical, light weight, and economic properties. *Macromol. Mater. Eng.* 2015;300:10–24.
17. Mahesh V, Joladarashi S, Kulkarni SM. Experimental study on abrasive wear behaviour of flexible green composite intended to be used as protective cladding for structures. *Int. J. Mod. Manuf. Technol.* 2019;XI:69–76.
18. Mahesh V, Joladarashi S, Kulkarni SM. An experimental study on adhesion, flexibility, interlaminar shear strength, and damage mechanism of jute/rubber-based flexible "green" composite. *J. Thermoplast. Compos. Mater.* 2019:1–28.
19. Mahesh V, Joladarashi S, Kulkarni SM. Damage mechanics and energy absorption capabilities of natural fiber reinforced elastomeric based bio composite for sacrificial structural applications. *Def. Technol.* [Internet]. 2020; Available from: Doi: 10.1016/j.dt.2020.02.013.

20. Mahesh V, Joladarashi S, Kulkarni SM. Development and mechanical characterization of novel polymer-based flexible composite and optimization of stacking sequences using VIKOR and PSI techniques. *J. Thermoplast. Compos. Mater.* 2019;34(8):1080–1102.

21. Mahesh V, Joladarashi S, Kulkarni SM. An experimental investigation on low-velocity impact response of novel jute/ rubber flexible bio-composite. *Compos. Struct.* 2019;225(111190):1–12.

22. Célino A, Fréour S, Jacquemin F, et al. The hygroscopic behavior of plant fibers: A review. *Front. Chem.* 2014;1:1–12.

23. Chittineni K. Functionally Gradient Syntactic Foams. Master of Science Thesis, Graduate Faculty of the Louisiana State University and Agricultural and Mechanical College, Louisiana; 2009.

24. Vinson J. *The Behavior of Sandwich Structures of Isotropic and Composite Materials.* 1st ed. CRC Press, Boca Raton, FL; 1999.

25. Gupta N. *Characterization of Syntactic Foams and their Sandwich Composites: Modelling and Experimental Approaches.* Graduate Faculty of the Louisiana State University and Agricultural and Mechanical College, Louisiana; 2003.

26. Mathivanan NR, Jerald J. Experimental investigation of woven E-glass epoxy composite laminates subjected to low-velocity impact at different energy levels. *J. Miner. Mater. Charact. Eng.* [Internet]. 2010;09:643–652. Available from: http://www.scirp.org/journal/PaperDownload.aspx?DOI=10.4236/jmmce.2010.97046.

27. Ansari MM, Chakrabarti A. Impact behavior of FRP composite plate under low to hyper velocity impact. *Compos. Part B Eng.* [Internet]. 2016;95:462–474. Available from: Doi: 10.1016/j.compositesb.2016.04.021.

28. Davies GAO, Hitchings D, Zhou G. Impact damage and residual strengths of woven fabric glass/polyester laminates. *Compos. Part A Appl. Sci. Manuf.* [Internet]. 1996,27.1147–1156. Available from: Doi: 10.1016/1359-835X(96)00083-8.

29. Belingardi G, Cavatorta MP, Duella R. Material characterization of a composite-foam sandwich for the front structure of a high speed train. *Compos. Struct.* 2003;61: 13–25.

30. Xia F, Wu X. Work on low-velocity impact properties of foam sandwich composites with various facesheets. *J. Reinf. Plast. Compos.* 2010;29:1045–1053.

31. Chai GB, Zhu S. A review of low-velocity impact on sandwich structures. *Proc. Inst. Mech. Eng. Part L J. Mater. Des. Appl.* 2011;225:207–230.

32. Daniel IM, Abot JL, Schubel PM, et al. Response and damage tolerance of composite sandwich structures under low velocity impact. *Exp. Mech.* 2012;52:37–47.

33. Feng D, Aymerich F. Experimental and numerical investigation into the damage response of composite sandwich panels to low-velocity impact. *Struct. Monit. Maint.* 2017;4:133–151.

34. Caprino G, Teti R. Impact and post-impact behavior of foam core sandwich structures. *Compos. Struct.* 1994;29:47–55.

35. Dogan A, Arikan V. Low-velocity impact response of E-glass reinforced thermoset and thermoplastic based sandwich composites. *Compos. Part B Eng.* 2017;127: 63–69.

36. Atas C, Sevim C. On the impact response of sandwich composites with cores of balsa wood and PVC foam. *Compos. Struct.* [Internet]. 2010;93:40–48 Available from: Doi: 10.1016/j.compstruct.2010.06.018.

37. Zhou J, Hassan MZ, Guan Z, et al. The low velocity impact response of foam-based sandwich panels. *Compos. Sci. Technol.* [Internet]. 2012;72:1781–1790. Available from: Doi: 10.1016/j.compscitech.2012.07.006.

38. Mahesh V, Joladarashi S, Kulkarni S. Investigation on effect of using rubber as core material in sandwich composite plate subjected to low velocity normal and oblique impact loading. *Sci. Iran. Trans. Mech. Eng.* 2019;26:897–907.

39. Leijten J, Bersee HEN, Bergsma OK, et al. Experimental study of the low-velocity impact behaviour of primary sandwich structures in aircraft. *Compos. Part A Appl. Sci. Manuf.* [Internet]. 2009;40:164–175. Available from: Doi: 10.1016/j.compositesa.2008.10.019.

40. Flores-Johnson EA, Li QM. Experimental study of the indentation of sandwich panels with carbon fibre-reinforced polymer face sheets and polymeric foam core. *Compos. Part B Eng.* 2011;42:1212–1219.

7 High-Velocity Impact Properties of the Composite Sandwich Panels

Hossein Ebrahimnezhad-Khaljiri
Faculty of Materials Science and Engineering
K.N. Toosi University of Technology

CONTENTS

7.1   Introduction 115
7.2   High-Velocity Impact Properties of Sandwich Panels...........................117
    7.2.1   Polymeric Core Sandwich Panels...117
    7.2.2   Metallic Core Sandwich Panels...119
    7.2.3   Bio-based Core Sandwich Panels 122
    7.2.4   Hybrid Core Sandwich Panels 122
    7.2.5   New Advanced Sandwich Panels..123
    7.2.6   Comparative Study 124
7.3   Conclusions...127
References..127

7.1   INTRODUCTION

The sandwich structures or panels are one of the new hybrid materials, which can be used in many applications. These materials are introduced as a subset of multilayer composite structures, optimized for use in various industries such as automotive, aerospace, marine and energy industries [1,2]. The sandwich panel consists of outer layers usually made up of composites or metals having high stiffness and a softer core material with lower density embedded between them [1,3]. Although the combination of these materials creates a proper structure for many engineering applications, sandwich panels are susceptible to impact damages, so the mechanical properties of these structures reduce after impact damaging. In this regard, to improve the resistance of sandwich panels against impact damaging, many efforts have been made by various research groups in the past years, especially in last decade. Selecting the materials with better impact performance and creating smart behaviors can be examples of these efforts [4,5].

DOI: 10.1201/9781003143031-7

To understand the performance of sandwich panels under impact damaging, it is better to introduce various types of impact tests. The impact tests are usually classified into three major groups, given the impactor speed. The low-impact test is defined as a test with an impactor speed lower than 10 m/s, which is commonly used for testing sandwich structures. In the medium-impact test, the initial velocity of the impactor can be 10–50 m/s, which is conventionally done by the drop weight impact test. The third group of impact tests have the highest initial velocity, around 50–1,000 m/s [2]. The high-velocity impact test shows the behavior of matter under high dynamic loadings or high strain rates [6]. Up to now, many efforts have been made to study the behavior of sandwich panels under high-velocity impact test. It is very important to understand how to classify these scientific reports.

As can be seen in Figure 7.1, the study on the impact behavior of sandwich panels can be categorized from the geometry point of view. In practice, honeycomb, cork and foam cores are the most common types of core geometries. The honeycomb cores are extremely used in the aerospace and automotive industries, due to the impressive impact resistance. Some new core geometries such as corrugated and Y-shape geometries have been investigated in past years, because of their better energy absorption as compared to the conventional core geometries. The research about the pyramidal cores shows that they have superior specific capability to disperse the impactor energy, as compared to honeycomb cores [3].

Although categorizing the sandwich panels from the point of view of geometry is reasonable, it cannot completely include all the sandwich panels, especially the new types of sandwich panels. Therefore, for studying the behavior of sandwich

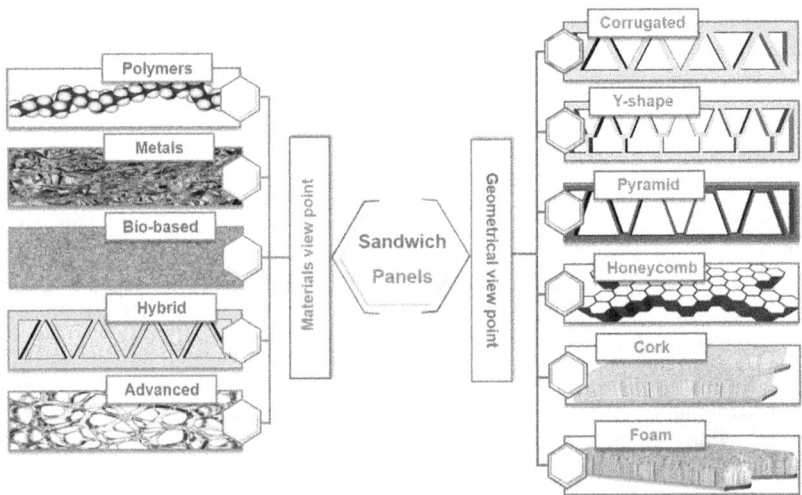

FIGURE 7.1 The classification of composite sandwich panels from the materials and geometrical points of view for subjecting under the high-velocity impact conditions.

panels under high-velocity impacts, the second classification is introduced in this chapter, which is the classification based on the materials used, especially the cores in the sandwich panels. The polymeric and metallic cores are conventional cores. The other group is the bio-based cores, which are balsa-, cork- and cellulose-based cores. The fiber metal laminates (FMLs) as face-sheets and 3D braided fibers as cores can be introduced as new advanced materials. Hybrid cores can be a separate group. Actually, in this group, two different kinds of materials can be simultaneously used as core or face-sheet. In light of the above, in this chapter, the impact behaviors of sandwich panels under high-velocity impacts are surveyed according to the second classification, aiming at achieving more comprehensive investigations.

7.2 HIGH-VELOCITY IMPACT PROPERTIES OF SANDWICH PANELS

7.2.1 POLYMERIC CORE SANDWICH PANELS

One of the best candidates for use as lightweight core materials in sandwich panels is the polymeric cores, especially polymeric foams. Therefore, comparing the behavior of various types of polymeric cores under high-velocity impacts can give proper view about their use in sandwich panels. To do so, the admirable research works are summarized in the following in order to obtain the appropriate perspective of the function of polymeric cores in the sandwich panels. One of the effective factors on the impact behaviors of foam core sandwich panels is the foam density. In the study of Nasirzadeh and Sabet [7], the polyurethane (PU) foam with various densities of 37, 49, 70, 95, 105 and 240 kg/m^3 was embedded between three-layered E-glass/epoxy composites. Then, the impact behavior of these composites was compared according to the obtained residual velocities, absorbed energies and specific absorbed energies. The lowest residual velocity and highest specific absorbed energy belonged to the sandwich panel containing PU core with the density of 49 kg/m^3 in various initial velocities. Also, the fracture pattern of this core showed extensive delamination between the foam core and back facing, with impactor yawing as the absorption mechanism. These researchers in another work [8] added various weight percentages of nanoclay (0, 0.25, 0.5, 1 and 3 wt.%) to the PU foam for enhancing the impact resistance of sandwich panels. They reported that the panel with 0.5 wt.% nanoclay added to the PU foam had the highest impact performance and energy absorption, as well as the lowest impactor residual velocity in various initial impact velocities. The results showed that the nanoclay simultaneously increased the number of cells per unit volume and the cell wall thickness, which reduced the foam density. Also, the scanning electron microscope (SEM) analysis depicted that the cells containing 0.25 and 0.5 wt.% nanoclay collapsed, whereas the foam with 1 and 3 wt.% nanoclay broke up to the lower region.

Another effective factor on the impact behavior of sandwich panels can be the thickness of the core materials. The simplest way of assessing that is increasing the number of foam layers in the sandwich panels. Also, if the density of each layer is different, the effect of this factor can be more effective. For example, Abbasi and Alavi Nia [9] used Al 1050 as skin and different layers of PU foam (1, 2, 3, 4 and 5) as core material in the sandwich panels. It should be noted that the density of each layer was different. The reported results showed that the panel with four layers of PU foam had the maximum limit velocity (as sign of the best performance), compared with other panels. Based on this work, it can be said that by increasing the density of foam core from the impact side of sandwich panel, the impact properties can be improved. With this view point, Alavi Nia and Kazemi [10] investigated the impact resistance of sandwich panels containing three- and five-layer PU foam with various foam densities. Similar results were reported in this work. The characterized fracture mechanisms were the shear plug on the impact side and petalling, plugging and dishing on the back face of the sandwich panels.

Although the PU foams are the most famous polymeric cores in sandwich panels, the polyvinyl chloride (PVC) foam and low-density polyethylene (LDPE) can also be used as core in sandwich panels [7,11]. For example, Ren et al. [11] used PVC foam with various configurations and core densities to investigate the high-velocity impact behavior of PVC–5A06 aluminum alloy, which were glued by the epoxy resin. They mentioned that increasing the density of PVC core could effectively improve the impact resistance. But the sequence of core layers had a negligible effect on the impact resistance of the sandwich panel. The noteworthy report in this work was the definition of energy absorption factor for comparing the impact resistance of different sandwich panels. This factor was calculated based on the energy absorption of panels per the final deflection of the center point of rear sheet. The calculation of the deflection was done by 3D digital image correlation (3D-DIC) method. This technique can give better comparable data than other normalized energy absorption methods about the impact behavior of sandwich panels under dynamic impact loadings.

The other important factor in the foam-based sandwich panels is the use of hybrid materials as skin or core. For example, Rolfe et al. [12] used PU foam as core and hybrid glass–carbon fibers/epoxy composites as skin of sandwich panels for studying the impact behavior by the 3D-DIC method. They discussed that the hybrid skins were more prone to delamination especially between the dissimilar materials. Therefore, because of that, the hybrid skin sandwich panels had the lower deflection and strain than the non-hybrid sandwich panels. This means that the hybrid skins have the higher capability to dissipate the energy of impactor. In the interesting study of Hassanpour Roudbeneh et al. [13], the effect of core density and hybrid cores on the impact resistance of sandwich panels was simultaneously compared. These researchers used PU foam with the different densities for filling the aluminum honeycomb as hybrid cores in the sandwich panels. The numerical and experimental results showed that the sandwich panel with the hybrid core had more than two times the impact resistance than the un-filled honeycomb sandwich panels. By increasing the foam density in the hybrid core, this resistance can be increased to about four times.

Although the foam-based core is more favorable for use in sandwich panels, some scientists have made many efforts to use another form of polymeric cores. Namely, Wang et al. [14] subjected the carbon fibers/epoxy composite sandwich panel including the carbon fibers/epoxy pyramidal lattice core to high-velocity impact ranging from 180 to 2,000 m/s. Then, they compared their sandwich panels with the aluminum and steel truss core sandwich structures. The comparison between the specific absorbed energy of these structures showed that the sandwich panel with pyramidal polymeric composite core had the higher impact resistance than the other mentioned structures. For a better understanding of the impact resistance of polymeric core sandwich panels, the results reported in the mentioned studies are listed in Table 7.1.

7.2.2 METALLIC CORE SANDWICH PANELS

The second group of sandwich panels can be metallic core-based panels. Given the research about these structures, it can be found that the metallic foam and honeycomb geometries are more favorable for use in impact conditions [15–17], as compared to other core geometries. Also, it was realized that the focus of scientists on the metallic composite structures has been more than on other composites. In other words, in sandwich panels with metallic core, most of the selected face-sheet was of metallic materials, especially aluminum alloys [18]. Some of the top researches about these structures under impact loading are introduced in the following. Also, the impact properties obtained in these researches are listed in Table 7.2. Khaire and Tiwari [19] fabricated sandwich panels with aluminum hexagonal honeycomb as core and AA-1100 as face-sheet and subjected them to impact loads. They investigated the effect of face-sheet thickness, cell wall thickness and cell size of honeycomb core on the impact response of sandwich panels. They mentioned that the face-sheet thickness had the dominant effect on the limit velocity and absorbed energy of metallic honeycomb core/metal shell panels. Also, by increasing the stiffness of these panels through enhancing the cell thickness or reducing the cell size, the impact resistance of sandwich panels was increased. In a similar work, Zhang et al. [20] investigated the effect of cell density, face-sheet thickness and impactor shape on the impact resistance of sandwich panels. Like previous work, the face-sheet thickness was more effective than the cell density on the energy absorption. Also, the deformation of sandwich panels under the impact of flat-nosed and conic-nosed impactor was pure shearing and totally tearing, respectively. This means that their panels were more protective against the conic-nosed impactor.

Hou et al. [21] used Al-5005 as face-sheet and aluminum foam as core for investigating the impact performance. The variations of this work were the face-sheet thickness (0.6, 1.0, 1.5 and 2.0 mm), foam density (5%, 10%, 15%, 18% and 20%) and impactor geometry (flat-ended, hemispherical and conical). It was found that the thicker face-sheet and cores with higher thickness and density were prone to producing higher limit velocity. Also, the panels with thicker skins tended to create a larger delamination area between the core and back face. The blunter

TABLE 7.1
The High-Velocity Impact Properties of Polymeric Core Sandwich Panels

Face-Sheet	Core	Impact Velocity (m/s)	Residual Velocity (m/s)	Limit Velocity (m/s)	Absorbed Energy (J)	Specific Absorbed Energy or Efficiency Energy Absorption	References
Glass fibers–polyester	PU foam	115.2	0	115	72	0.9 J/g	[7]
		151.4	85		87	1.15 J/g	
Glass fibers–polyester	PU foam–nanoclay	129	50	116	76	-	[8]
		136	65		78	-	
Aluminum	PU foam–aluminum honeycomb	90	37.9	82	-	815 J/kg	[13]
Carbon fibers–epoxy	Pyramidal carbon fibers–epoxy	500	400	-	-	0.022 m²/kg	[14]
		1,300	1,200	-	-	0.018 m²/kg	

TABLE 7.2
The High-Velocity Impact Properties of Metallic Core Sandwich Panels

Face-Sheet	Core	Impact Velocity (m/s)	Residual Velocity (m/s)	Limit Velocity (m/s)	Absorbed Energy (J)	Specific Absorbed Energy or Efficiency Energy Absorption	References
Aluminum	Aluminum honeycomb	191.3	133.2	137.4	98.1	-	[20]
Aluminum	Aluminum foam	186.4	143.9	109.6	49.3	-	[21]
Aluminum	Aluminum–SiC$_p$	181	104	148	153.6	529.3 J/kg	[22]
Ti	Aluminum honeycomb	141	0	141	238.6	1028 J/kg	[25]
CFRP	Aluminum honeycomb	191	174	-	66.7	-	[31]

impactor created a larger petalling area and had the tendency to enhance the limit velocity and energy dissipation. Golestanipour et al. [22] fabricated Al 1100 face-sheet/Al A356–SiC$_p$ sandwich panels. In other words, they used metallic composite core in their study. The effects of face-sheet thickness, density and thickness of aluminum composite foam were investigated in this work. The obtained results showed that these sandwich panels had similar behavior like those in the previous work, except that increasing the foam thickness had no noteworthy effect on the function of energy absorption.

Although using aluminum face-sheet and core is more favorable in sandwich panels, other metals such as steels have rarely been used as core and face-sheet in these structures. Namely, Liang et al. [23] fabricated steel face-sheet/aluminum foam core sandwich panels. The interesting idea in this work was the use of polyvinylidene fluoride (PVDF) transducers for measuring the stress variations after the high-velocity impact test. The results showed that about 90% of stress was reduced, after impact testing. This means that this structure had an excellent energy absorption. Using steel alloys in sandwich panels is not very common, due to their high density, which in turn results in an increase in the weight of the sandwich panels. But it seems that the use of austenitic steels is reasonable compared to others because they are tougher than other steel alloys [24]. Titanium alloys are one of the new candidates that can be used in sandwich panels, due to their low weight, high strength and excellent corrosion resistance. Although using titanium as face-sheet in sandwich panels can be easy, using the same as core is rare because of the need of highly advanced technology for fabricating titanium cores [25]. Novak et al. [26] fabricated aluminum face-sheet/titanium core sandwich panels and subjected them to impact tests. They used selective electron beam melting for producing the core with chiral auxetic geometry.

In new research works, some researchers used polymeric composite face-sheet instead of metallic face-sheet in sandwich panels with metallic core because polymeric composite face-sheets have lower weight and higher mechanical properties than metallic face-sheets. The glass fibers/epoxy and carbon fibers/epoxy composites are examples of proper composite face-sheets in sandwich panels [27,28]. Namely, Tang et al. [29] fabricated honeycomb core/carbon fibers-epoxy composite face-sheet sandwich panels and investigated their impact properties by using the high-velocity impact test. The microscopic investigation showed that the carbon fibers with the cracking, delamination and pulling out mechanisms improved the impact resistance of sandwich panels, which are the similar mechanisms previously reported in other sandwich structures [30]. In a similar work, Tang et al. [31] used aluminum foam instead of aluminum honeycomb in the sandwich panel. In this work, the effects of density and velocity of impactor on the impact resistance of these sandwich panels were investigated. They reported that by increasing the impactor velocity, the difference between the peak values of impact pressure of aluminum foam was increased. This means that the aluminum foam core had the attenuation effect on strong shock weave.

7.2.3 Bio-based Core Sandwich Panels

As we know, the face-sheet acts as the load-bearing structure, whereas the core structure serves to absorb the energy of impactor. Therefore, the core structures have an important role in the impact properties of sandwich panels. The new requirements and environmental issues have caused the use of bio-based cores in sandwich panels [32]. For example, Jover et al. [33] used balsa as core in sandwich panels with the carbon fibers/epoxy face-sheet and subjected them to high-velocity impact conditions. The obtained limit velocity in this work was about 96 m/s, which means that this structure had the proper impact resistance and could absorb the energy of impactor. In another work, Najafi et al. [34] fabricated sandwich panels with the glass fibers/epoxy and FML face-sheet, and balsa core. The obtained results showed that the sandwich panel with FML face-sheet had the higher energy absorption capacity than the sandwich panel with composite shell.

The second bio-based core with proper impact properties for use in sandwich panels can be agglomerated cork. These materials can be produced by cork grains with polymeric binders, which have an important role in the dissipation of energy in the sandwich panels under impact situations. The results reported by Ivañez et al. [35] indicated that the sandwich panel with agglomerated cork had 30% higher energy absorption, as compared to monolithic laminates. Sarasini et al. [36] used flax/epoxy face-sheet/agglomerated cork sandwich panel for investigating the impact behavior. The obtained results showed that the use of agglomerated cork with flax/epoxy composite made the limit velocity of this sandwich panel comparable with the sandwich panels with synthetic or conventional foams.

7.2.4 Hybrid Core Sandwich Panels

Using two cores simultaneously in sandwich panels is one of the ways to enhance the impact properties. In other words, some researchers have developed hybrid cores in sandwich panels. The hybridization can be performed in the face-sheet and core. Given the researches performed, it can be said that the hybridization in the core part can be geometrical/materials or materials/materials forms. In the first group, the geometrical structures such as pyramid with the second materials such as PU foam can be simultaneously used as hybrid core in sandwich panels for improving the impact performance. In the second group, two different materials such as foam with polymeric or ceramic materials or even nanomaterials can be used at the same time in the core of sandwich panels. Investigating the literature can give a better view about the impact properties of these sandwich panels; hence, some of the interesting research works are reviewed in the following.

Hassanpour Roudbeneh et al. [37] filled aluminum honeycomb structure with PU foam to use as core in sandwich panels with aluminum skins. The variations of this work were the foam density. By comparing the un-filled and filled cores in the sandwich panels, about 35% improvement in the limit velocity was seen in this work. O'Masta et al. [38] first designed a corrugated Al 6061-T6 core and then filled it with prismatic alumina inserts to use as a hybrid core in sandwich panels with the face-sheet of ultrahigh molecular weight polyethylene (UHMWPE). This means that these researchers used the polymeric, metallic and ceramic materials at the same time to block the impactor. The obtained results indicated that this structure could block the impactor with an initial velocity of 2.7 km/s, whereas the impactor with an initial velocity of 1.71 km/s had a residual velocity of 0.43 km/s, after crushing the pristine corrugated aluminum sandwich structure.

A comparative study about the impact properties of non-hybrid and hybrid cores in the sandwich panels was done by Yungwirth et al. [39,40]. They first fabricated aluminum pyramidal core/steel face-sheet sandwich panels. Then, they filled the core with PU and alumina as ceramic prisms. In the next step, they used Kevlar instead of stainless steel face-sheet. The obtained data indicated that changing the face-sheets or filling the pyramidal core with PU had no noteworthy effect on improving the impact resistance of sandwich panels. Combining pyramidal core with the alumina prisms caused the limit velocity of hybrid core sandwich panel be twice that of an equivalent monolithic steel plate. Jung et al. [41] first coated aluminum foam with Ni particles and then filled that with pourable silicone. After that, this foam was adhesively bonded to the aluminum to complete the sandwich structure, which was subsequently subjected to impact test with an initial velocity of 300 m/s. The reported results showed that this structure could absorb about 97% of impactor initial velocity.

Ahmadi et al. [42] first produced syntactic foam by adding ceramic microballoons to the epoxy resin with various volume fractions of 0%, 20%, 40% and 60%. To study the effect of nanoclay on the impact resistance, they fabricated nanoreinforced syntactic foam with 40 vol.% and six different weight percentages of nanoclay (0, 1, 2, 3, 5 and 7 wt.%). These nanoreinforced foams were sandwiched by 8 plies of glass fibers/epoxy composite. The maximum increment in the perforation energy was about 10% for foam containing 3 wt.% nanoclay.

7.2.5 New Advanced Sandwich Panels

To achieve higher impact resistance, the researchers have made many efforts, which resulted in the creation of new sandwich panels. One of the simplest ways for improving the impact resistance of sandwich panels is improving the energy dissipation of impactor by the face-sheets. For this reason, some researchers used FML structures instead of metal or polymeric composite skins in the sandwich panels [43]. The FMLs have the features of both metallic materials and polymeric composites at the same time [44,45]. In other words, by using the ductile behavior of metals and high mechanical performance of composites, the capability of sandwich panels to dissipate the energy of impactor or blocking that can be improved. To understand the performance of sandwich panels with FML face-sheets, reviewing some studies in the literature can give a better view about the impact resistance of those under a high-speed impactor. Villanueva and Cantwell [46] compared the influence of FLM and conventional composite face-sheets on the impact resistance of sandwich panels. They mentioned that the FML face-sheets offered about 23% higher specific absorbed energy compared to composite face-sheets, when the sandwich panels were subjected to high-velocity impact test. Also, the fracture of aluminum alloys; longitudinal splitting and fiber fractures in the FML skins; and indentation, crushing and delamination of aluminum foam were the characterized energy absorption mechanisms in the sandwich panel with FML skin.

Ghalami-Choobar et al. [47] fabricated FML face-sheet/PU core sandwich panels and investigated their impact resistance. The used FML was the different configurations between glass fibers/epoxy composite and aluminum 1050. They mentioned that the dominate fracture in the impact touch face was the shear off, whereas delamination, debonding and dishing were seen on the back face-sheet. In their admirable research work, Liu et al. [48] fabricated sandwich panels with FML (Al 5005/glass fibers-epoxy composites) face-sheet and aluminum foam core for studying the effect of face-sheet and core thickness on the impact resistance of these sandwich structures. The obtained results showed that the thickness of face-sheet had the higher influence on the impact resistance of the sandwich panels than the core thickness, so with a slight increase in the thickness of face-sheet, the structure could block an impactor with an initial velocity of 215 m/s.

The other new advanced materials introduced for improving the impact properties of sandwich panels are the 3D braided fibers, which can be used as core in the sandwich panels. Chatterjee et al. [49] improved the impact resistance of sandwich panels by the incorporation of shear-thickening fluids (STFs) into the 3D braided glass fibers. Also, the face-sheets of this sandwich structure were the Kevlar/epoxy composite. This means that these researchers have fabricated sandwich panels with proper impact properties without using metals. In other words, this work can be a new trend toward producing lighter sandwich panels with a higher impact resistance. The reported results showed that the Kevlar-epoxy/STF/3D braided fibers sandwich panel could absorb 96.3% of the impactor energy with an initial velocity of 157.5 m/s. By using the impactor with a velocity of 404 m/s, this structure could dissipate 20% of that [50]. Substituting glass/Kevlar hybrid composite with Kevlar composite in the face-sheets caused the structure to absorb 94.5% of the energy of impactor with an initial velocity of 173 m/s [51].

7.2.6 COMPARATIVE STUDY

To compare the impact performance of sandwich panels with different cores, Figures 7.2–7.4 are drawn. Figure 7.2 shows the residual velocity versus the impact (initial) velocity of impactor after crossing the sandwich panels. Based on this figure, it can be said that the bio-based core sandwich panels can block an impactor with a velocity lower than 120 m/s. If the initial velocity of the impactor is more than 120 m/s, these sandwiches cannot block the impactor, but they can absorb some part of its energy by crushing the bio-based core. The most important advantage of bio-based cores is the reduction in the environmental issues, as compared to other cores. Based on Figure 7.2, it can be realized that the panels with new advanced core materials have the highest performance, when the velocity

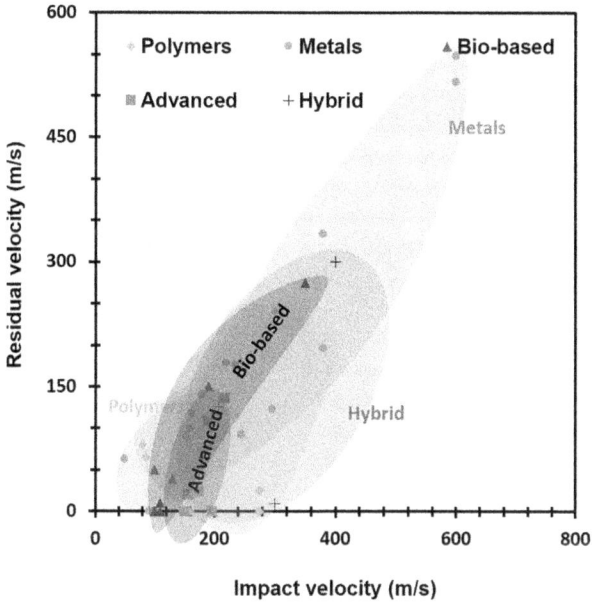

FIGURE 7.2 The comparison of impact performance between sandwich panels with different cores by using the residual velocity of impactor at various impact velocities.

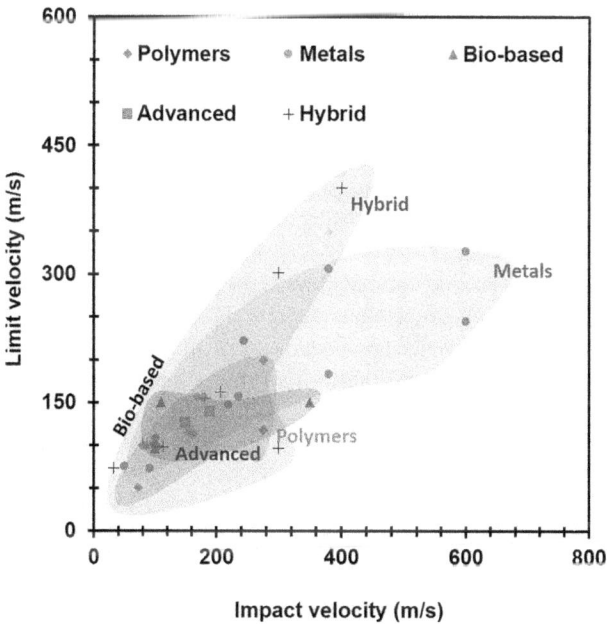

FIGURE 7.3 The comparison of impact performance between sandwich panels with different cores by using the limit velocity at various impact velocities.

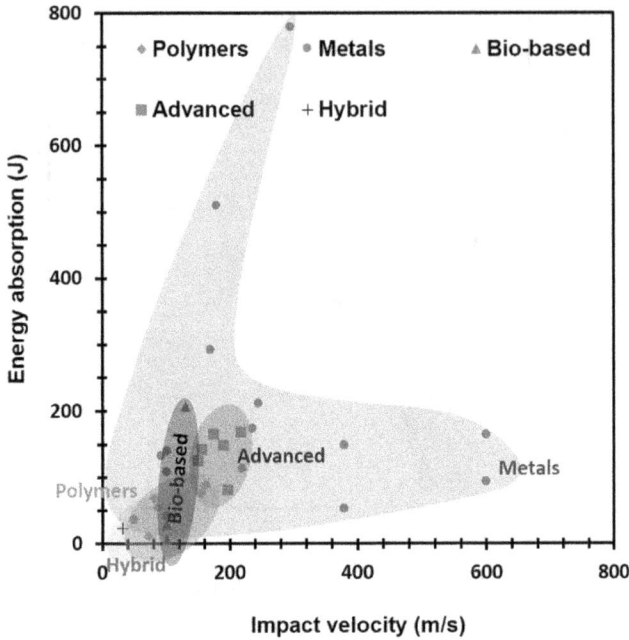

FIGURE 7.4 The comparison of impact performance between sandwich panels with different cores by using the energy absorption capability at various impact velocities.

of impactor is between 120 and 200 m/s. This means that more researches for developing this type of cores should be performed. The polymeric and hybrid cores can be used for the moderate-impact applications. By observing Figure 7.2, it can be said that under the impactor with a velocity of more than 400 m/s, the sandwich panels with metallic core have a higher resistance against the impactor. But, as previously mentioned, using metallic cores increases the weight of sandwich structures.

The limit velocity can be another important factor for comparing the impact performance of sandwich panels with various cores under high-velocity impacts. Figure 7.3 summarizes the general capability of impact resistance of sandwich panels with various cores according to the limit velocity in the different impact velocities. Based on Figure 7.3, the range of limit velocity of new advanced core structures is so limited. It can be said that these core structures have recently been introduced, and in future, they can be used instead of metallic cores in the sandwich panels. As can be seen in Figure 7.3, the metallic cores can be absorb impactors with higher impact velocities. But, they can increase the weight of the structure. So, for reducing the weight of these structures, reducing the thickness of metallic cores and putting another absorbing material can improve the impact resistance of sandwich structures. The obtained data about the limit velocity of hybrid cores in Figure 7.3 confirm this theory. The other parameter that can be used for comparing the impact properties of these structures can be absorbed energy in different impact velocities. Figure 7.4 gives a point of view about the capability of energy absorption of the sandwich panels. Based on this figure, although the energy absorption can give data about the impact resistance of different sandwich panels, the effect of other important factors such as weight and thickness of the panels is not considered. Therefore, normalizing the energy absorption can give a better view about the impact resistance of sandwich panels. Given the researches, various methods have been developed for normalizing the absorbed energy, which are energy absorption per one of the influencing factors such as density, areal density, deflection and damaged area in the perpendicular or parallel direction [52]. However, a comprehensive method for showing a better comparison between sandwich panels with different cores has not yet been reported.

7.3 CONCLUSIONS

In this chapter, the impact resistance of sandwich panels was investigated from the materials point of view. The first class of these sandwich structures was the polymeric core panels. Polyurethane foams as polymeric cores were the most famous cores that can be subjected to high-velocity impacts. The second group was the panels with metallic cores in the form of pyramid, foam, corrugated, etc. This group of sandwich panels can be used for blocking impactors with various initial impact velocities. To reduce the weight of structures, production costs and environmental issues, the bio-based core sandwich panels were introduced. Given the researches, it can be said that the balsa and agglomerated cork can be good candidates for use as core in these structures for dissipating the impact energies. The impact resistances of hybrid cores with geometrical/materials or materials/materials forms were studied as a new progressive trend. Also, the 3D braided fibers and FMLs were introduced as new advanced core and face-sheet for enhancing the impact resistance of lighter sandwich panels.

REFERENCES

1. Birman, V., and G. A. Kardomateas. 2018. Review of current trends in research and applications of sandwich structures. *Composites Part B: Engineering* 142: 221–40.
2. Susainathan, J., F. Eyma, E. De Luycker, A. Cantarel, and B. Castanie. 2018. Experimental investigation of impact behavior of wood-based sandwich structures. *Composites Part A: Applied Science and Manufacturing* 109: 10–19.
3. Khan, T., V. Acar, M. R. Aydin, B. Hülagü, H. Akbulut, and M. Ö. Seydibeyoğlu. 2020. A review on recent advances in sandwich structures based on polyurethane foam cores. *Polymer Composites* 41: 2355–2400.
4. Chai, G. B., and S. Zhu. 2011. A review of low-velocity impact on sandwich structures. *Proceedings of the Institution of Mechanical Engineers, Part L: Journal of Materials: Design and Applications* 225: 207–30.
5. Eslami Farsani, R., and H. Ebrahimnezhad-Khaljiri. 2021. Smart epoxy composites. In *Epoxy Composites: Preparation, Characterization and Applications*, eds. J. Parameswaranpillai, H. Pulikkalparambil, S. M. Rangappa, and S. Siengchin, 349–394. Weinheim: Wiley-VCH Verlag GmbH & Co.

6. Bratton, K. R., K. J. Hill, C. Woodruff, L. L. Campbell, C. B. Cagle, M. L. Pantoya, J. Magallanes, J. Abraham, and C. Meakin. 2020. High velocity impact testing of intermetallic projectiles. *Journal of Dynamic Behavior of Materials* 6: 236–45.
7. Nasirzadeh, R., and A. R. Sabet. 2014. Study of foam density variations in composite sandwich panels under high velocity impact loading. *International Journal of Impact Engineering* 63: 129–39.
8. Nasirzadeh, R., and A. R. Sabet. 2016. Influence of nanoclay reinforced polyurethane foam toward composite sandwich structure behavior under high velocity impact. *Journal of Cellular Plastics* 52: 253–75.
9. Abbasi, M., and A. Alavi Nia. 2020. High-velocity impact behavior of sandwich structures with AL faces and foam cores-experimental and numerical study. *Aerospace Science and Technology* 105: 106039.
10. Alavi Nia, A., and M. Kazemi. 2020. Experimental study of ballistic resistance of sandwich targets with aluminum face-sheet and graded foam core. *Journal of Sandwich Structures and Materials* 22: 461–79.
11. Ren, P., Q. Tao, L. Yin, Y. Ma, J. Wu, W. Zhao, Z. Mu, Z. Guo, and Z. Zhao. 2020. High-velocity impact response of metallic sandwich structures with PVC foam core. *International Journal of Impact Engineering* 144: 103657.
12. Rolfe, E., C. Kaboglu, R. Quinn, P. A. Hooper, H. Arora, and J. P. Dear. 2018. High velocity impact and blast loading of composite sandwich panels with novel carbon and glass construction. *Journal of Dynamic Behavior of Materials* 4: 359–72.
13. Hassanpour Roudbeneh, F., G. Liaghat, H. Sabouri, and H. Hadavinia. 2020. High-velocity impact loading in honeycomb sandwich panels reinforced with polymer foam: A numerical approach study. *Iranian Polymer Journal* 29: 707–21.
14. Wang, B., G. Zhang, S. Wang, L. Ma, and L. Wu. 2014. High velocity impact response of composite lattice core sandwich structures. *Applied Composite Materials* 21: 377–89.
15. Sun, G., D. Chen, H. Wang, P. J. Hazell, and Q. Li. 2018. High-velocity impact behaviour of aluminium honeycomb sandwich panels with different structural configurations. *International Journal of Impact Engineering* 122: 119–36.
16. Liu, C., Y. X. Zhang, Q. H. Qin, and R. Heslehurst. 2014. High velocity impact modelling of sandwich panels with aluminium foam core and aluminium sheet skins. *Applied Mechanics and Materials* 553: 745–50.
17. Sibeaud, J. M., L. Thamié, and C. Puillet. 2008. Hypervelocity impact on honeycomb target structures: experiments and modeling. *International Journal of Impact Engineering* 35: 1799–1807.
18. Yang, S., C. Qi, D. Wang, R. Gao, H. Hu, and J. Shu. 2013. A comparative study of ballistic resistance of sandwich panels with aluminum foam and auxetic honeycomb cores. *Advances in Mechanical Engineering* 2013: 589216.
19. Khaire, N., and G. Tiwari. 2020. Ballistic response of hemispherical sandwich shell structure against ogive nosed projectile. *Thin-Walled Structures* 154: 106869.
20. Zhang, Q. N., X. W. Zhang, G. X. Lu, and D. Ruan. 2018. Ballistic impact behaviors of aluminum alloy sandwich panels with honeycomb cores: An experimental study. *Journal of Sandwich Structures and Materials* 20: 861–84.
21. Hou, W., F. Zhu, G. Lu, and D. Ning Fang. 2010. Ballistic impact experiments of metallic sandwich panels with aluminium foam core. *International Journal of Impact Engineering* 37: 1045–55.
22. Golestanipour, M., A. Babakhani, and S. M. Zebarjad. 2020. High-velocity perforation behaviour of sandwich panels with Al/SiCp composite foam core. *Journal of Composite Materials* 54: 1483–95.

23. Liang, X., H. Luo, Y. Mu, L. Wu, and H. Lin. 2017. Experimental study on stress attenuation in aluminum foam core sandwich panels in high-velocity impact. *Materials Letters* 203: 100–102.

24. Dean, J., A. S. Fallah, P. M. Brown, L. A. Louca, and T. W. Clyne. 2011. Energy absorption during projectile perforation of lightweight sandwich panels with metallic fibre cores. *Composite Structures* 93: 1089–95.

25. Rahimijonoush, A., and M. Bayat. 2020. Experimental and numerical studies on the ballistic impact response of titanium sandwich panels with different facesheets thickness ratios. *Thin-Walled Structures* 157: 107079.

26. Novak, N., M. Vesenjak, G. Kennedy, N. Thadhani, and Z. Ren. 2020. Response of chiral auxetic composite sandwich panel to fragment simulating projectile impact. *Physica Status Solidi (B) Basic Research* 257: 6–11.

27. Odac, L. K., C. Klaslan, A. Tademirci, and M. Güden. 2012. Projectile impact testing of glass fiber-reinforced composite and layered corrugated aluminium and aluminium foam core sandwich panels: A comparative study. *International Journal of Crashworthiness* 17: 508–18.

28. Gao, G. w., E. L. Tang, M. H. Feng, Y. F. Han, Y. Li, M. Liu, Y. L. Xu, et al. 2018. Research on dynamic response characteristics of CFRP/Al HC SPs subjected to high-velocity impact. *Defence Technology* 14: 503–12.

29. Tang, E., W. Li, and Y. Han. 2020. Research on the interacting duration and microscopic characteristics created by high-velocity impact on CFRP/Al HC SP structure. *Journal of Materials Research and Technology* 9: 1640–51.

30. Rahmani, H., R. Eslami-Farsani, and H. Ebrahimnezhad-Khaljiri. 2020. High velocity impact response of aluminum- carbon fibers-epoxy laminated composites toughened by nano silica and zirconia. *Fibers and Polymers* 21: 170–78.

31. Tang, E., X. Zhang, and Y. Han. 2019. Experimental research on damage characteristics of CFRP/aluminum foam sandwich structure subjected to high velocity impact. *Journal of Materials Research and Technology* 8: 4620–30.

32. Senthilkumar, K., R. Nagarajan, S. M. K. Thiagamani, and S. Siengchin. 2021. *Mechanical and Dynamic Properties of Biocomposites.* Weinheim: John Wiley and Sons. ISBN: 9783527346264, 3527346260.

33. Jover, N., B. Shafiq, and U. Vaidya. 2014. Ballistic impact analysis of balsa core sandwich composites. *Composites Part B: Engineering* 67: 160–69.

34. Najafi, M., R. Ansari, and A. Darvizeh. 2019. Experimental characterization of a novel balsa cored sandwich structure with fiber metal laminate skins. *Iranian Polymer Journal* 28: 87–97.

35. Ivañez, I., S. Sánchez-Saez, S. K. Garcia-Castillo, E. Barbero, A. Amaro, and P. N.B. Reis. 2020. High-velocity impact behaviour of damaged sandwich plates with agglomerated cork core. *Composite Structures* 248: 112520.

36. Sarasini, F., J. Tirillò, L. Lampani, E. Barbero, S. Sanchez-Saez, T. Valente, P. Gaudenzi, and C. Scarponi. 2020. Impact behavior of sandwich structures made of flax/epoxy face sheets and agglomerated cork. *Journal of Natural Fibers* 17: 168–88.

37. Hassanpour Roudbeneh, F., G. Liaghat, H. Sabouri, and H. Hadavinia. 2019. Experimental investigation of impact loading on honeycomb sandwich panels filled with foam. *International Journal of Crashworthiness* 24: 199–210.

38. O'Masta, M. R., B. G. Compton, E. A. Gamble, F. W. Zok, V. S. Deshpande, and H. N.G. Wadley. 2015. Ballistic impact response of an UHMWPE fiber reinforced laminate encasing of an aluminum-alumina hybrid panel. *International Journal of Impact Engineering* 86: 131–44.

39. Yungwirth, C. J., J. O'Connor, A. Zakraysek, V. S. Deshpande, and H. N.G. Wadley. 2011. Explorations of hybrid sandwich panel concepts for projectile impact mitigation. *Journal of the American Ceramic Society* 94: s62–75.

40. Yungwirth, C. J., H. N.G. Wadley, J. H. O'Connor, A. J. Zakraysek, and V. S. Deshpande. 2008. Impact response of sandwich plates with a pyramidal lattice core. *International Journal of Impact Engineering* 35: 920–36.

41. Jung, A., E. Lach, and S. Diebels. 2014. New hybrid foam materials for impact protection. *International Journal of Impact Engineering* 64: 30–38.

42. Ahmadi, H., G. Liaghat, and S. C. Charandabi. 2020. High velocity impact on composite sandwich panels with nano-reinforced syntactic foam core. *Thin-Walled Structures* 148: 106599.

43. Liu, C., Y. X. Zhang, and J. Li. 2017. Impact responses of sandwich panels with fibre metal laminate skins and aluminium foam core. *Composite Structures* 182: 183–90.

44. Eslami-Farsani, R., H. Aghamohammadi, S. M. R. Khalili, H. Ebrahimnezhad-Khaljiri, and H. Jalali. 2020. Recent trend in developing advanced fiber metal laminates reinforced with nanoparticles: A review study. *Journal of Industrial Textiles* (August 26). Sage. Doi:10.1177/1528083720947106.

45. Ebrahimnezhad-khaljiri, H., R. Eslami-farsani, and S. Talebi. 2020. Investigating the high velocity impact behavior of the laminated composites of aluminum / jute fibers- epoxy containing nanoclay particles. *Fibers and Polymers* 21: 2607–13.

46. Reyes, V. G., and W. J. Cantwell. 2004. The high velocity impact response of composite and FML-reinforced sandwich structures. *Composites Science and Technology* 64: 35–54.

47. Ghalami-Choobar, M., and M. Sadighi. 2014. Investigation of high velocity impact of cylindrical projectile on sandwich panels with fiber-metal laminates skins and polyurethane core. *Aerospace Science and Technology* 32: 142–52.

48. Liu, C., Y. X. Zhang, and L. Ye. 2017. High velocity impact responses of sandwich panels with metal fibre laminate skins and aluminium foam core. *International Journal of Impact Engineering* 100: 139–53.

49. Chatterjee, V. A., S. K.R Verma, D. Bhattacharjee, I. Biswas, and S. Neogi. 2019. Enhancement of energy absorption by incorporation of shear thickening fluids in 3D-Mat sandwich composite panels upon ballistic impact. *Composite Structures* 225: 111148.

50. Chatterjee, V. A., R. Saraswat, S. K. Verma, D. Bhattacharjee, I. Biswas, and S. Neogi. 2020. Embodiment of dilatant fluids in fused-double-3D-Mat sandwich composite panels and its effect on energy-absorption when subjected to high-velocity ballistic impact. *Composite Structures* 249: 112588.

51. Chatterjee, V. A., S. K. Verma, D. Bhattacharjee, I. Biswas, and S. Neogi. 2021. Manufacturing of dilatant fluid embodied kevlar-glass-hybrid-3D-fabric sandwich composite panels for the enhancement of ballistic impact resistance. *Chemical Engineering Journal* 406: 127102.

52. Bigdilou, M. B., R. Eslami-Farsani, H. Ebrahimnezhad-Khaljiri, and M. A. Mohammadi. 2020. Experimental assessment of adding carbon nanotubes on the impact properties of kevlar-ultrahigh molecular weight polyethylene fibers hybrid composites. *Journal of Industrial Textiles* (April 22). Sage. Doi:10.1177/1528083720921483.

8 Investigation of Blast Loading Response of the Composite Sandwich Panels

Hukum Chand Dewangan and Subrata Kumar Panda
National Institute of Technology

Nitin Sharma
KIIT University

Chetan Kumar Hirwani
National Institute of Technology Patna

CONTENTS

8.1 Introduction 131
8.2 Theoretical and Geometrical Development 133
8.3 Boundary Conditions... 135
8.4 Blast Load... 135
8.5 Simulation Modeling 136
8.6 Results and Discussion 136
8.7 Concluding Remarks 144
References.. 145

8.1 INTRODUCTION

The modern engineering industries (aeronautical, marine, space, automotive, civil, etc.) require high-performance, lightweight structural components without compromising their strength and stiffness. The utilization of the laminated composite in place of monolithic materials can fulfill the requirement of structural elements. The laminated composite structure's strength and stiffness can significantly be increased by laminating a thick layer of lightweight low-density material between two thin, stiff, strong sheets of dense material, which is called sandwich structure. The sandwich composite has materialized as a most capable

DOI: 10.1201/9781003143031-8

type of structural material to form enhanced fatigue life, more excellent energy absorption that produces higher resistance to impact load, reduced proneness to corrosion, and excellent acoustic and thermal isolation. However, any material's structure/structural component may not experience a fixed amount of loading. The structural deformation behavior is a function of the load amplitude and time. The magnitude of the load varies at different instances of time. These time-dependent loads, i.e., dynamic load, can cause severe damage to the structure/structural components and must be taken into account to achieve a reliable design.

The investigation of the transient response of the laminated and sandwich structures subjected to the various types of time-dependent pulse loading, including blast load, has been reported for a long time (Gupta et al. 1987; Kant and Mallikarjuna 1989; Librescu and Nosier 1990; Wang, Lam, and Liu 2001) via different solution methods (simulation, analytical, and numerical). The three-dimensional (3D) displacement field model is presented for the transient analysis of sandwich panels subjected to blast loading (Librescu, Oh, and Hohe 2004; Hause and Librescu 2005, 2007). The higher-order shear deformation theory-based finite element (FE) model has been presented for the transient analysis of sandwich panels (Patel et al. 2005; Park et al. 2008). The enhancement of the resistance against the sandwich plate's blast load is presented (Bahei-El-Din and Dvorak 2008). The sandwich panels with different core materials (cellular, micro-architectural, and macro-architectural) subjected to three types of blast loading (air, water, and simulated blast loads) have been considered for experimental analysis (Yuen et al. 2009). The experimental investigation has been carried out for the sandwich structures' blast response (Shukla et al. 2009; Wang and Shukla 2011; Arora, Hooper, and Dear 2011). The optimized thickness of face sheets via fiber orientation of the laminated sandwich exposed to blast loading has been calculated using the refined zigzag model (Icardi and Ferrero 2009, 2010). The radial basis function has been employed to predict the transient behavior of composite and sandwich plates (Roque et al. 2011). The Galerkin method, in association with the finite difference method, has been implemented to predict the dynamic response of sandwich panels exposed to time-dependent loads (Kazancı 2017; Kazancı and Mecitoğlu 2008; Kazancı 2011). The time-dependent transverse deflection behavior of laminated and sandwich plates subjected to different pulse loading types is evaluated (Marjanović and Vuksanović 2013, 2014) using Reddy's generalized laminated plate theory (GLPT). The non-polynomial shear deformation theory-based isogeometric analysis (Gupta and Ghosh 2019) has been reported to predict the transient deflection behavior of laminated and sandwich plates.

Numerous literature studies present the transient behavior of sandwich panels subjected to several types of time-dependent pulse loading via analytical, numerical, and experimental techniques. The authors found that the work analyzing the transient characteristics of the sandwich panel subjected to blast loading via developing a simulation model using any commercial FE tools is limited. Based on the literature survey, this work aims to fill this gap and present a methodology to analyze the blast load effect on the sandwich plate's time-dependent deflection by developing a FE simulation model in a commercially available FE tool (Ansys Mechanical APDL). In this context, a flat sandwich panel's model is considered to have the isotropic soft core in between the orthotropic layered face sheets. The validity and accuracy of the developed simulation have been established by comparing the results with the published data in the literature. Additionally, some examples are solved to evaluate the influence of blast load on the transient deflection of the flat sandwich panel for several types of geometry parameters (side–length ratio, ratio of core to face thickness, ratio of side to thickness, and end boundary restrictions) as well as various time-dependent loading conditions.

8.2 THEORETICAL AND GEOMETRICAL DEVELOPMENT

A multilayered laminated sandwich composite plate is assumed to have a soft core between the hard and stiff laminates having 'N' number of orthotropic layers stacked in a particular orientation with length 'l', width 'b', and overall thickness 'h' for the analysis. The top and bottom laminated face sheets consist of an equal number of laminae and are of equal thickness. The sandwich core and face sheet thickness is denoted as 'h_c' and 'h_f', respectively. The schematic view of the sandwich plate in the Cartesian coordinate system (X, Y, Z) is shown in Figure 8.1. The plate's mid-plane overlaps with XY plane, and the origin is situated at the left corner of the plate. The global displacement (u, v, and w) of an arbitrary point at the plate's mid-plane is presented about X-, Y-, and Z-axes, respectively. The displacement is approximated based on the first-order shear deformation theory (FSDT), and the assumed polynomial function is given as:

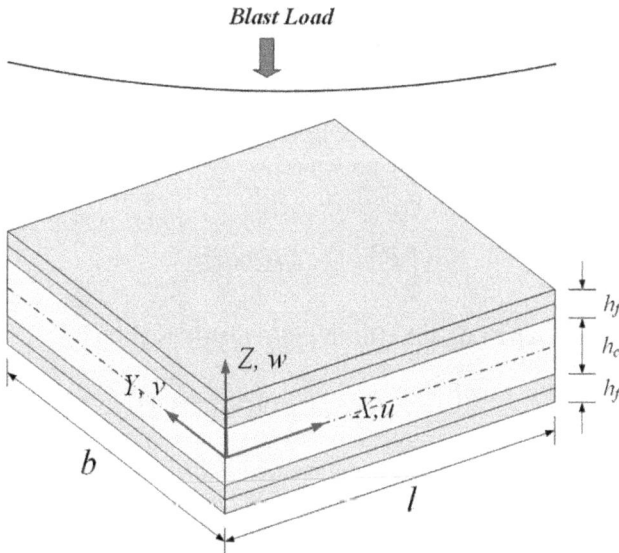

FIGURE 8.1 Geometry and contour of the laminated sandwich plate subjected to the blast load.

$$u = u^* + z\phi_X$$

$$v = v^* + z\phi_Y \tag{8.1}$$

$$w = w^* + z\phi_Z$$

where u^* and v^* are in-plane displacements; w^* is out-of-plane displacement about XY plane; ϕ_X, ϕ_Y, and ϕ_Z are rotations of normal to mid-plane about Y, X, and Z-axes, respectively, during the deformation of the panel.

The general form of strain–displacement relationship utilized for the analysis is given as:

$$\epsilon = \left\{ \begin{array}{c} \epsilon_X \\ \epsilon_Y \\ \gamma_{XY} \\ \gamma_{YZ} \\ \gamma_{ZX} \end{array} \right\} = \left\{ \begin{array}{c} \partial u / \partial X \\ \partial v / \partial Y \\ \partial u / \partial Y + \partial v / \partial X \\ \partial v / \partial Z + \partial w / \partial Y \\ \partial u / \partial Z + \partial w / \partial X \end{array} \right\} \tag{8.2}$$

The transient response solution is calculated using the static equilibrium, considering the effect of acceleration at a particular time t. Initially, a finite number of time sub-steps 'n' is considered in the total time domain such that $\Delta t = T/n$. The time interval is deemed to be such that $t \in [0, T]$ within the period 'T'. Newmark's integration method is used to acquire the solution in the time domain.

$$[m]\{\ddot{\delta}\} + [K]\{\delta\} = \{f\} \tag{8.3}$$

where $[m]$, $[K]$, and $\{f\}$ are the mass, stiffness, and externally applied force. The acceleration and displacement are presented as $\{\ddot{\delta}\}$ and $\{\delta\}$, respectively.

For a specific time step 't', the effective stiffness is given by:

$$\left[\hat{K}\right] = [K] + \alpha_0 [m] \tag{8.4}$$

For the next time step $t + \delta t$, the effective load matrix is given by:

$$^{t+\Delta t}\left[\hat{f}\right] = {}^{t+\Delta t}[f] + [m]\left(\alpha_0^t \delta + \alpha_1^t \ddot{\delta}\right) \tag{8.5}$$

Further, the displacement and acceleration are calculated using the equations:

$$\left[\hat{K}\right]^{t+\Delta t}\delta = {}^{t+\Delta t}\{f\}$$

$$^{t+\Delta t}\ddot{\delta} = \alpha_0\left({}^{t+\Delta t}\delta - {}^t\delta\right) - \alpha_1^t\ddot{\delta} \tag{8.6}$$

TABLE 8.1
Different Edge Support Conditions

Support Type		Descriptions
BC-(a)	$u^* = v^* = w^* = \phi_X = \phi_Y = \phi_Z = 0$ at $X = 0, l$ and $Y = 0, b$	All edge clamped
BC-(b)	$v^* = w^* = \phi_y = \phi_z = 0$ at $X = 0, l$;	All edge simply supported
	$u^* = w^* = \phi_x = \phi_z = 0$ at $Y = 0, b$	
BC-(c)	$v^* = w^* = \phi_y = \phi_z = 0$ at $X = 0, l$	Two opposite edges clamped
BC-(d)	$u^* = v^* = w^* = \phi_X = \phi_Y = \phi_Z = 0$ at $X = 0$	Only one edge clamped
BC-(e)	$u^* = v^* = w^* = \phi_X = \phi_Y = \phi_Z = 0$ at $X = 0, l$ and $Y = 0$	Three consecutive edges clamped
BC-(f)	$u^* = v^* = w^* = \phi_X = \phi_Y = \phi_Z = 0$ at $X = 0$ and $Y = 0$	Two consecutive edges clamped

8.3 BOUNDARY CONDITIONS

The transient analysis of structural panels that has been carried out considers edge support conditions as given in Table 8.1.

8.4 BLAST LOAD

The explosion creates very high pressure movement with approximately in triangular profile tailed by decay is known as blast. The simulation of this blast load can be simulated accurately by using Friedlander exponential decay equation (Dewey 2018) with respect to time. The structural components may be exposed to time-dependent loading such as blast load during their service life. Therefore, the structural analysis and design requires the study of blast and dynamic responses. In the past, different categories of blast pulse models have been studied in the literature. The blast load can be categorized into the following three types:

1. Air blast load: the sudden release of energy in the air due to the explosion.
2. Underwater blast load: the resultant of the explosion in water which can cause fluid–structural interaction.
3. Simulated blast load: it is equivalent to blast pulse load due to explosion for research purposes.

This study assumes that source the blast is situated at a sufficient distance from the plate; the blast pressure is considered uniform and can be expressed as:

$$f(t) = f_0(1 - t/T)e^{-\alpha t/l} \tag{8.7}$$

There are three independent appearances specified to describe the blast load. The magnitude of the initial shock is represented as the blast pressure f_0, and α is a decay coefficient.

8.5 SIMULATION MODELING

This work proposes a FE simulation model-based FSDT using Ansys Mechanical APDL to analyze the transient responses of a laminated sandwich plate subjected to blast load. The necessary steps involved in developing the simulation model is presented in Figure 8.2, and the detailed explanation is provided in the following lines:

1. Preprocessing: It involves the type of element and material selection. The eight-noded shell element (SHELL 281) has been considered having six degrees of freedom for each node. The sandwich material consists of isotropic soft core between the hard and stiff multilayered orthotropic laminates stacked in a particular orientation.
2. Modeling: The input data for the sandwich panel's layup scheme, and geometrical and boundary parameters have been provided, and the model has been discretized (meshing). The real-time stacking sequence of the laminated sandwich and element distribution along x- and y-axes are presented in Figures 8.3 and 8.4.
3. Solution and analysis: Analysis type and required input data for Newmark's integration are provided and submitted for the solution. Then, the time history graph is plotted against the loading conditions. The real-time plot of deflection vs time history is shown in Figure 8.5.

8.6 RESULTS AND DISCUSSION

The results' convergence behavior by developed FE simulation model for the laminated panel's transient responses is computed for altered element densities and the number of time steps. A four-layered $(30/-30)_s$ square laminate ($l = b = 1.27$ m, $h = 0.0254$ m) with edge support condition BC-(a) is considered for the analysis. The laminated plate is subjected to blast load $p = p_0(1 - T/t_1)e^{-\alpha T/t_1}$, where $p_0 = 68.95 \times 10^{-3}$ N/m^2, $t_1 = 0.004s$, $\alpha = 1.98$. The material properties considered for the analysis are: $\rho = 1{,}610$ kg/m^3, $E_1 = 131.69 \times 10^9$ N/m^2, $E_2 = 8.55 \times 10^9$ N/m^2, $G_{12} = G_{13} = G_{23} = 6.67 \times 10^9$, N/m^2 and $v_{12} = 0.3$. The central deflection of the plate is normalized using $W_{nd} = 100W(E_2h^3/p_0l^4)$. Figure 8.6 shows the transient response of the laminate for different mesh densities with a constant time step = 300. Figure 8.7 shows the transient response of the laminate with a mesh density = 6×6 and different numbers of time steps. The normalized deflection values follow a similar behavior from the mesh refinement = 6×6 and the number of time steps = 100 onward, so it is considered for further analysis. Subsequently, the results are compared with the previously published results (Maleki, Tahani, and Andakhshideh 2012), and the comparison shows excellent coherence.

Additionally, one more problem is solved for the square laminated sandwich plate with edge support condition BC-(b). The sandwich plate $((0/90)_2/\text{core}/(0/90)_2)$ is subjected to triangular pulse loading $p = p_0(1 - T/t)$, where $p_0 = 68.946 \times 10^6$ N/m^2. The following geometrical and material parameters are considered for the analysis:

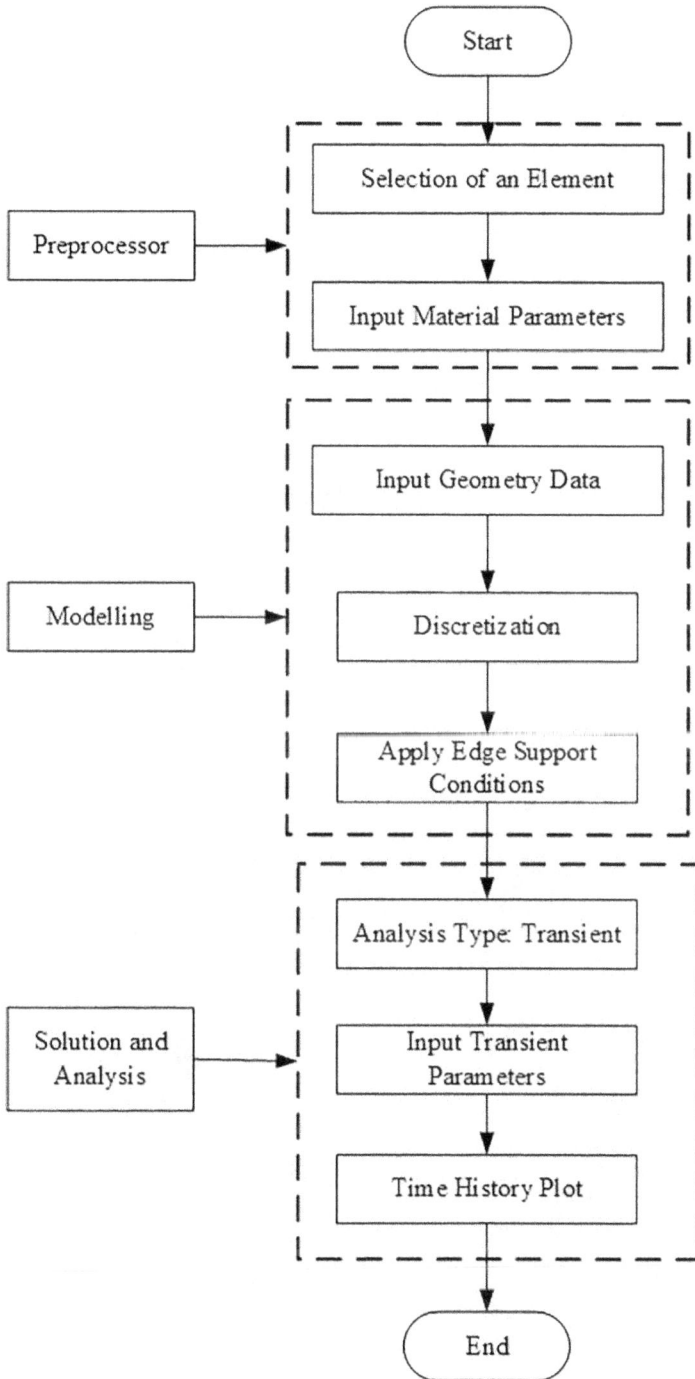

FIGURE 8.2 Modeling steps for the sandwich panel for transient analysis.

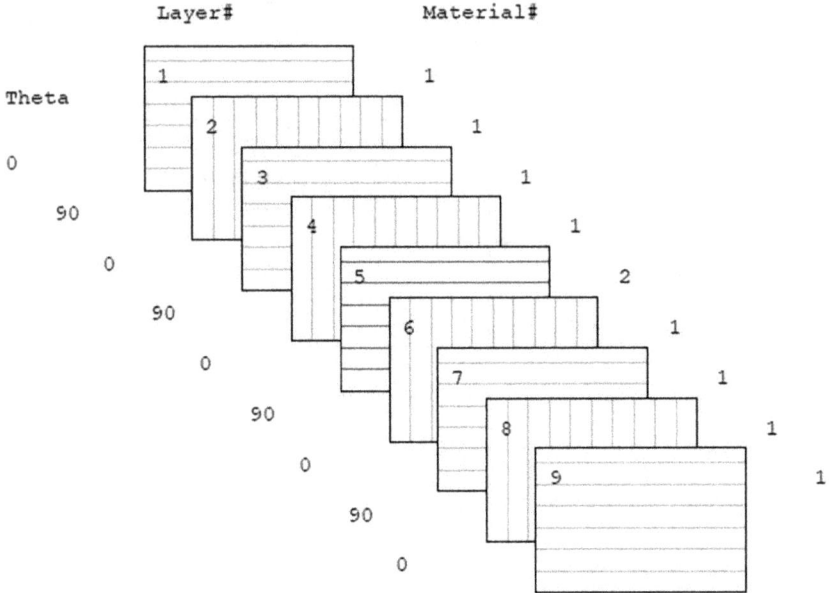

FIGURE 8.3 Example layup scheme of the sandwich panel provided in Ansys.

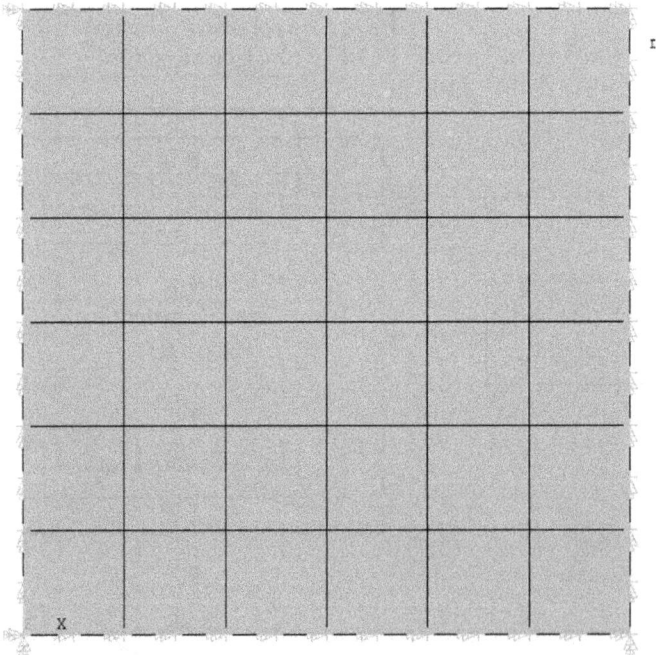

FIGURE 8.4 Example discretization of the sandwich panel in Ansys.

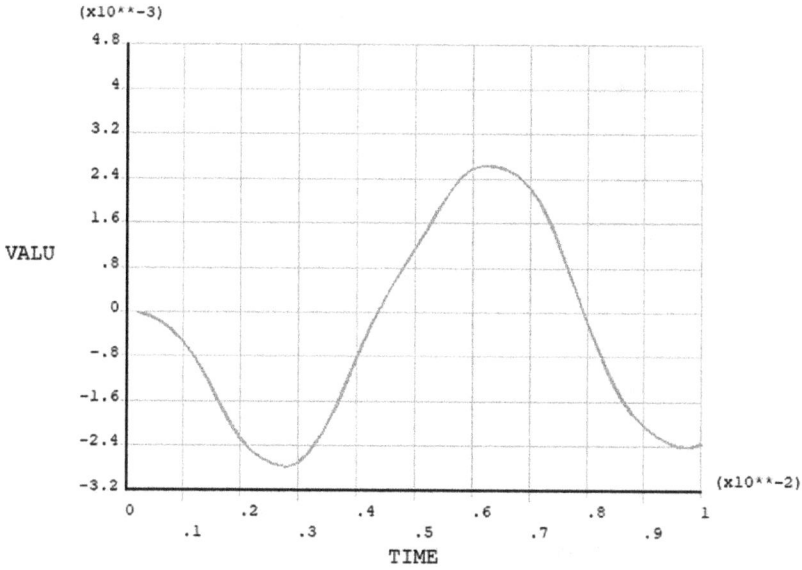

FIGURE 8.5 Real-time results of the transient deflection of a laminated plate subjected to blast load in Ansys Mechanical APDL.

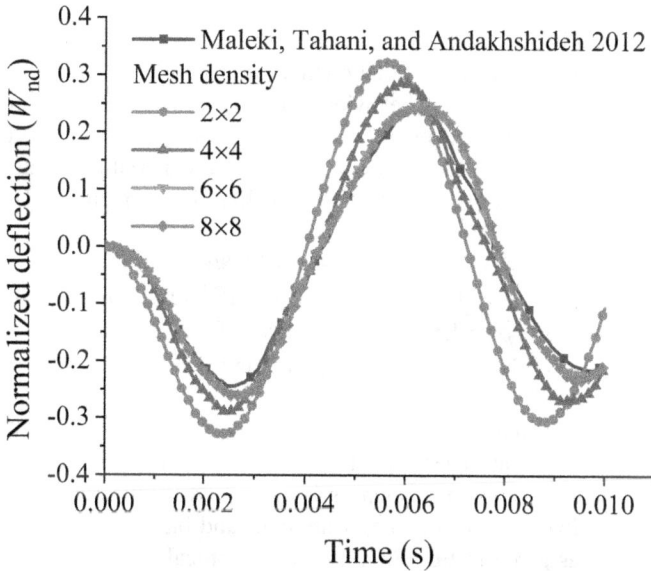

FIGURE 8.6 Transient deflection of a square laminated plate $(30/-30)_s$ subjected to blast load for different mesh densities ($l = b = 1.27$ m, $h = 0.0254$ m, BC-(a)).

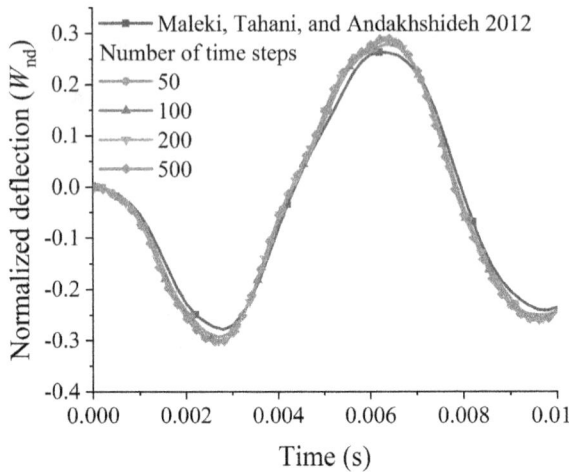

FIGURE 8.7 Transient deflection of a square laminated plate $(30/-30)_s$ subjected to blast load for different numbers of time steps $(l = b = 1.27$ m, $h = 0.0254$ m, BC-(a)).

$l = b = 1$ m, $h = 0.1$ m, $h_f/h = 0.025$.

Face sheet: $E_1 = 128 \times 10^9$ N/m^2, $E_2 = 11 \times 10^9$ N/m^2, $G_{12} = G_{23} = 4.48 \times 10^9$ N/m^2, $G_{23} = 1.53 \times 10^9$ N/m^2, $\rho_f = 1{,}500$ kg/m^3 and $v_{12} = 0.25$.

Core: $E_c = 103.63 \times 10^6$ N/m^2, $G_c = 50 \times 10^6$ N/m^2, $\rho_c = 130$ kg/m^3 and $v_{12} = 0.32$.

The obtained time-dependent central deflection values are presented in Figure 8.8 and compared with the literature results (Nayak, Shenoi, and Moy 2004), which shows the presently developed FE simulation model's versatility and robustness.

Hereafter, the developed FE simulation model has been extended to examine the square sandwich plates' transient behavior. The sandwich plate consists of isotropic core material in between cross-ply (0/90) laminated face sheets, and the final form can be written as 0/90/core/90/0. The geometrical and material parameters considered for the analysis are as given in the following lines unless mentioned otherwise:

$l = b = 0.5$ m, $h = 0.05$ m, $h_c/2h_f = 1$, $p_0 = 100$ N/m^2, $\alpha = 1.98$.

Face sheet: $E_1 = 139 \times 10^9$ N/m^2, $E_2 = 9.86 \times 10^9$ N/m^2, $G_{12} = G_{13} = G_{23} = 5.24 \times 10^9$ N/m^2, $\rho_f = 1{,}590$ kg/m^3, and $v_{12} = 0.3$.

Core: $E_c = 90 \times 10^6$ N/m^2, $\rho_c = 170.6$ kg/m^3, and $v_c = 0.45$.

The sandwich panel may be exposed to several types of time-dependent loads during their service life. Therefore, the influence of different time-dependent loads on the panel's transient responses is examined in this case. A variety of time-dependent loads (stepped, sine, triangular, and blast pulse) are considered for the analysis as given in Eq. (8.8), and the graphical representation is shown in Figure 8.9, where (f_0) is the magnitude of the load intensity and is assumed to be uniformly distributed over the panel's surface area. The obtained transient

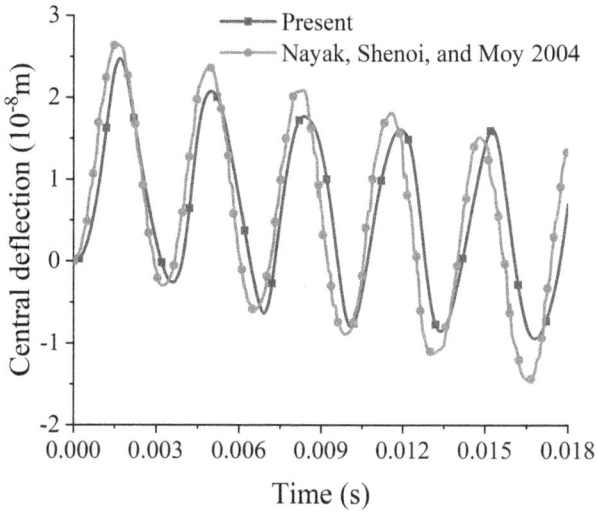

FIGURE 8.8 Time-dependent central deflection of a square laminate subjected to triangular pulse load ($l = b = 1.27$ m, $h = 0.0254$ m, BC-(a)).

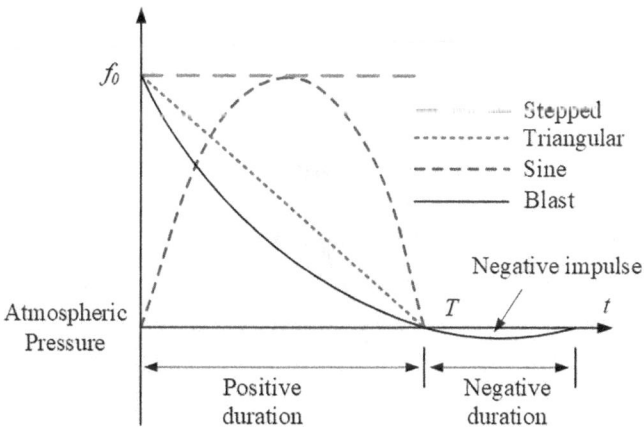

FIGURE 8.9 Graphical representation of different pulse loads.

deflection values of the sandwich plate with edge support condition BC-(b) are presented in Figure 8.10.

$$\left.\begin{array}{ll} \text{Stepped pulse:} & f(t) = f_0 \\[6pt] \text{Sine pulse:} & f(t) = f_0 \sin(\pi T / t) \\[6pt] \text{Triangular pulse:} & f(t) = p_0(1 - T / t) \\[6pt] \text{Blast pulse:} & f(t) = f_0(1 - T / t)e^{-\alpha T/t} \end{array}\right\} \quad (8.8)$$

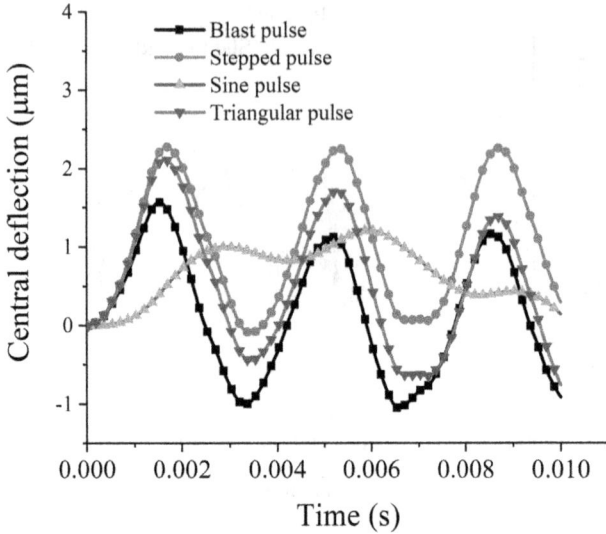

FIGURE 8.10 The transient central deflection of the square laminate plate exposed to different types of pulse load ($l = b = 0.5$ m, $h = 0.05$ m, hc/2hf = 1, BC-(b)).

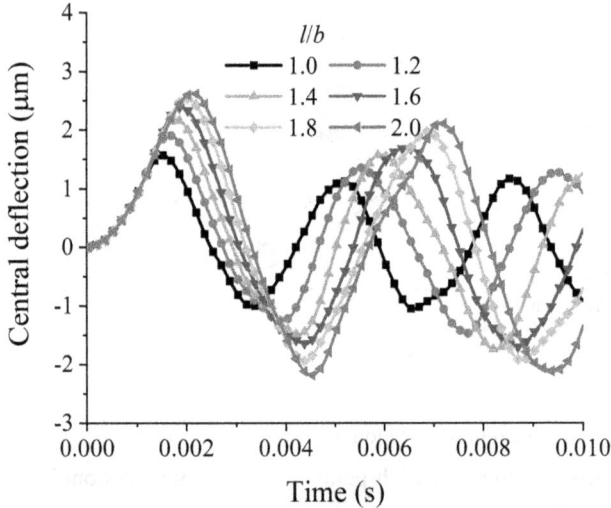

FIGURE 8.11 Transient deflection of a laminated plate subjected to blast pulse load for different side–length ratios ($h = 0.05$ m, hc/2hf = 1, BC-(b)).

The sandwich plate's time-dependent deflection values are obtained for different side–length ratios ($l / b = 1, 1.2, 1.4, 1.6, 1.8, 2$) having edge support condition BC-(b) subjected to the blast pulse loading. The results in Figure 8.11 show that the transient central deflection values increase with an increase in the length-to-width ratio due to the increase in the structure's slenderness ratio. The total cycle time increases with the increase in the side–length ratio. It can also be noticed that the deflection values are higher for the stepped pulse loading.

Now, the effect of the edge support conditions on the time-dependent deflection responses of sandwich panels is investigated. The deflection values are computed for five different edge support conditions BC-(a, c, d, e, and f) and presented in Figure 8.12. The results show a decline in the deflection values with an increase in the number of constraints due to higher stiffness.

This numerical example is solved to demonstrate the influence of the length-to-thickness ratio of the sandwich panel with edge support condition BC-(b). Figure 8.13 represents the results obtained for the time-dependent deflection values for different length-to-thickness ratios (l/h = 10, 25, 50, 75, 100). The results show that the deflection values decrease with an increase in the length-to-thickness ratio due to the reduction in the structure's thickness. The total cycle time increases with an increase in the said parameter.

The influence of the ratio of core to face thickness $\left(h_c \,/\, 2h_f\right)$ on the time-dependent deflection responses of a sandwich panel with edge support condition BC-(b) is examined for six different core-to-face thickness ratios ($h_c \,/\, 2h_f$ =1, 1.2, 1.4, 1.6, 1.8, and 2). From the results presented in Figure 8.14, it can be seen that the deflection values increase with an increase in the core-to-face thickness ratio while the total cycle time reduces. The reduction in the laminated face sheets' thickness causes the decline in the final structural stiffness and shows higher deflection values.

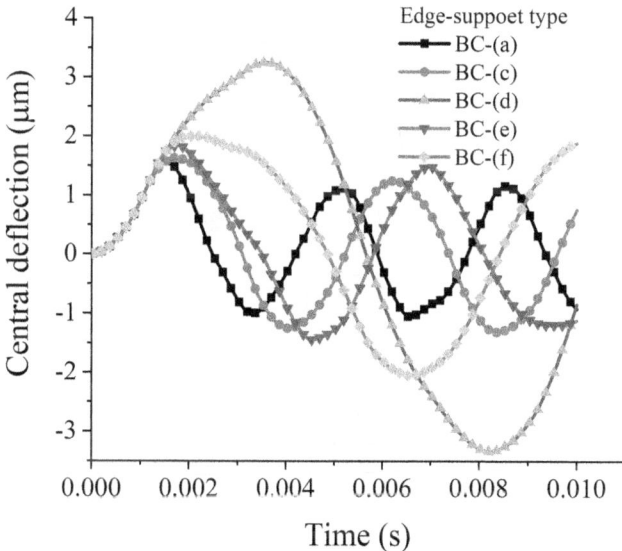

FIGURE 8.12 Transient deflection of a laminated plate subjected to blast pulse load for different end support conditions ($l = b = 0.5$ m, $h = 0.05$ m, hc/2hf = 1).

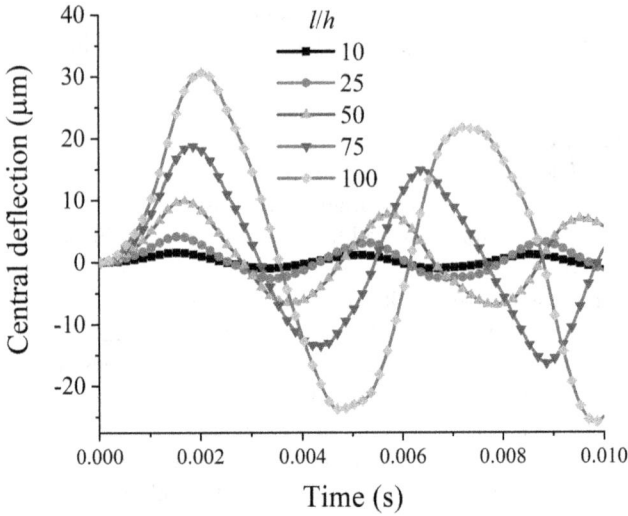

FIGURE 8.13 Transient deflection of a laminated plate subjected to blast pulse load for different side–thickness ratios ($l = b = 0.5$ m, hc/2hf = 1, BC-(b)).

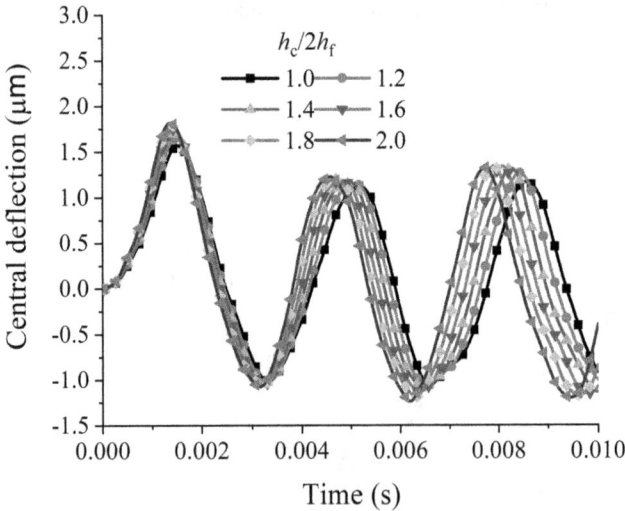

FIGURE 8.14 Transient deflection of the laminated plate subjected to blast pulse loads for different core-to-face thickness ratios ($l = b = 0.5$ m, $h = 0.05$ m, BC-(b)).

8.7 CONCLUDING REMARKS

The time-dependent deflection behavior of a flat sandwich plate exposed to blast load is addressed. The effects of the sandwich plate's structural geometry characteristics on the transient response have been evaluated, and the related conclusions have been underlined. The implemented FE simulation model is based on FSDT and developed using Ansys Mechanical APDL. The developed simulation method's excellent performances have been shown in different conditions to provide accurate solutions. The study shows that the proposed FE simulation model can solve time-dependent problems with low computational cost and acceptable accuracy. The parameters such as loading type, side-to-thickness ratio, aspect ratio, end support condition, and core-to-face thickness ratio significantly affect the structure's final responses.

REFERENCES

Arora, H., P.A. Hooper, and J.P. Dear. 2011. "Dynamic response of full-scale sandwich composite structures subject to air-blast loading." *Composites Part A: Applied Science and Manufacturing* 42 (11): 1651–62. Doi: 10.1016/j.compositesa.2011.07.018.

Bahei-El-Din, Y.A., and G.J. Dvorak. 2008. "Enhancement of blast resistance of sandwich plates." *Composites Part B: Engineering* 39 (1): 120–27. Doi: 10.1016/j.compositesb.2007.02.006.

Dewey, J.M. 2018. "The Friedlander equations." In *Blast Effects: Physical Properties of Shock Waves*, 37–55. Cham: Springer International Publishing. Doi: 10.1007/978-3-319-70831-7_3.

Gupta, A., and A. Ghosh. 2019. "Isogeometric static and dynamic analysis of laminated and sandwich composite plates using nonpolynomial shear deformation theory." *Composites Part B: Engineering* 176 (December 2018): 107295. Doi: 10.1016/j.compositesb.2019.107295.

Gupta, A.D., F.H. Gregory, R.L. Bitting, and S. Bhattacharya. 1987. "Dynamic analysis of an explosively loaded hinged rectangular plate." *Computers & Structures* 26 (1–2): 339–44. Doi: 10.1016/0045-7949(87)90263-X.

Hause, T., and L. Librescu. 2005. "Dynamic response of anisotropic sandwich flat panels to explosive pressure pulses." *International Journal of Impact Engineering* 31 (5): 607–28. Doi: 10.1016/j.ijimpeng.2004.01.002.

Hause, T., and L. Librescu. 2007. "Dynamic response of doubly-curved anisotropic sandwich panels impacted by blast loadings." *International Journal of Solids and Structures* 44 (20): 6678–6700. Doi: 10.1016/j.ijsolstr.2007.03.006.

Icardi, U., and L. Ferrero. 2009. "Laminated and sandwich panels subject to blast pulse loading." *Journal of Mechanics of Materials and Structures* 4 (9): 1573–94. Doi: 10.2140/jomms.2009.4.1573.

Icardi, U., and L. Ferrero. 2010. "Optimization of sandwich panels to blast pulse loading." *Journal of Sandwich Structures & Materials* 12 (5): 521–50. Doi: 10.1177/1099636209106255.

Kant, T., and B.N. Mallikarjuna. 1989. "Transient dynamics of composite sandwich plates using 4-, 8-, 9-noded isoparametric quadrilateral elements." *Finite Elements in Analysis and Design* 5 (4): 307–18. Doi: 10.1016/0168-874X(89)90010-3.

Kazancı, Z. 2011. "Dynamic response of composite sandwich plates subjected to time-dependent pressure pulses." *International Journal of Non-Linear Mechanics* 46 (5): 807–17. Doi: 10.1016/j.ijnonlinmec.2011.03.011.

Kazancı, Z. 2017. "Computational methods to predict the nonlinear dynamic response of blast loaded laminated composite plates." In *Explosion Blast Response of Composites*, 85–112. Elsevier. Doi: 10.1016/B978-0-08-102092-0.00004-2.

Kazancı, Z., and Z. Mecitoğlu. 2008. "Nonlinear dynamic behavior of simply supported laminated composite plates subjected to blast load." *Journal of Sound and Vibration* 317 (3–5): 883–97. Doi: 10.1016/j.jsv.2008.03.033.

Librescu, L., and A. Nosier. 1990. "Response of laminated composite flat panels to sonic boom and explosive blast loadings." *AIAA Journal* 28 (2): 345–52. Doi: 10.2514/3.10395.

Librescu, L., S.Y. Oh, and J. Hohe. 2004. "Linear and non-linear dynamic response of sandwich panels to blast loading." *Composites Part B: Engineering* 35 (6–8): 673–83. Doi: 10.1016/j.compositesb.2003.07.003.

Maleki, S., M. Tahani, and A. Andakhshideh. 2012. "Transient response of laminated plates with arbitrary laminations and boundary conditions under general dynamic loadings." *Archive of Applied Mechanics* 82 (5): 615–30. Doi: 10.1007/s00419-011-0577-1.

Marjanović, M.S., and D.M. Vuksanović. 2013. "Linear transient analysis of laminated composite plates using GLPT." *Acta Technica Napocensis: Civil Engineering & Architecture* 56 (2): 58–71.

Marjanović, M.S., and D.M. Vuksanović. 2014. "Transient analysis of laminated composite and sandwich plates with embedded delaminations using GLPT." In Proceedings of the 9th International Conference on Structural Dynamic, EURODYN, edited by A. Chuna, E. Caetano, P. Ribeiro, and G. Muller, 3373–80. Porto, Portugal. https://grafar.grf.bg.ac.rs/handle/123456789/643.

Nayak, A.K., R.A. Shenoi, and S.S.J. Moy. 2004. "Transient response of composite sandwich plates." *Composite Structures* 64 (3–4): 249–67. Doi: 10.1016/S0263-8223(03)0015-1.

Park, T., S.-Y. Lee, J.W. Seo, and G.Z. Voyiadjis. 2008. "Structural Dynamic Behavior of Skew Sandwich Plates with Laminated Composite Faces." *Composites Part B: Engineering* 39 (2): 316–26. Doi: 10.1016/j.compositesb.2007.01.003.

Patel, B.P., S.S. Gupta, M. Joshi, and M. Ganapathi. 2005. "Transient response analysis of bimodulus anisotropic laminated composite plates." *Journal of Reinforced Plastics and Composites* 24 (8): 795–821. Doi: 10.1177/0731684405047768.

Roque, C.M.C., A.J.M. Ferreira, A.M.A. Neves, C.M.M. Soares, J.N. Reddy, and R.M.N. Jorge. 2011. "Transient analysis of composite and sandwich plates by radial basis functions." *Journal of Sandwich Structures & Materials* 13 (6): 681–704. Doi: 10.1177/1099636211419132.

Shukla, A., S.A. Tekalur, N. Gardner, M. Jackson, and E. Wang. 2009. "Performance of Novel Composites and Sandwich Structures Under Blast Loading." In *Major Accomplishments in Composite Materials and Sandwich Structures*, 503–40. Dordrecht: Springer Netherlands. Doi: 10.1007/978-90-481-3141-9_20.

Wang, E., and A. Shukla. 2011. "The blast response of sandwich composites with biaxial in-plane compressive loading." In *Conference Proceedings of the Society for Experimental Mechanics Series*, 1:383–91 Doi: 10.1007/978-1-4614-0216-9_53.

Wang, Y.Y., K.Y. Lam, and G.R. Liu. 2001. "A strip element method for the transient analysis of symmetric laminated plates." *International Journal of Solids and Structures* 38 (2): 241–59. Doi: 10.1016/S0020-7683(00)00035-4.

Yuen, S.C.K., G.N. Nurick, M.D. Theobald, and G.S. Langdon. 2009. "Sandwich Panels Subjected to Blast Loading." In *Dynamic Failure of Materials and Structures*, edited by A. Shukla, G. Ravichandran, and Y.D.S. Rajapakse, 297–325. Boston, MA: Springer US. Doi: 10.1007/978-1-4419-0446-1_10.

9 Flexural Behavior of Reinforced Concrete Sandwich Wall Panels Enabled by Fiber-Reinforced Polymer (FRP) Connectors

Junqi Huang
Hefei University of Technology

Sushil Kumar and Jian-Guo Dai
The Hong Kong Polytechnic University

CONTENTS

9.1 Introduction 148
9.2 Development and Performance Characterization of GFRP Connectors 149
    9.2.1 Details of the Proposed Connectors 149
    9.2.2 Test Specimens 149
    9.2.3 Test Results 150
9.3 Structural Performance of RC One-Way Slabs with Geopolymer Concrete 151
    9.3.1 Test Specimens 152
    9.3.2 Test Results 153
    9.3.3 Comparison with the Code of Practice and Guidelines 154
9.4 Flexural Performance of the Sandwich Panel with Geopolymer Concrete 156
    9.4.1 Test Specimens 156
    9.4.2 Test Results 157
    9.4.3 Degree of Composite Action 157
    9.4.4 Simplified Approach for Stiffness Prediction of the PCSWP 158

DOI: 10.1201/9781003143031-9

9.5 Fire and Post-Fire Residual Performance of the Sandwich Panel
with Geopolymer Concrete.. 160
9.5.1 Test Specimens 161
9.5.2 Test Procedure 161
9.5.3 Test Results.. 162
9.6 Conclusions.. 163
References.. 165

9.1 INTRODUCTION

Precast concrete sandwich wall panel (PCSWP), a typical structural element used in the precast industry, has widely been adopted in official and residential buildings as the facade or load-bearing walls. It consists of inner and outer reinforced concrete (RC) wythes, a core insulation layer, and connectors that link the two RC wythes. The PCSWPs can be divided into three types based on the out-of-plane stiffness and strength performance: (1) fully composite type, which means the two RC wythes work together as one panel (i.e., the interface slip between the two wythes is zero); (2) non-composite type, which means the two RC wythes work independently; and (3) partially composite type, which lies between the two above-mentioned extreme mechanical behaviors (PCI committee 2011). Traditionally, concrete blocks and steel bent-up bars have been used as the connectors. However, owing to the high thermal conductivity of concrete and steel, the thermal bridge effect occurs (Bush and Stine 1994; Benayoune et al. 2008; Kinnane et al. 2020). Although these traditional connectors provide a high degree of composite action, the energy efficiency of the PCSWP is reduced. In order to avoid such effects, fiber-reinforced polymer (FRP) materials have recently been adopted to manufacture the connectors because of their low thermal conductivity and high strength (Salmon et al. 1997; Frankl et al. 2011; Choi et al. 2015a, 2015b; Chen et al. 2015; O'Hegarty et al. 2019; Tomlinson and Fam 2015). However, most of the existing FRP connectors are designed to transfer one-directional shear force. The so-formed PCSWPs using these existing connectors are usually either non-composite or having low composite action due to the lower strength and stiffness of the connectors.

This chapter aims to develop an innovative PCSWP system. In this system, environmental-friendly geopolymer concrete is adopted to replace ordinary Portland cement (OPC) concrete for use in the two RC wythes. The geopolymer concrete is produced by the alkaline activation of industrial by-products (i.e., fly ash and slag) to realize ambient-temperature curing and achieve comparable performance to OPC concrete (Qian et al. 2020; Ding et al. 2016, 2019). FRP rebar is used to replace steel rebar to improve durability and minimize the thickness of the entire panel. In addition, a tubular glass FRP (GFRP) connector is developed to enhance the composite action of the PCSWP. As a result, the proposed PCSWP system can retain the energy efficiency of all the existing PCSWPs, but is entitled to a low carbon footprint (Davidovits 1991; Nath and Sarker 2014), high durability, and superior structural efficiency. This chapter consists of investigation studies on the behavior of the proposed PCSWP system, including (1) the development and performance characterization of GFRP connectors, (2) the structural performance of RC one-way slab with geopolymer concrete, (3) the flexural behavior of the proposed PCSWP system, and (4) the fire and post-fire residual performance of the proposed PCSWP system.

9.2 DEVELOPMENT AND PERFORMANCE CHARACTERIZATION OF GFRP CONNECTORS

Huang and Dai (2019) conducted in-plane direct shear tests on the three types of GFRP connectors to obtain the strength, stiffness, and deformability of the connectors. A brief introduction and major findings of this study are presented in this section.

9.2.1 DETAILS OF THE PROPOSED CONNECTORS

In total, three types of GFRP connectors were developed for potential use in PCSWPs, including flat plate, corrugated plate, and hexagonal tube connectors (see Figure 9.1). The flat plate connector may fail in shear buckling. Therefore, the corrugated plate connector was explored, aiming to delay or avoid buckling. Both the flat plate and corrugated plate could only transfer one-way in-plane shear (i.e., one-way connector). Thus, the hexagonal tube connector was developed, aiming to transfer two-way shear force (i.e., two-way connector). Besides, for all the proposed connectors, steel rebars were inserted into the predrilled holes on the ends of the GFRP laminate to form a reliable anchorage system.

9.2.2 TEST SPECIMENS

The performance of the proposed connectors was investigated through the direct shear test. In total, 25 specimens were prepared and tested. For the one-way connectors, the investigated parameters were the thickness of the GFRP laminate (i.e., two and four plies of glass fiber sheet) and the projected length (i.e., 100 and 160 mm). For the two-way connectors, the investigated parameters were the GFRP laminate thickness and the loading direction. For each parameter combination, two identical specimens were tested. The specimen designation was in the form of A-B-C, in which "A", "B", and "C" represented the connector type,

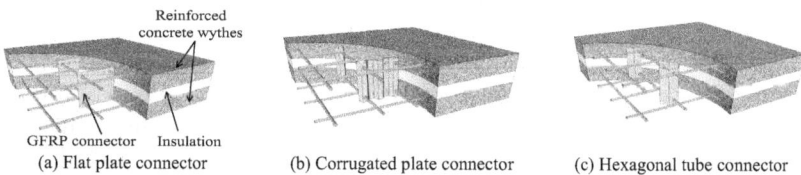

(a) Flat plate connector   (b) Corrugated plate connector   (c) Hexagonal tube connector

FIGURE 9.1 Configuration of the proposed connectors (Huang and Dai 2019).

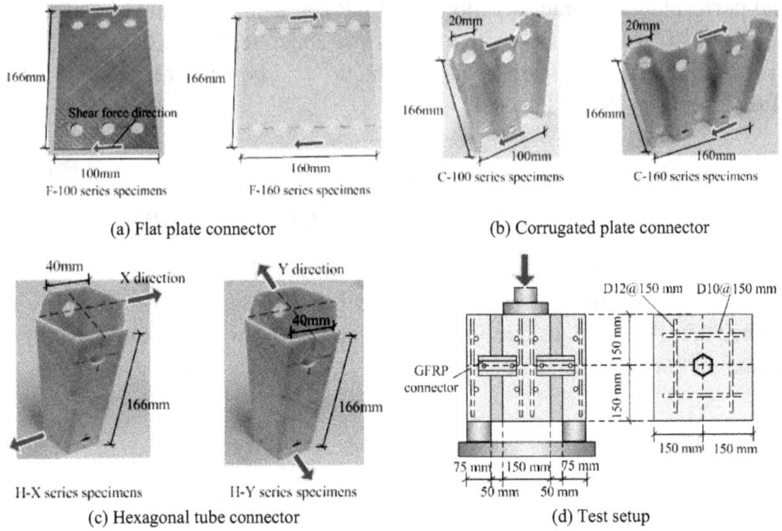

(a) Flat plate connector

(b) Corrugated plate connector

(c) Hexagonal tube connector

(d) Test setup

FIGURE 9.2 Geometry of the investigated connectors and test setup (Huang and Dai 2019).

the projected length, or shear force direction (X- or Y-direction; see Figure 9.2), and the laminate thickness, respectively. Here, the flat plate, corrugated plate, and hexagonal tube connectors were marked by "F", "C", and "H" in the specimen designation, respectively. For the final specimen (H-Y-4-NB), the bond between concrete and insulation was removed by covering the insulation with a plastic sheet. Other design parameters of H-Y-4-NB were the same as those of H-Y-4. The length, width, and height of all the specimens were 400, 300, and 300 mm, respectively (see Figure 9.2d). Two 50 mm thick extruded polystyrene (XPS) foam boards were adopted as the insulations. The relative slip between the outer and core RC wythes was recorded by the installed linear variable differential transformers (LVDTs).

9.2.3 Test Results

The shear force versus relative slip relationships of these specimens are shown in Figure 9.3. Here, the dashed curves represent the response of the individual specimen and the solid curves represent the average values of the two identical specimens. Generally, the failure of all the proposed connectors happened in a progressive manner. For the one-way connector, when the GFRP laminate was thinner (i.e., two plies), the flat plate connector reflected significantly lower deformability than the corrugated plate connector. This was probably because of the premature buckling of the GFRP laminate of the flat plate connector. The buckling of the flat plate connector could be avoided with an increase in thickness to four plies, although their deformability was still slightly different. For the

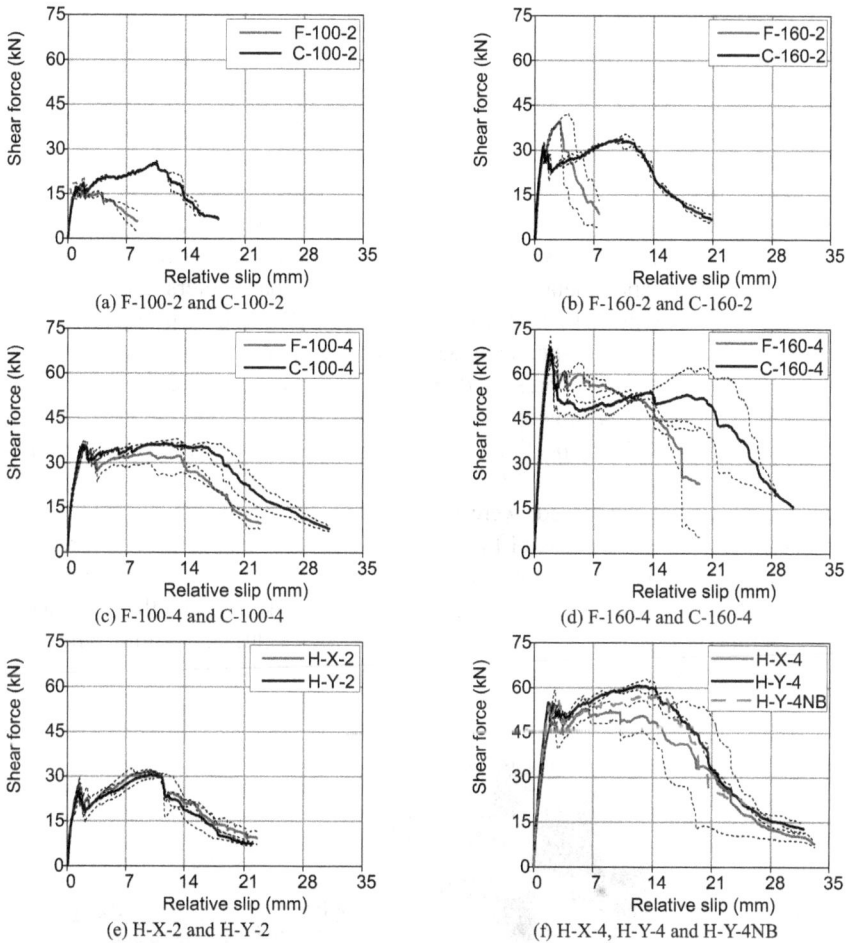

FIGURE 9.3 Shear force versus relative slip relationships of the direct shear test specimens (Huang and Dai 2019).

two-way connector, H-X specimens reflected a similar response with the H-Y specimens, which confirmed the similar performance of the hexagonal tube connector in two orthogonal shear force directions. In addition, H-Y-4-NB had the same response as H-Y-4, which confirmed that the bond between concrete and insulation had no effect on the shear resistance and stiffness of the connectors.

9.3 STRUCTURAL PERFORMANCE OF RC ONE-WAY SLABS WITH GEOPOLYMER CONCRETE

Huang and Dai (2018, 2021) investigated the out-of-plane performance of steel- and basalt FRP (BFRP)-reinforced geopolymer concrete one-way slabs. A brief introduction and major findings of this study are presented in this section.

9.3.1 TEST SPECIMENS

The out-of-plane flexural performance of steel-reinforced geopolymer concrete one-way slab and the out-of-plane shear performance of the BFRP-reinforced geopolymer concrete one-way slabs were investigated experimentally. In total, 18 specimens were prepared and tested. Twelve of these were steel-reinforced specimens, and the investigated parameters included the concrete type (OPC and geopolymer concrete), concrete compressive strength level (30, 40, and 50 MPa), and reinforcement ratio (0.82% and 1.20%). The other six specimens were the BFRP-reinforced specimens, and the investigated parameters included concrete compressive strength level (30, 40, and 50 MPa) and reinforcement ratio (1.20% and 2.18%). The specimen designation was in the form of A-B-C, in which "A", "B", and "C" represented concrete type, reinforcement, and concrete strength, respectively. The geopolymer concrete and OPC concrete were marked as "GCS" and "OCS," respectively. The specimens with the reinforcement ratio of 0.82% (steel-reinforced), 1.20% (steel-reinforced), 1.20% (BFRP-reinforced), and 2.18% (BFRP-reinforced) were marked as "S10," "S12," "B12," and "B16," respectively. The dimensions of the specimens were $2,100 \times 500 \times 120$ mm^3 (i.e., length $\times$ width $\times$ depth), with a concrete cover of 20 mm. All the specimens were tested under four-point loading condition (see Figure 9.4). The lengths of the shear span and the pure bending region were equal, with a value of 640 mm. The midspan deflection of the slabs was recorded by the installed LVDTs.

FIGURE 9.4　Test setup of the steel- and BFRP-reinforced slabs (Huang 2019).

9.3.2 Test Results

The load versus mid-span deflection relationships of the specimens are presented in Figures 9.5 and 9.6, and the failure modes of the specimens are presented in Figure 9.7. For the steel-reinforced specimens, the GCS specimens reflected similar flexural performance to their OCS counterparts, including the initial cracking load, yielding load, and load-carrying capacity (Figure 9.5). Also, the failure modes of both GCS and OCS specimens were governed by steel yielding and concrete crushing in the top part of the pure bending region (Figure 9.7a and b). BFRP-reinforced specimens were governed by a shear compression failure. The specimens with a lower reinforcement ratio could keep good integrity until the end of the test (Figure 9.7c). However, specimens with a higher reinforcement ratio had severe concrete cover splitting (Figure 9.7d). The

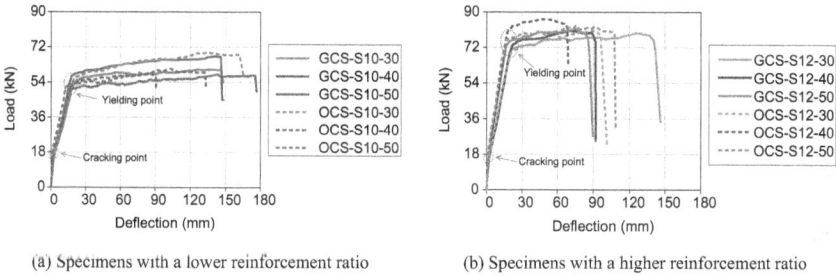

(a) Specimens with a lower reinforcement ratio

(b) Specimens with a higher reinforcement ratio

FIGURE 9.5 Load versus mid-span deflection relationships of the steel-reinforced slabs (Huang 2019).

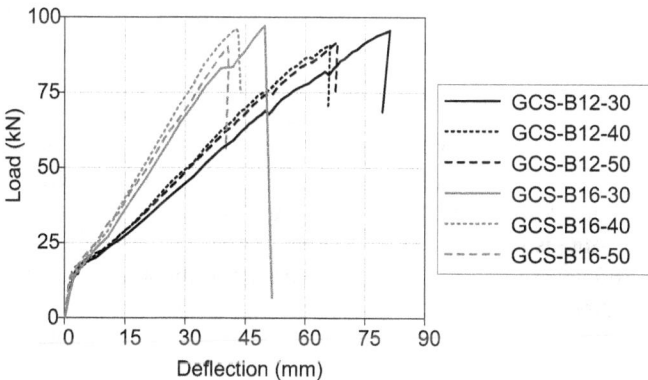

FIGURE 9.6 Load versus mid-span deflection relationships of the BFRP-reinforced slabs (Huang 2019).

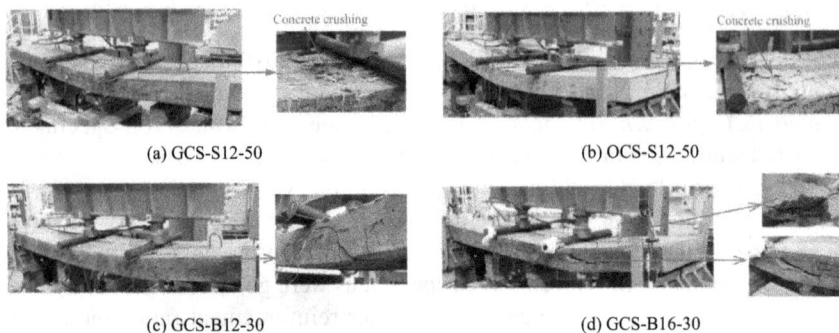

(a) GCS-S12-50   (b) OCS-S12-50

(c) GCS-B12-30   (d) GCS-B16-30

FIGURE 9.7 Typical failure mode of the test steel- and BFRP-reinforced slabs (Huang 2019).

peak load of all specimens was 90.6~96.9 kN. Besides, increasing the reinforcement ratio from 1.20% to 2.18% increased the average post-cracking stiffness by 60%.

9.3.3 COMPARISON WITH THE CODE OF PRACTICE AND GUIDELINES

The design formulae given in the code of practice and guidelines were used to assess the performance of the specimens. For steel-reinforced specimens, the design formulae of GB 50010-2010 (2010) and ACI 318-05 (2005) were used to calculate the cracking load and the load-carrying capacity, and the predicted values were compared with the test results and are presented in Table 9.1. The loads at the service limit state are also listed in Table 9.1. The deflection and the crack width limits as recommended by GB 50010-2010 (2010) are $l/200$ (l is the length between two supports) and $\omega = 0.2$ mm, respectively. ACI 318-05 (2005) recommends the deflection limit of $l/180$. At cracking load, the average ratio between the calculated and tested values is 0.89 and 0.95 for ACI 318-05 (2005) and GB 50010-2010 (2010), respectively (see Table 9.1). The average ratio between the calculated and tested ultimate load values is 0.82 for both GB 50010-2010 (2010) and ACI 318-05 (2005). In addition, the GCS specimen could maintain a level of around half of the load-carrying capacity when the service limit state is reached.

For the BFRP-reinforced specimens, the design formulae given in ACI 440.1R-06 (2006), CAN/CSA-S806-12 (2012), and JSCE (1997) were used to calculate the shear resistance, and the obtained values were compared with the test results, which is summarized in Table 9.2. The average ratio of the calculated and tested results is 0.54 (0.43 – 0.66), 0.39 (0.31 – 0.47), and 0.83 (0.69 – 0.98), respectively. Here, ACI 440.1R-06 (2006) and CAN/CSA-S806-12 (2012) lead to a conservative prediction, and JSCE (1997) shear design method predicts the capacity with a reasonable accuracy.

TABLE 9.1
Comparison of the Cracking Load, Load at Service Limit State, and Load-Carrying Capacity

Specimen ID	Cracking Load (kN)			Load at Service Limit State (kN)			Load-Carrying Capacity (kN)		
	ACI	GB	Test	l/200	l/180	ω = 0.2 mm	ACI	GB	Test
GCS-S10-30	13.0 [0.85]	12.8 [0.84]	15.3	30.1 (0.50)	33.1 (0.54)	37.7 (0.62)	46.5 [0.76]	47.0 [0.77]	60.8
GCS-S12-30	13.0 [0.77]	12.8 [0.76]	16.8	38.3 (0.48)	41.1 (0.52)	48.9 (0.61)	70.8 [0.89]	72.1 [0.90]	79.8
GCS-S10-40	15.1 [0.85]	16.9 [0.95]	17.7	35.3 (0.52)	37.7 (0.56)	40.2 (0.60)	47.4 [0.70]	47.7 [0.71]	67.5
GCS-S12-40	15.1 [0.92]	16.9 [1.03]	16.4	37.4 (0.46)	40.5 (0.50)	45.2 (0.56)	73.0 [0.91]	73.8 [0.82]	80.5
GCS-S10-50	16.7 [1.01]	18.6 [1.12]	16.6	32.8 (0.57)	36.0 (0.62)	34.5 (0.59)	45.0 [0.78]	45.2 [0.78]	58.0
GCS-S12-50	16.7 [0.92]	18.6 [1.03]	18.1	46.1 (0.57)	50.6 (0.62)	44.1 (0.54)	70.4 [0.86]	70.9 [0.87]	81.5

Note: The number written inside the round brackets represents the ratio between the load at service limit state and the tested load-carrying capacity; the number written inside the square brackets represents the ratio between the predicted and tested cracking loads or load-carrying capacities.

TABLE 9.2
Comparisons of the Shear Resistance between the Test and Calculated Results

Specimen Designation	Test (kN)	ACI 440.1R-06		CAN/CSA-S806-02		JSCE	
		Predict (kN)	Predict/test	Predict (kN)	Predict/Test	Predict (kN)	Predict/Test
GCS-B12-30	47.7	20.5	0.43	15.0	0.31	32.7	0.69
GCS-B12-40	45.3	22.4	0.50	16.7	0.37	36.5	0.81
GCS-B12-50	45.8	24.0	0.52	18.2	0.40	39.7	0.87
GCS-B16-30	48.5	25.2	0.52	17.4	0.36	36.5	0.75
GCS-B16-40	48.0	27.7	0.58	19.4	0.40	40.7	0.85
GCS-B16-50	45.0	29.7	0.66	21.1	0.47	44.3	0.98

9.4 FLEXURAL PERFORMANCE OF THE SANDWICH PANEL WITH GEOPOLYMER CONCRETE

Huang (2019) and Huang and Dai (2020) investigated the out-of-plane flexural performance of precast geopolymer concrete sandwich panels enabled with FRP connectors. A brief introduction and major findings of this study are presented in this section.

9.4.1 TEST SPECIMENS

Eight PCSWP specimens with geopolymer concrete were prepared and tested to evaluate the out-of-plane flexural performance. The length, width, and thickness of the specimens were 2,340, 300, and 200 mm, respectively. The investigated parameters consisted of the reinforcement type (steel and BFRP rebar), the reinforcement ratio (0.30% and 0.43%), the type of connector (commercial plate-type GFRP connector and the developed tubular GFRP connector), and spacing between connectors (300 and 525 mm). The plate-type GFRP connector was provided by Thermomass. Co. Ltd. The shear force versus relative slip relationship of each type of connector is shown in Figure 9.8a. The specimen designation was in the form of A-B-C-D, in which "A", "B", "C", and "D" represented the type of connector, the spacing between connectors (300 and 525 mm), the type of rebar (steel and BFRP rebar), and the diameter of the longitudinal rebar (10 and 12 mm). Here, for the type of connector, the plate-type and tubular connectors were marked as "P" and "H", respectively. For the type of rebar, the steel and BFRP rebars were marked as "S" and "B", respectively. Four-point loading condition was applied to all the specimens (see Figure 9.8b). The midspan deflection and the relative slip between the two RC wythes were recorded by the installed LVDTs.

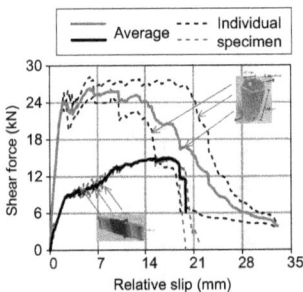

(a) Direct shear response of the adopted connectors

(b) Test setup

FIGURE 9.8 Direct shear response of the adopted connectors and flexural test setup (Huang and Dai 2020).

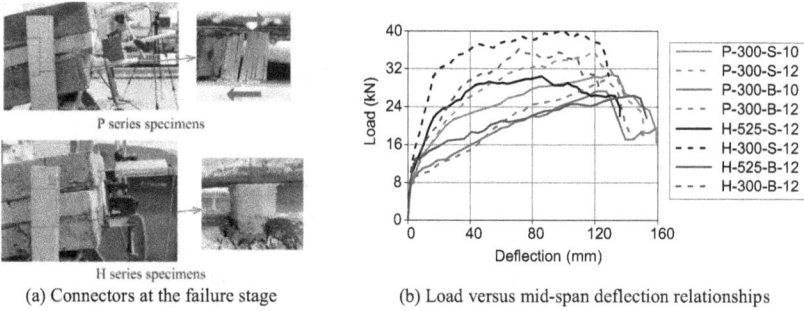

P series specimens

H series specimens

(a) Connectors at the failure stage   (b) Load versus mid-span deflection relationships

FIGURE 9.9 Flexural test results of the sandwich panel with geopolymer concrete (Huang and Dai 2020).

9.4.2 Test Results

The typical failure modes of the specimens are shown in Figure 9.9a. For the specimens enabled by the plate-type GFRP connector, the GFRP laminates were broken into pieces and the specimens were dominated by the connector failure. In contrast, the failure of specimens with hexagonal GFRP tube connector was governed by the concrete local failure, although considerable damage to the GFRP laminate also occurred.

Figure 9.9b presents the load versus mid-span deflection relationships of the eight specimens. It can be seen that the post-cracking stiffness and the load-carrying capacity of the specimens depend on the reinforcement type, the type of connector, and the spacing between connectors. The two above-mentioned performance parameters (i.e., the post-cracking stiffness and the load-carrying capacity) could be enhanced by using a higher-stiffness rebar (i.e., replacing the BFRP rebar with steel rebar) and a higher-stiffness connector (i.e., replacing the commercial FRP connector with the hexagonal tube connector), and reducing the spacing of the connectors.

9.4.3 Degree of Composite Action

The composite action of the tested PCSWPs at the initial loading stage and the ultimate stage was evaluated through the degree of composite action in terms of initial stiffness and ultimate strength (Choi et al. 2015a), respectively (DCA_i and DCA_u). Here, the degree of composite action in terms of initial stiffness can be calculated through $DCA_i = \left[\left(I_t - I_n \right) / \left(I_f - I_n \right) \right] \times 100\%$, in which I_t is the moment of inertia of the test PCSWP, and this value can be obtained from the experimental load–deflection curve. I_n and I_f are the theoretical moments of inertia of the non-composite and fully composite PCSWPs, respectively. The degree of composite action in terms of ultimate strength can be calculated through $DCA_u = \left[\left(P_t - P_n \right) / \left(P_f - P_n \right) \right] \times 100\%$, in which P_t is the load-carrying capacity

TABLE 9.3

Degree of Composite Action in Terms of Initial Stiffness and Ultimate Strength

Specimen ID	I_{test} ($\times 10^7$ mm^4)	I_f ($\times 10^7$ mm^4)	I_n ($\times 10^7$ mm^4)	DCA_i (%)	P_t (kN)	P_f (kN)	P_n (kN)	DCA_u (%)
P-300-S-10	3.12	19.69	2.11	5.76	30.80	28.12	11.27	115.92
P-300-S-12	3.05	19.69	2.11	5.37	35.50	41.09	15.99	77.75
P-300-B-10	3.03	19.69	2.11	5.23	27.30	71.75	14.53	22.33
P-300-B-12	3.05	19.69	2.11	5.37	29.80	87.52	17.16	17.96
H-525-S-12	3.44	19.69	2.11	7.54	30.30	41.08	15.98	57.04
H-300-S-12	4.46	19.69	2.11	13.38	40.20	41.08	15.98	96.48
H-525-B-12	3.51	19.69	2.11	7.99	29.40	87.44	17.14	17.44
H-300-B-12	4.33	19.69	2.11	12.66	36.20	87.44	17.14	27.11

of the test PCSWP. P_n and P_f are the theoretical moments of inertia of the non-composite and fully composite PCSWPs, respectively. The theoretical values of moments of inertia were obtained according to ACI 318-05 (2005) and ACI 440.1R-06 (2006) for steel- and BFRP-reinforced specimens, respectively. The calculated values of the degree of composite action in terms of initial stiffness and ultimate strength are presented in Table 9.3. Here, it can be noted that the values of both DCA_i and DCA_u can be increased by using a higher-stiffness connector and/or designing the connector with closer spacing.

9.4.4 SIMPLIFIED APPROACH FOR STIFFNESS PREDICTION OF THE PCSWP

In order to predict the deflection of the PCSWPs during both pre-cracking and post-cracking stages, a simplified approach was proposed, which is a closed-form solution based on the continuum method. Herein, three loading conditions were considered, as shown in Figure 9.10.

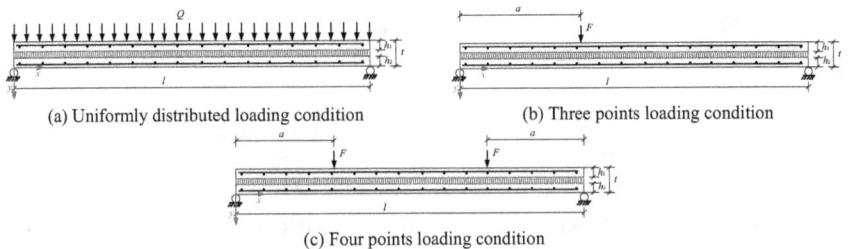

(a) Uniformly distributed loading condition   (b) Three points loading condition

(c) Four points loading condition

FIGURE 9.10 Loading conditions considered in the closed-form solutions (Huang 2019).

a. For uniformly distributed loading condition, the mid-span deflection (y_{UDL}) and the cracking load (Q_{cru}) were calculated as:

$$y_{UDL} = \frac{1}{E_c(I_1+I_2)} \left(\frac{2Q\left(e^{-0.5\lambda l} - e^{0.5\lambda l}\right)}{\lambda^4\left(e^{-\lambda l} - e^{\lambda l}\right)} + \frac{Q}{8\lambda^2}l^2 - \frac{Q}{\lambda^4} \right) \quad (9.1)$$

$$Q_{cru} = \left(\frac{h_2}{2(I_1+I_2)f_r} \left(-\frac{2\left(e^{-0.5\lambda l} - e^{0.5\lambda l}\right)}{\lambda^2\left(e^{-\lambda l} - e^{\lambda l}\right)} + \frac{1}{\lambda^2} \right) + \frac{1}{f_r A_2} \left(\frac{2\left(e^{-0.5\lambda l} - e^{0.5\lambda l}\right)}{\lambda^2 r\left(e^{-\lambda l} - e^{\lambda l}\right)} + \frac{1}{8r}l^2 - \frac{1}{\lambda^2 r} \right) \right)^{-1} \quad (9.2)$$

b. For three-point loading condition, the deflection at the loading point (y_{TPL}) and the cracking load (F_{crt}) were calculated as:

$$y_{TPL} = \frac{F}{\lambda^2 E_c(I_1+I_2)} \left[\frac{\left(e^{\lambda a} - e^{2\lambda l - \lambda a}\right)\left(e^{\lambda a} - e^{-\lambda a}\right)}{2\lambda\left(e^{2\lambda l} - 1\right)} - \frac{(a-l)a}{l} \right] \quad (9.3)$$

$$F_{crt} = \left(\frac{-h_2}{2f_r(I_1+I_2)} \left(\frac{\left(e^{\lambda a} - e^{2\lambda l - \lambda a}\right)}{2\lambda\left(e^{2\lambda l} - 1\right)}\left(e^{\lambda a} - e^{-\lambda a}\right) \right) + \frac{1}{f_r A_2} \left(\frac{\left(e^{\lambda a} - e^{2\lambda l - \lambda a}\right)}{2\lambda r\left(e^{2\lambda l} - 1\right)}\left(e^{\lambda a} - e^{-\lambda a}\right) + \frac{a}{r} - \frac{a^2}{rl} \right) \right)^{-1} \quad (9.4)$$

c. For four-point loading condition, the mid-span deflection (y_{FPL}) and the cracking load (F_{crf}) were calculated as:

$$y_{FPL} = -\frac{F\left(e^{\lambda a} - e^{-\lambda a}\right)e^{0.5\lambda l}}{\lambda^3 E_c(I_1+I_2)\left(1+e^{\lambda l}\right)} + \frac{Fa}{\lambda^2 E_c(I_1+I_2)} \quad (9.5)$$

$$F_{crf} = \left(\frac{h_2\left(e^{\lambda a} - e^{-\lambda a}\right)e^{\lambda 0.5 l}}{2f_r(I_1+I_2)\lambda\left(1+e^{\lambda l}\right)} + \frac{a}{A_2 r f_r} - \frac{\left(e^{\lambda a} - e^{-\lambda a}\right)e^{\lambda 0.5 l}}{A_2 \lambda r f_r\left(1+e^{\lambda l}\right)} \right)^{-1} \quad (9.6)$$

where I_1 and I_2 are the gross moments of inertia of the top and bottom RC wythes, respectively; h_1 and h_2 are the section thickness of the top and bottom RC wythes, respectively; l is the length between two supports of the panel; a is the length of the shear span; r is the distance between the center of the two RC wythes; E_c is the concrete elastic modulus; λ is a parameter which is calculated as $\lambda = \left[kr^2 / E_c(I_1+I_2) \right]^{0.5}$, in which $k = K/S$. K is the stiffness of a connector, and S is the connector spacing.

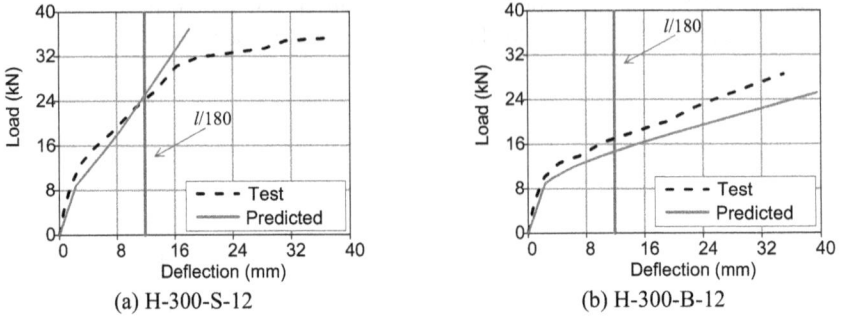

FIGURE 9.11   Comparisons of load versus mid-span deflection relationships between test and simplified approach (Huang 2019).

In the proposed approach, an equivalent moment of inertia (I_e) was introduced to calculate the deflection of the PCSWP during the post-cracking stage, as shown in the following:

$$I_e = \left(\frac{M_{cr}}{M}\right)^3 I_g + \left[1 - \left(\frac{M_{cr}}{M}\right)^3\right]\beta I_{cr} \tag{9.7}$$

where M_{cr} is the cracking moment of the PCSWP; M is the applied load in the moment; I_g and I_{cr} are the gross moment of inertia and the cracked moment of inertia of the PCSWP, respectively. Both I_g and I_{cr} were calculated based on the developed closed-form solution. Also, β is a reduction factor that is equal to 1.0 for steel-reinforced PCSWP and 0.27 for BFRP-reinforced PCSWP.

The test load–deflection relationships of the mentioned PCSWP specimens were used to validate the simplified approach. Here, the predicted load versus mid-span deflection relationships of H-300-S-12 and H-300-B-12 are compared to those obtained from tests in Figure 9.11. The closed-form solution reasonably predicted the load–deflection relationship during the pre-cracking stage. For steel-reinforced PCSWP, by adopting the equivalent moment of inertia, a reasonable prediction of the load–deflection relationship was achieved before $l/180$ (mid-span deflection). However, after this point, the differences between the predicted and the test results were larger. These differences were caused due to the damaged connector and yielded steel rebar. For BFRP-reinforced PCSWP, the predicted results using the equivalent moment of inertia agreed well with the tests.

9.5   FIRE AND POST-FIRE RESIDUAL PERFORMANCE OF THE SANDWICH PANEL WITH GEOPOLYMER CONCRETE

Huang et al. (2021) studied the structural performance of precast geopolymer concrete sandwich panels subjected to one-side fire and also investigated the post-fire residual performance. A brief introduction and major findings of this study are presented in this section.

9.5.1 Test Specimens

In this section, the fire and post-fire residual performance of the PCSWPs was evaluated. Six PCSWP specimens were fabricated. The dimensions of the panels were $1,600 \times 900 \times 200$ mm^3 or $1,600 \times 900 \times 250$ mm^3 (height $\times$ width $\times$ thickness). The RC wythes were 75 or 100 mm thick, and the insulation was 50 mm thick. Solid RC end-capping beams with the dimensions of $300 \times 300 \times 900$ mm^3 (height $\times$ width $\times$ length) were fabricated at the top and bottom of the PCSWP for transferring the load. The investigated parameters included the concrete type (geopolymer and OPC concrete), the reinforcement type (steel and BFRP rebar), the connector type (plate-type and hexagonal tube GFRP), and the RC wythe thickness (75 and 100 mm). The specimen designation was in the form of A-B-C-D, in which "A", "B", "C", and "D" represented the concrete type, connector type, reinforcement type, and the RC wythe thickness (75 and 100 mm). The geopolymer concrete and OPC concrete were marked as "G" and "O", respectively. The plate-type and hexagonal tube connectors were marked as "P" and "H", respectively. The steel and BFRP reinforcements were marked as "S" and "B", respectively. The final specimen G-P-B-75a had the same design parameter as G-P-B-75. However, it was only tested under axial load with ambient temperature conditions for comparison.

9.5.2 Test Procedure

The test program consisted of two stages. First, a one-side fire exposure scheme was designed in a furnace built for testing the structural members (see test setup in Figure 9.12a). During the fire test, a constant axial compressive load was applied at the top RC beam of each PCSWP specimen by the MTS hydraulic

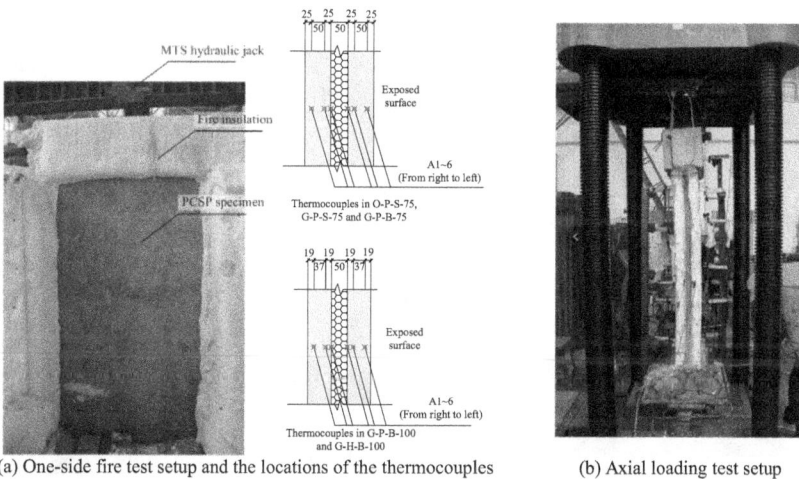

(a) One-side fire test setup and the locations of the thermocouples   (b) Axial loading test setup

FIGURE 9.12 Test setup of the one-side fire test and the post-fire axial loading test.

jack, reaching a nominal axial load ratio of 0.15. LVDTs were placed at the top RC beams to measure the axial deformation. Thermocouples were placed to monitor the temperature history of concrete (see Figure 9.12a). The fire exposure test was terminated after a fire duration of 240 min. The furnace temperature was controlled following the ISO-834 standard fire curve. After the fire test was completed, an axial loading test was conducted on the fire-exposed PCSWPs to evaluate their post-fire strength while taking the unexposed G-P-B-75a as a reference (see test setup in Figure 9.12b). Here the axial load was applied in displacement control by a hydraulic jack. The test was terminated when the load decreased to 85% of the peak load. The axial deformations of the specimens were recorded by the installed LVDTs.

9.5.3 TEST RESULTS

At the fire duration of 234 min, the OPC concrete PCSWP (O-P-S-75) buckled toward the heating side and the vertical load could no longer be withstood by the specimen. The concrete near the top and bottom of the exposed RC wythe was crushed. However, all PCSWP specimens made of geopolymer concrete had no obvious failure on the exposed RC wythe, although a residual deformation that formed a 20 mm eccentricity could be seen after the fire test (see Figure 9.13).

The temperature–time relationships of O-P-S-75 and G-P-S-75 are compared and shown in Figure 9.14. Here, O-P-S-75 presented a similar temperature curve to that of G-P-S-75, including the shape of the curve and the temperature value, indicating that the PCSWP made by geopolymer concrete reflected similar thermal property with that made by OPC concrete.

The load versus axial deformation relationships of the post-fire loading test specimens are shown in Figure 9.14, and the failure modes are shown in Figure 9.15. The failure of exposed geopolymer PCSWP specimens was due to

(a) O-P-S-75   (b) G-P-S-75

FIGURE 9.13   Failure of specimens after the fire test.

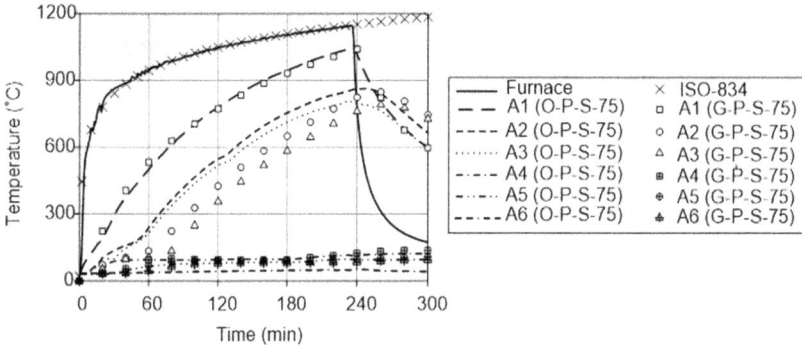

FIGURE 9.14 Relationship between temperature and time in concrete.

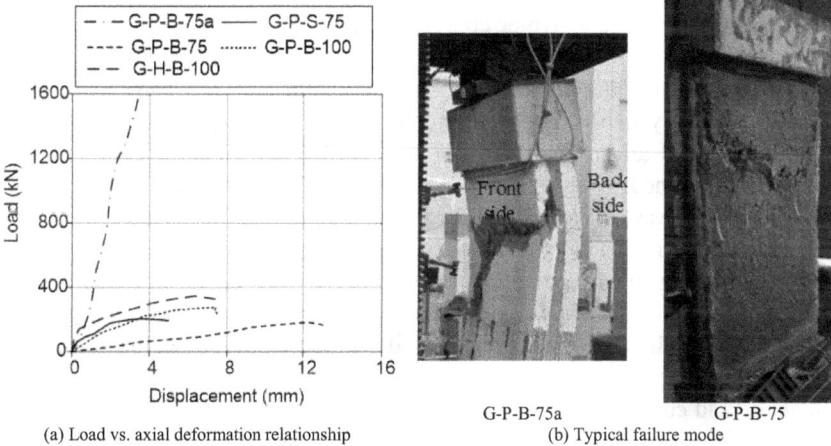

(a) Load vs. axial deformation relationship

G-P-B-75a   G-P-B-75

(b) Typical failure mode

FIGURE 9.15 Post-fire axial loading test results.

the crushing of the exposed RC wythe and the tensile failure of the unexposed RC wythe. Here, G-P-B-75a achieved a strength of 1,500 kN. However, after the fire exposure of 240 min, the residual strength of the counterpart specimen (i.e., G-P-B-75) was only 196 kN.

9.6 CONCLUSIONS

This chapter presented the experimental and theoretical studies conducted on a proposed BFRP-bar-reinforced geopolymer concrete sandwich/solid wall panel enabled with FRP connectors. In addition, in-plane direct shear performance of three types of FRP connectors was also studied. The performance of the proposed wall panels was compared with steel-reinforced OPC concrete sandwich/solid wall panels. A closed-form solution, which was obtained following a simplified approach, to predict the stiffness of the proposed sandwich wall panel was developed. Further, the major findings from the tests conducted to study the fire performance of the proposed sandwich wall are also presented.

The study consisting of direct shear test on the proposed GFRP connectors can be concluded as: (1) the failure of the connector occurred in a progressive manner; (2) owing to the shear buckling of the GFRP laminate, the plate-type connector had less deformability than the corrugated one. However, increasing the laminate thickness avoided the shear buckling and enhanced the deformability; and (3) the hexagonal tube connector offered a good potential to be used as a two-way connector because of the similar and satisfactory performance along with two orthogonal directions.

For the structural performance of steel- and BFRP-reinforced geopolymer concrete one-way slabs, it was concluded that: (1) the flexural performances of the steel-reinforced geopolymer concrete one-way slabs and the OPC concrete counterparts were similar; (2) the existing design provisions given in the code of practice had the potential to be used for predicting the cracking load and the load-carrying capacity of the steel-reinforced geopolymer concrete one-way slabs; (3) the failure of all the tested BFRP-reinforced geopolymer concrete one-way slabs was governed by shear compression; and (4) the equation in JSCE shear design method was recommended for use in calculating the shear resistance of the BFRP-reinforced geopolymer concrete one-way slabs.

The following conclusions were drawn from the study on the structural performance of PCSWPs, which also includes the development of a simplified approach for stiffness prediction: (1) the failure modes of the PCSWP specimens with the plate-type and hexagonal tube GFRP connectors were governed by the connector breakage and concrete local failure, respectively; (2) using a high-stiffness connector and/or reduced connector spacing, the stiffness and load-carrying capacity of the PCSWP could be enhanced and the degree of composite action in terms of both initial stiffness and ultimate strength could be improved; and (3) the simplified approach gave an accurate prediction of the load–deflection relationship for a PCSWP under out-of-plane load, indicating a good potential for use in the design of PCSWP.

For the fire and post-fire residual performance of the PCSWP, it was concluded that (1) the geopolymer concrete PCSWP survived with structural integrity after 240 min of standard fire exposure, whereas the OPC concrete PCSWP failed due to the buckling of the exposed wythe; (2) the geopolymer concrete PCSWP reflected temperature field similar to that reflected by OPC concrete PCSWP; and (3) a significant strength deterioration (i.e., by 88.5%) of PCSWP specimens was observed after the 240 min exposure as compared to the ambient strength in spite of the well-kept structural integrity.

REFERENCES

ACI 318-05. 2005. *Building Code Requirements for Reinforced Concrete and Commentary*, American Concrete Institute, Farmington Hills, MI.

ACI 440.1R-06. 2006. *Guide for the Design and Construction of Structural Concrete Reinforced with FRP Bars*, American Concrete Institute, Farmington Hills, MI.

Benayoune, A., Samad, A. A., Trikha, D. N., Ali, A. A., and Ellinna, S. H. M. 2008. Flexural behaviour of pre-cast concrete sandwich composite panel–experimental and theoretical investigations. *Construction and Building Materials*, 22(4), 580–592.

Bush, T. D., and Stine, G. L. 1994. Flexural behavior of composite precast concrete sandwich panels with continuous truss connectors. *PCI Journal*, 39(2), 112–121.

CAN/CSAS806-12. 2012. *Design and Construction of Building Components with Fiber Reinforced Polymers. Canadian Standards Association*, Rexdale, ON.

Chen, A., Norris, T. G., Hopkins, P. M., and Yossef, M. 2015. Experimental investigation and finite element analysis of flexural behavior of insulated concrete sandwich panels with FRP plate shear connectors. *Engineering Structures*, 98, 95–108.

Choi, I., Kim, J., and Kim, H. R. 2015a. Composite behavior of insulated concrete sandwich wall panels subjected to wind pressure and suction. *Materials*, 8(3), 1264–1282.

Choi, K. B., Choi, W. C., Feo, L., Jang, S. J., and Yun, H. D. 2015b. In-plane shear behavior of insulated precast concrete sandwich panels reinforced with corrugated GFRP shear connectors. *Composites Part B: Engineering*, 79, 419–429.

Davidovits, J. 1991. Geopolymers: Inorganic polymeric new materials. *Journal of Thermal Analysis*, 37, 1633–1656.

Ding, Y., Dai, J. G., and Shi, C. J. 2016. Mechanical properties of alkali-activated concrete: A state-of-the-art review. *Construction and Building Materials*, 127, 68–79.

Ding, Y., Dai, J. G., and Shi, C. J. 2019. Fracture properties of alkali-activated slag and ordinary Portland cement concrete and mortar. *Construction and Building Materials*, 165, 310–320.

Frankl, B. A., Lucier, G. W., Hassan, T. K., and Rizkalla, S. H. 2011. Behavior of precast, prestressed concrete sandwich wall panels reinforced with CFRP shear grid. *PCI Journal*, 56(2), 42–54.

GB 50010-2010. 2010. *Code for Design of Concrete Structures*, China Structural Science Academe, Beijing, China.

Huang, J.-Q. 2019. Structural performance of precast reinforced geopolymer concrete sandwich panels enabled by FRP connectors. PhD thesis, The Hong Kong Polytechnic University. https://theses.lib.polyu.edu.hk/handle/200/10015

Huang J.-Q., and D. Jian-Guo. 2020. Flexural performance of precast geopolymer concrete sandwich panel enabled by FRP connector. *Composite Structures*, 248: 112563.

Huang J.-Q., and D. Jian-Guo. 2018. Experimental investigation on flexural performance of steel reinforced geopolymer concrete one-way slabs, The 8th International Conference of Asian Concrete Federation (ACF2018), 4–7 November, 2018, Fuzhou, China.

Huang J.-Q., and D. Jian-Guo. 2019. Direct shear tests of glass fiber reinforced polymer connectors for use in precast concrete sandwich panels. *Composite Structures*, 207: 136–147.

Huang J.-Q., and D. Jian-Guo. 2021. Shear performance of BFRP reinforced geopolymer concrete one-way slab, 10th International Conference on FRP Composites in Civil Engineering (CICE2020), 8–10 December, 2021, İstanbul.

Huang J.-Q., X. Yu-Ye., H. Hao., and D. Jian-Guo. 2021. Structural behavior of FRP connector enabled precast geopolymer concrete sandwich panels subjected to one-side fire exposure. *Fire Safety Journal*. (Revised submission).

JSCE (Japan Society of Civil Engineers). 1997. *Recommendation for Design and Construction of Concrete Structures using Continuous Fiber Reinforcing Materials.* Tokyo.

Kinnane, O., O'Hegarty, R., and West, R. 2020. Structural shear performance of insulated precast concrete sandwich panels with steel plate connectors. *Engineering Structures*, 215, 110691.

Nath, P., and Sarker, P. K. 2014. Effect of GGBFS on setting, workability and early strength properties of fly ash geopolymer concrete cured in ambient condition. *Construction and Building Materials*, 66, 163–171.

O'Hegarty, R., West, R., Reilly, A., and Kinnane O. 2019. Composite behaviour of fibre-reinforced concrete sandwich panels with FRP shear connectors. *Engineering Structures*, 198, 109475.

PCI Committee. 2011. State of the art of precast/prestressed concrete sandwich wall panels. *PCI Journal*, 56(2), 131–176.

Qian, L. P., Wang, Y. S., Alrefaei, Y., and Dai, J. G. 2020. Experimental study on full-volume fly ash geopolymer mortars: Sintered fly ash versus sand as fine aggregates. *Journal of Cleaner Production*, 263, 121445.

Salmon, D. C., Einea, A., Tadros, M. K., and Culp, T. D. 1997. Full scale testing of precast concrete sandwich panels. *ACI Structural Journal*, 94, 239–247.

Tomlinson, D., and Fam, A. 2015. Flexural behavior of precast concrete sandwich wall panels with basalt FRP and steel reinforcement. *PCI Journal*, 50(6), 51–71.

10 Axial Behavior of Reinforced Concrete Sandwich Wall Panels Enabled by Fiber-Reinforced Polymer (FRP) Connectors

Jian-Guo Dai and Sushil Kumar
The Hong Kong Polytechnic University

Junqi Huang
Hefei University of Technology

CONTENTS

10.1 Introduction 168
10.2 Materials 168
    10.2.1 Geopolymer Concrete.. 168
    10.2.2 BFRP Bar/Grid 169
    10.2.3 GFRP Tubular Connectors 171
    10.2.4 Insulation 172
10.3 Axial Compression Behavior – Experimental Investigation 172
10.4 Eccentric Compression Behavior – Experimental Investigation 176
10.5 Role of Connectors 184
10.6 Simplified Theoretical Investigation... 186
    10.6.1 SWPs Subjected to Eccentric Axial Load................................. 186
10.7 Prediction of Load Capacity from the Existing Empirical Design
    Formulae.. 187
10.8 Concluding Remarks 188
References.. 189

DOI: 10.1201/9781003143031-10

10.1 INTRODUCTION

Precast concrete sandwich wall panels (SWPs), also known as insulated concrete walls, are thermally efficient prefabricated elements. The higher environmental and economic cost of heating and cooling has led researchers to further improve the thermal resistance of the buildings with an enhanced level of structural efficiency. The use of prefabricated elements ensures higher quality control and reduces the in situ construction time. Therefore, prefabricated SWPs have successfully been used as an exterior facade or load-bearing walls in building envelopes for decades. SWPs are composed of two reinforced concrete wythes, which surround an insulation layer and are linked by connectors. Traditionally, steel ties or concrete blocks have been used as connectors. However, their higher conductivity led to thermal bridging. Therefore, traditionally used connectors have been replaced with the ones produced from nonmetallic fiber-reinforced polymer (FRP) due to their high strength and low thermal conductivity. The shape, material, and number of connectors depend on the required level of composite action, cost, and thermal conductivity of the connector.

SWPs have been used in several residential and commercial structures around the globe. The total thickness of a SWP generally varies from 130 to 305 mm, with the thickness of insulation varying from 25 to 100 mm (PCI Committee 2011). A SWP can be composite, partially composite, or non-composite. Non-composite walls are designed with a small number of connector ties to minimize thermal bowing of the SWP (Einea et al. 1991, Frankl et al. 2011). The shear connectors for composite or partially composite walls are designed to provide significant shear transfer between concrete wythes and allow the wythes to deflect together such that the entire SWP behaves as one integral unit. The composite SWPs have added strength and stiffness as compared to the non-composite ones. However, the potential end movements from thermal bowing should be considered. Although theoretically a SWP can be designed as a fully composite or non-composite system, in reality, most of the SWPs tend to behave as partially composite systems.

A SWP is often subjected to in-plane concentric or eccentric gravity loads (e.g., dead and live loads), in-plane lateral cyclic loads (e.g., under seismic action), out-of-plane transverse loads (e.g., under wind action), or a combination of these. This chapter presents the investigation findings of a research work carried out to understand the in-plane axial load behavior of reinforced concrete SWP enabled with FRP connectors.

10.2 MATERIALS

10.2.1 GEOPOLYMER CONCRETE

Geopolymer concrete is clinker-free concrete. It is prepared by the alkali activation of aluminosilicate materials (e.g., industrial by-products – fly ash and slag) and mixing the resulting alkali-activated binders with fine and coarse aggregates and curing under proper conditions. Alkali-activated binders can be divided into two subclasses: high calcium (e.g., slag-based binders) and low calcium or no calcium (e.g., low-calcium fly ash-based binders). The main hydration product of alkali-activated slag-based binder is a C-S-H gel having a lower Ca/Si ratio than traditional Portland cement-based concrete; in contrast, the main hydration product of low-calcium or calcium-free binders is a three-dimensional N-A-S-H gel. The physical and chemical properties of geopolymer concrete depend mainly on the raw materials used and the curing condition. A typical mix design and the corresponding compressive strength properties are shown in Table 10.1.

Ding et al. (2016) conducted a state-of-the-art review of the mechanical properties of geopolymer concrete (i.e., alkali-activated concrete). The compressive strength of ambient-cured fly ash-based geopolymer concrete develops very slowly; however, partial replacement of fly ash with slag leads to early strength. As mentioned earlier, the primary hydration product of slag-based geopolymer concrete is C-S-H. In contrast, it is N-A-S-H for fly ash-based concrete. The elastic modulus of concrete mainly depends on the type and content of coarse aggregates and the elastic modulus of paste. The Young's modulus of the N-A-S-H paste is much lower than that of the C-S-H paste. Lee and Lee (2013) stated that the elastic modulus of alkali-activated fly ash-slag concrete was 20%–40% lower than that predicted using ACI 318-14 (2014). However, there is a significant discrepancy in the reported test data due to the sensitivity of mechanical properties to the source of raw materials (i.e., fly ash and slag). The bond stress–slip curves of steel reinforcement with geopolymer concrete have been found similar to that with OPC concrete. Alkali-activated fly ash paste has been reported to have better temperature resistance than OPC paste; however, slag-based alkali-activated paste has a similar temperature resistance performance to OPC paste for the temperature range of 400°C–800°C. Ding et al. (2018) conducted an experimental study and reported the fracture properties of geopolymer concrete.

The contribution of concrete to the axial load capacity of a squat column/wall panel (P_c) is given by $k_3 \times f'_c \times A_c$, where f'_c is the cylinder compressive strength and A_c is the sectional area of concrete. The empirical parameter k_3 is always less than 1. This difference in the strength of a concrete structural element and a concrete cylinder is due to the change in shape, size, and other factors (e.g., curing conditions and strain rate). The recommended value of k_3 is 0.85 for OPC concrete (ACI318-19 2019). The value of k_3 has been reported as 0.7 for ambient-cured geopolymer concrete columns (Tran et al. 2019, Albitar et al. 2017) and 0.9 for heat-cured geopolymer concrete columns (Tran et al. 2019, Sarker et al. 2008). However, more research studies need to be conducted in the future to verify the above observations.

10.2.2 BFRP BAR/GRID

FRP bar has been used for decades as internal reinforcement to mitigate corrosion and enhance the durability of reinforced concrete structures. The most widely used FRP bars are made from carbon (CFRP), glass (GFRP), and aramid (AFRP). CFRP is expensive for regular RC structures, while GFRP and

TABLE 10.1
Typical Mix Design of Geopolymer Concrete

	Slag	Fly Ash	Sand	Stone	Water	Sodium Silicate Solution	Sodium Hydroxide	w/b	n (%)	M_s	fc	fst	E
						kg/m³					MPa		GPa
Mix-1	200	200	716	1,074	145	56	9	0.45	3	1.5	31.7	2.88	22.7
Mix-2	300	100	712	1,068	133	74	12	0.45	4	1.5	52.1	3.56	23.8
Mix-3	200	200	716	1,074	102	93	15	0.40	5	1.5	65.7	4.32	25.7
Mix-4	200	200	712	1,068	139	62	19	0.45	5	1.0	44.4	3.58	27.4
Mix-5	400	0	712	1,068	133	74	12	0.45	4	1.5	63.1	3.97	22.3

Source: Data from Ding (2017).

Note: M_s, the modulus of alkali activator; n, the alkali concentration; fc, fst, and E are the concrete compressive strength, splitting tensile strength, and elastic modulus, respectively; w/b, water-to-binder ratio.

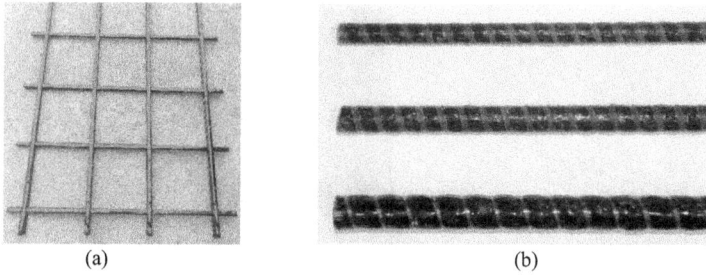

FIGURE 10.1   BFRP reinforcement: (a) BFRP grid and (b) BFRP bar.

AFRP are sensitive to the alkaline environment within the concrete (Sim et al. 2005). Basalt FRPs (BFRPs) are non-hazardous and environmentally friendly fibers produced from basalt rock by drawing and winding fibers from the melt. BFRP bars have recently been introduced as reinforcing material due to their lower cost, excellent resistance to acid and freeze–thaw cycle, better performance under fire, improved corrosion resistance, high accessibility, lightweight, and non-magnetic properties (Wang et al. 2017). Therefore, there is an incentive to use BFRP as reinforcement in RC structures. Several studies have been conducted using BFRP as flexural and shear reinforcement (Tomlinson and Fam 2015) and as main reinforcement in columns (Ibrahim et al., 2016). ACI 440.1R-15 (2015) provides the design guidelines for FRP-reinforced concrete section; however, it covers design guidelines only for the reinforcement produced from AFRP, CFRP, and GFRP.

BFRP reinforcement has been used as sand-coated bars, deformed surface (e.g., ribbed or helically wound) bars, and grids (see Figure 10.1). The material constitutive relationship of the BFRP bar/grid is linear elastic. Figure 10.2 shows a chord of the BFRP grid after failure in tension. The elastic modulus has mostly been reported to be varying from around 48,000 to 70,000 MPa (Huang and Dai 2020, Tomlinson et al. 2016). The ultimate strength and ultimate strain have been reported to be nearly 1,100 MPa and 2%, respectively.

10.2.3   GFRP TUBULAR CONNECTORS

Traditionally, steel ties or concrete blocks have been used as connectors. Owing to the thermal bridging effect, these have been replaced by connectors produced from FRP. The connector material is desired to have low cost and high thermal and corrosion resistance. These connectors can be produced in different shapes and configurations, such as pin, rod, grids, plates, and trusses. Naito et al. (2012) performed a comprehensive study to determine the strength, stiffness, and deformability of 14 types of connectors having different shapes and produced from different FRP materials. Connectors in truss girder configuration have been found very effective in resisting shear. However, these existing types of connectors are effective only to resist one-way shear. Huang and Dai (2019) proposed a

FIGURE 10.2   A chord of the BFRP grid after failure in tension.

GFRP tubular connector (see Figure 10.3) with excellent strength, stiffness, and deformability in two orthogonal directions. Figure 10.4 represents the idealized shear force–slip relationship of the GFRP tubular connector.

10.2.4   INSULATION

SWPs are provided with closed-cell rigid foams to make them thermally insulated. The two most commonly used foams are expanded polystyrene (EPS) and extruded polystyrene (XPS). Naito et al. (2012) conducted an experimental study and reported that the shear strength of a SWP with EPS insulation was 21% higher than the one with XPS when used in conjunction with a CFRP truss shear tie.

10.3   AXIAL COMPRESSION BEHAVIOR – EXPERIMENTAL INVESTIGATION

A few experimental studies have been conducted to investigate the in-plane axial compression load behavior of the SWPs. These studies have been carried out for the following two types of SWP systems:

FIGURE 10.3   GFRP connector.

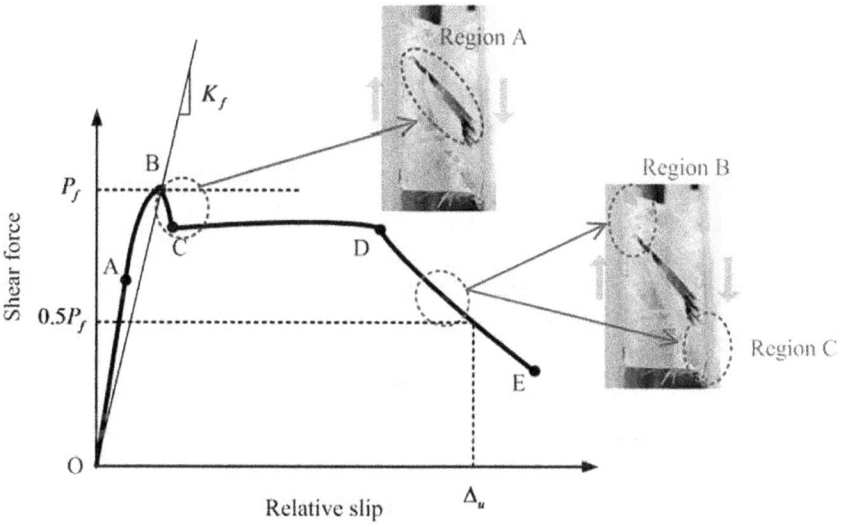

FIGURE 10.4   Idealized shear force vs relative slip relationships of the test specimens. (Adapted from Huang and Dai (2019).)

a. traditional form of SWPs in which the applied load is primarily sustained by the inner wythe (i.e., axial load is applied to inner wythe). The entire cross section of the SWP (mostly being partially composite) only comes into play to resist the out-of-plane first-order and second-order bending moments; and

b. both the concrete wythes are load bearing. Axial load is applied to the end-capping beams, which connect the two wythes at the top and bottom ends.

Kumar et al. (2021) conducted an experimental study to investigate the axial load behavior (load eccentricity = 0) of BFRP-grid-reinforced geopolymer concrete SWPs. Two 75 mm thick concrete wythes were connected laterally with GFRP tubular connectors. Smooth-surface XPS foam of 50 or 100 mm thickness was used as an insulation layer. The test parameters considered for the study were the slenderness ratio of the SWP (λ, defined as the effective height divided by total sectional thickness; 8, 11, 14, 17), longitudinal spacing of connectors ($S = 200, 300, 600$ mm), and the ratio of insulation thickness to wythes' thickness ($t_i/t_w = 0.33, 0.66$). High-strength OPC concrete end-capping beams were provided at both the ends of the SPWs to avoid the local crushing failure and simulate the floor beam's effect. Figure 10.5 shows the fabricated SWPs, which are being cured in ambient conditions.

SWPs were tested with both ends hinged. All the SWPs had material failure by crushing of concrete. No cracks were spotted on the concrete surface during the entire course of loading. There was no spalling of concrete until failure. The failure modes of typical squat and slender SWPs are shown in Figure 10.6. The typical lateral deflection profile of a slender SWP is shown in Figure 10.7. Figure 10.8 shows the strain variation across the sectional thickness of a typical slender SWP. The strain is almost linearly varying initially. However, as the applied load approached the ultimate axial load, the curvature of two individual wythes becomes different, possibly due to local bending of the wythes. The typical load–strain relationship is shown in Figure 10.9. The maximum recorded compressive strain in concrete and the corresponding value in reinforcement at the ultimate axial load were 2,973 and 2,142 $\mu\varepsilon$, respectively.

The axial load capacity of the SWP was reduced by 26% when the slenderness ratio increased from 8 to 17 (Figure 10.10). The spacing of connectors had a marginal impact on the axial load capacity of the SWPs. Increasing the t_i/t_w increased the bending stiffness (EI/H_{eff}) of the SWPs. For squat SWPs, t_i/t_w did not have any significant impact on axial load capacity as the second-order effect was minimal. However, the axial load capacity of the slender SWPs increased by 21% as t_i/t_w increased from 0.33 to 0.66 (Figure 10.11).

Benayoune et al. (2007) conducted a full-scale experimental study to investigate the axial load behavior of steel-mesh-reinforced OPC concrete SWPs using the continuous steel truss connectors. Axial load was applied to the centroid of the entire cross section of SWPs. The slenderness ratio (H_{eff}/t_i) of SWPs was varied from 10 to 20, and the two t_i/t_w values were 0.5 and 0.63. The panels were tested

FIGURE 10.5 Fabricated sandwich wall panel specimens.

with the bottom end fixed and the top end hinged. The load capacity reduced by 21% as slenderness increased from 10 to 20. Benayoune et al. (2007) proposed an empirical design formulation to predict the load capacity of axially loaded SWPs by amending the ACI design formula and adding a term for the contribution of reinforcement to the total capacity (Eq. 10.1).

$$P = 0.4 f_c' A_g \left[1 - \left(\frac{H}{40t} \right)^2 \right] + 0.67 f_y A_{sc} \tag{10.1}$$

FIGURE 10.6   Failure modes of the SWPs.

10.4   ECCENTRIC COMPRESSION BEHAVIOR – EXPERIMENTAL INVESTIGATION

In a real scenario, the SWPs are subjected to gravity and live loads acting eccentric to the centroid of SWP's cross section. Axial load–moment interaction comes into the picture to analyze and design an SWP when an eccentric axial load is applied. Due to load eccentricity and slenderness, SWPs deform laterally. The applied axial load, load eccentricity, and lateral deflection result in the development of bending moment in SWP.

The N–M curve of an SWP may differ from a solid wall panel owing to the partial interaction between the two concrete wythes (i.e., partially composite action) in a SWP. Research studies have been conducted for the eccentric axial load applied to SWP in two different ways.

1. In the first type of SWP system, the wythes are not connected at the ends and the load is applied to the inner wythe. For example, in traditional SWPs, round ties spaced at a considerable distance were provided to transfer the dead load of the outer wythe (facade) to the load-bearing inner wythe. However, nowadays, FRP connectors (e.g., truss girder shape), which effectively provide a certain level of composite actions, are used. The axial load is applied directly to the inner wythe and is mainly resisted by the inner wythe, which may result in a high value

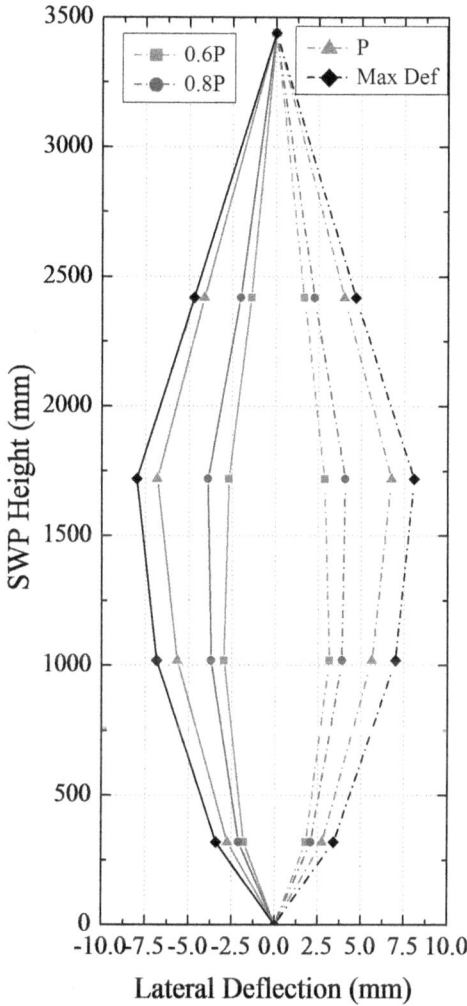

FIGURE 10.7 Typical lateral deflection profile of a concentrically loaded SWP. (Adapted from Kumar et al. (2021).)

of the axial load ratio. It implies that a squat wall panel subjected to axial load at the centroid of the inner wythe is always non-composite as the load is resisted by the inner wythe only. The full cross section comes into play only to resist the bending due to load eccentricity and second-order effect, and the degree of composite action depends upon the effectiveness of the connectors.

2. The wythes are connected at the top and bottom ends by the end-capping beams, and the eccentric load is applied to the capping beam. The presence of the end-capping beams suppresses the interface slip between the two wythes. The primary role of the connectors is to prevent the lateral

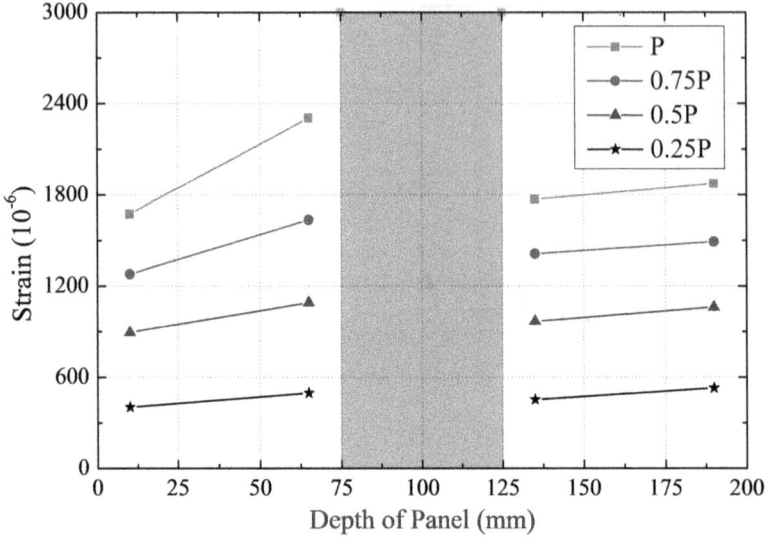

FIGURE 10.8   Typical strain variation across the sectional thickness.

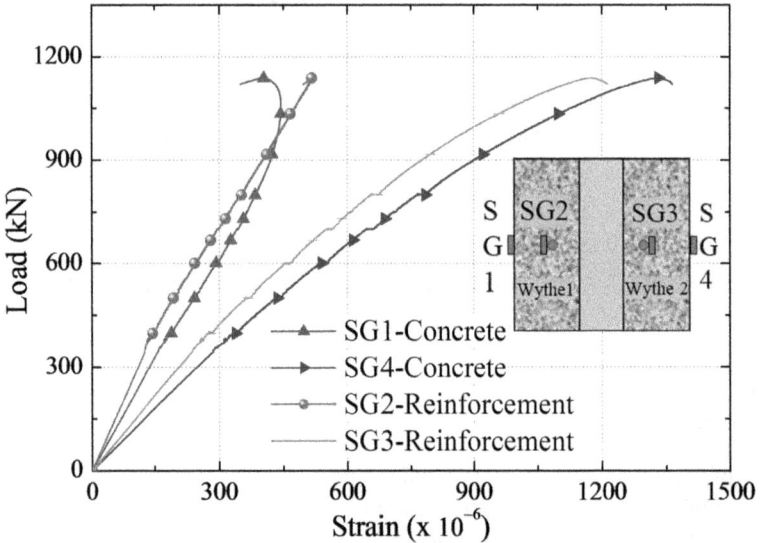

FIGURE 10.9   Typical load–strain relationship.

separation of the two wythes. These SWPs have been found both structurally and thermally efficient. The section depth of such SWPs can be
up to two-thirds of that of equivalent non-composite ones. The reduction
in self-weight leads to improved spacing optimization and reduced seismic loading.

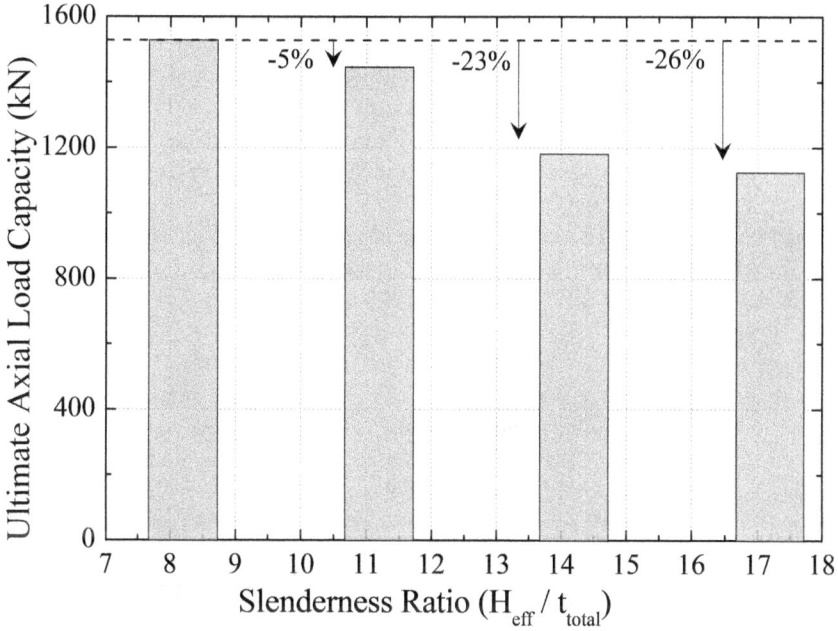

FIGURE 10.10 Variation in load capacity with slenderness ratio. (Adapted from Kumar et al. (2021).)

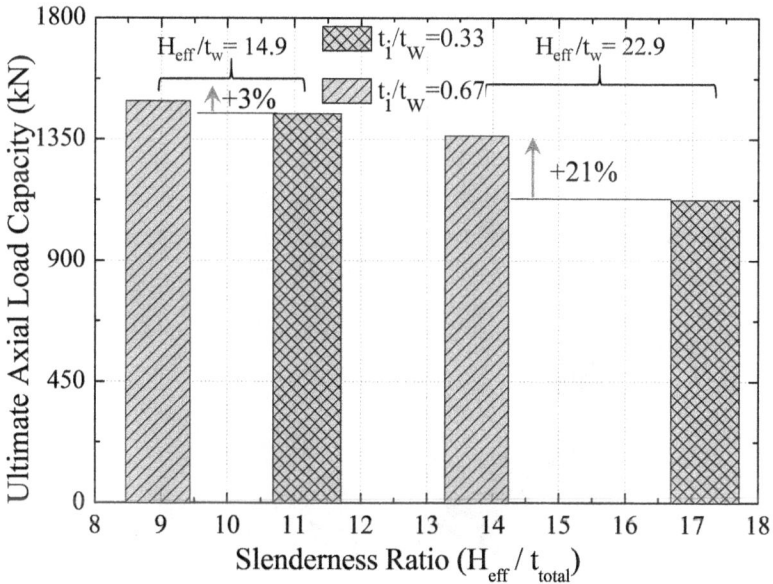

FIGURE 10.11 Effect of insulation thickness on load capacity. (Adapted from Kumar et al. (2021).)

An experimental study has recently been conducted to investigate the in-plane eccentric load behavior of BFRP-grid-reinforced geopolymer concrete SWPs. The wythes were connected by high-strength OPC end-capping beams. The two 75 mm thick concrete wythes were connected laterally with GFRP tubular connectors. Smooth-surface XPS foam of 50 mm thickness was used as an insulation layer. The test parameters considered included the slenderness of SWP, load eccentricity, and type of reinforcement (BFRP bar or grid). The sand-coated-type BFRP bar was used. The ultimate tensile strength and ultimate tensile strain of the BFRP grid/bar were 1,100 MPa and 2%, respectively. The elastic modulus of the BFRP grid and BFRP bar was 58,000 and 48,000 MPa, respectively. The percentage of longitudinal reinforcement was 0.55%.

Figure 10.12 shows the test SWP mounted on the machine. The SWPs were tested under the load eccentricities of 0, $0.15t_t$ (referred to as small-e-SWPs), and $0.45t_t$ and $0.63t_t$ (referred to as large-e-SWPs). The slenderness ratios of the tested SWPs were 11.2 (squat SWP) and 17.2 (slender SWP). All the SWPs had material failure by crushing of concrete except for the slender large-e-SWP, which underwent stability failure. The small-e-SWPs had a compression-controlled failure without any visible flexural–tensile cracks. The large-e-SWPs developed several cracks on the tensile face of the SWP. The average of the maximum recorded

FIGURE 10.12 SWP mounted on test machine.

compressive strain in concrete at the failure of the tested SWPs was 2,424 $\mu\varepsilon$, with the largest recorded strain value of 3,478 $\mu\varepsilon$. The maximum recorded tensile strain in BFRP reinforcement (BFRP grid) at the failure of the tested SWPs was 8,560 $\mu\varepsilon$, which was nearly half of the ultimate tensile strain capacity of BFRP reinforcement. Strain values in the BFRP bar were found lower than those in the BFRP grid. This was probably because the load transfer from concrete to reinforcement is much better in the BFRP-grid-reinforced SWPs than in the BFRP-bar-reinforced SWPs.

Figure 10.13 shows the strain variation across the section thickness of a slender small-e-SWP. The strain variation is initially linear from the surface of one wythe to another wythe. However, the curvature of two wythes at the mid-height of SWP is different (i.e., non-linear variation) at a higher load level, probably due to the local bending of the wythes. The typical lateral deflection profile of the SWP is given in Figure 10.14. Both the wythes of SWPs deflected by almost the same magnitude in the same direction, indicating connectors were effective in keeping the wythes together. At the ultimate axial load, the maximum lateral deflection was $H/470$ and $H/178$ for squat small-e-SWPs and slender small-e-SWPs, respectively, while $H/100$ and $H/65$ for squat large-e-SWPs and slender large-e-SWPs. Here, H is the height of the SWP.

Figure 10.15 shows the variation of ultimate axial load with load eccentricity. The axial load capacities of BFRP-grid-reinforced squat and slender large-e-SWPs ($e = 0.45t_t$) were reduced to 27% and 23% of the counterpart concentrically loaded SWPs, respectively. ACI 318-19 (2019) recommends an upper limit of 1.4 on the ratio of second-order to first-order moment when calculated using factored forces and moments. In this study, the second-order moments were found to be

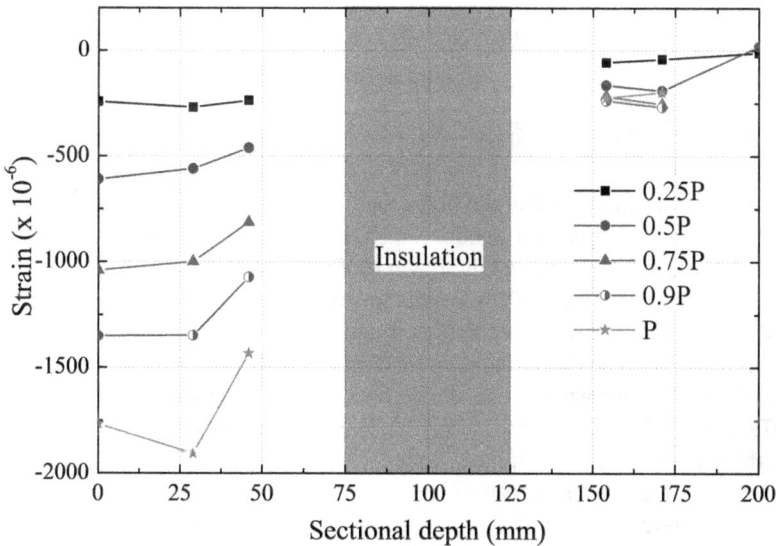

FIGURE 10.13 Typical strain variation across the sectional thickness.

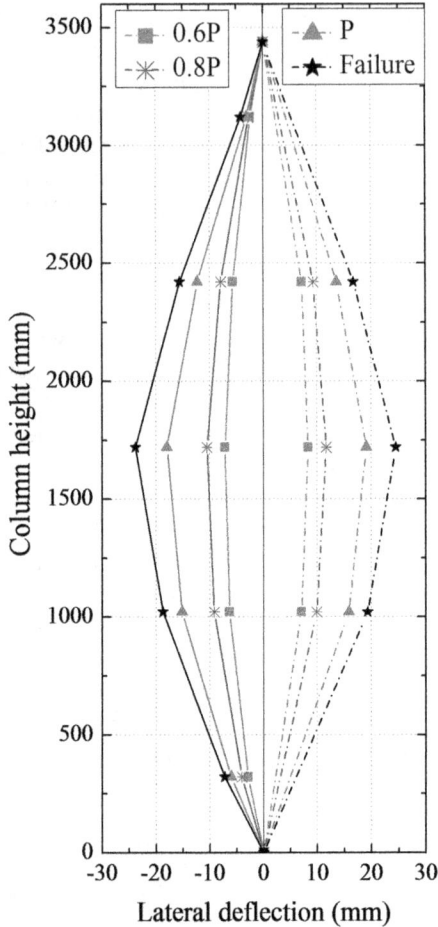

FIGURE 10.14 Typical lateral deflection profile an eccentrically loaded SWP.

17% and 59% at an average of the first-order moments, calculated at their ultimate axial load, for the squat and slender SWPs, respectively. The experimental axial load–moment interaction curve is shown in Figure 10.16.

Tomlinson and Fam (2018) conducted an experimental study and developed axial load–moment interaction diagrams of partially composite SWPs. Traditionally used SWPs without end-capping beams were tested. The load was applied to the inner wythe. The slenderness ratio (i.e., ratio of the height to the total sectional thickness of SWP) was 10. The N–M interaction curve was divided into three regions based on failure mode – (1) yielding and rupture of outer wythe reinforcement at low axial loads, (2) excessive slip or connector yielding failure, and (3) concrete crushing of inner wythe at high axial loads. The strength-based composite action decreased from 87% to 11% as axial load increased from 0% to 31% of axial load capacity of the inner wythe. This is because the outer wythe

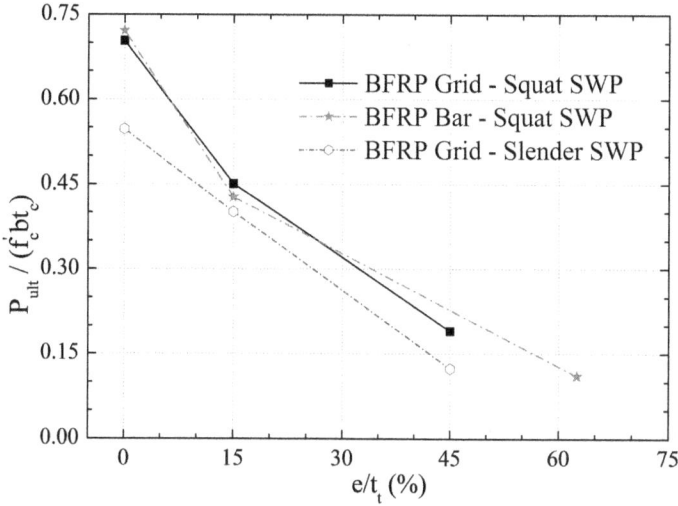

FIGURE 10.15 Variation of load capacity with load eccentricity.

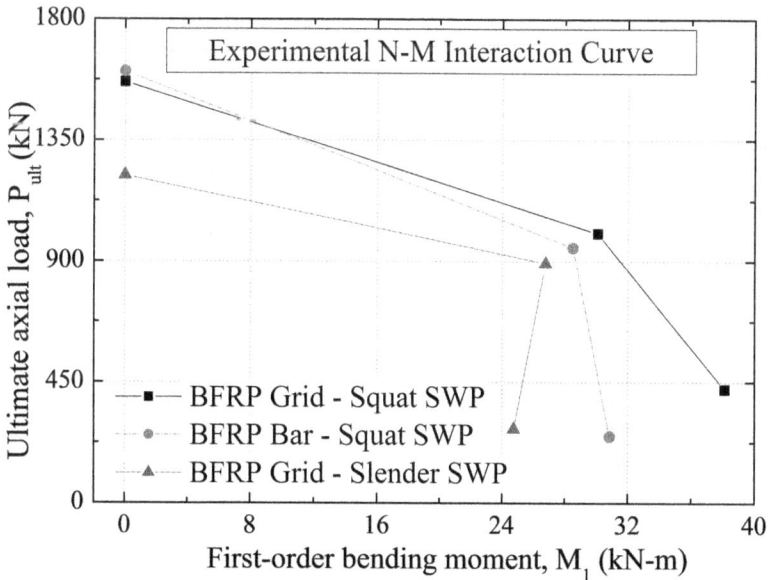

FIGURE 10.16 Experimental load–moment interaction curve.

contributes to the load capacity only when a significant moment is acting on the SWP, which corresponds to the region of the N–M curve under a small axial load.

Gara et al. (2011) conducted an experimental investigation on in situ SWPs with non-shear connectors. The slenderness ratios of the SWPs were 13, 15, and 20, and the values of t_l/t_w were 1.1, 1.7, and 2.3. Both the wythes were connected at the ends. SWPs were tested for concentric and eccentric axial loads. The eccentric axial load was applied at the centroid of one of the concrete wythes. The panels were tested with the bottom end hinged and top end restrained. The axial load capacity of concentrically and eccentrically loaded SWPs reduced by 16% and 33%, respectively, when the slenderness ratio increased from 13 to 20. The axial load capacity of eccentrically loaded SWPs with slenderness ratios of 13, 15, and 20 was 65%, 61%, and 52% of the counterpart concentrically loaded SWPs.

Benayoune et al. (2006) conducted a full-scale experimental study to investigate the eccentric axial load behavior of steel-mesh-reinforced OPC concrete SWPs using the continuous steel truss connectors. Axial load was applied at an eccentricity of 40mm to the centroid of the entire cross section of SWPs. The slenderness ratio (H/t_t) of SWPs was varied from 10 to 20, and the t_i/t_w values were 0.5 and 0.63. The panels were tested with the bottom end fixed and the top end hinged. All the specimens failed in compression by crushing of concrete. The load capacity reduced by 38% as slenderness increased from 10 to 20. The stresses in steel connectors were well below the yield stress. The strain variation across the cross section of SWP from one surface of the wythe to the other wythe was linear, indicating nearly fully composite action was achieved.

10.5   ROLE OF CONNECTORS

Connectors are used in a SWP to connect the concrete wythes laterally. The primary purpose of the connector is to allow the two wythes to deflect laterally together and restrict the relative longitudinal movement between the two wythes. In the SWPs with wythes connected at the top and bottom ends, the interface slip between the two wythes is restricted by the solid RC end-capping beams. Theoretically, the connectors in concentrically or eccentrically loaded SWPs are subjected to axial stress, bending, and transverse shear. The connectors proposed by Huang and Dai (2019) are tubular-shaped, which makes them effective in bending. Connectors in eccentrically loaded SWPs are subject to additional stresses due to (1) second-order effect resulting from load eccentricity and slenderness and (2) unequal post-cracking stiffness of the two wythes. In order to prevent the separation of the two wythes, connectors may be subject to significant tensile stress. In addition, the difference in the curvature of the two connected wythes at the location between two longitudinal connectors may result in the development of additional bending and transverse shear in connectors. The absence of sufficient anchorage to connectors may lead connectors to slip out of the wythes, causing the difference in the local curvature of the two wythes. Figure 10.17 shows the strain values at the top and bottom of a connector measured at a connector's mid-length, located at the mid-height of a SWP. Figure 10.18 shows similar values for a connector located at the mid-height of another SWP. It can be seen in Figure 10.17 that the connector is subjected to significant bending, which is probably due to a local change in the curvature of the two wythes. Figure 10.18 shows that the strain in the connector drops suddenly at a load much lower than the ultimate axial load, which indicates the slip of the connector out of the wythe due to insufficient anchorage.

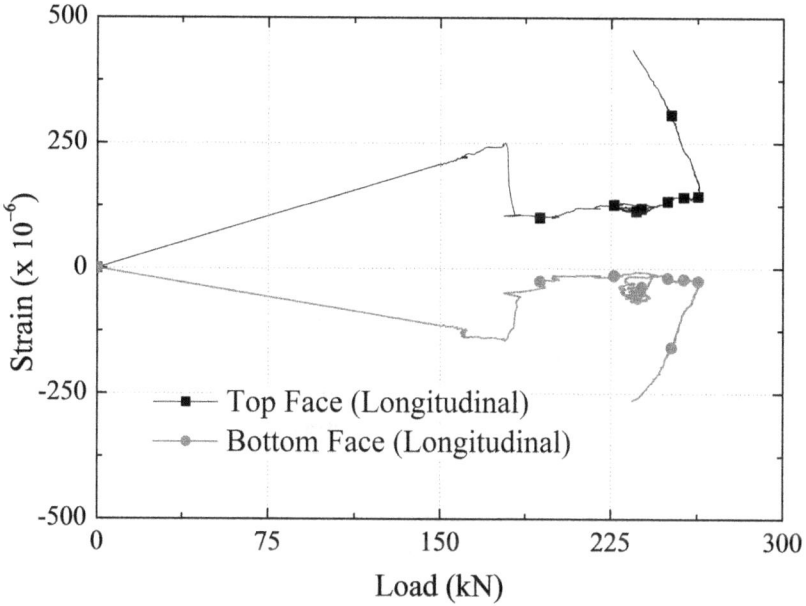

FIGURE 10.17   Load vs strain curve of a connector located at mid-height of a SWP.

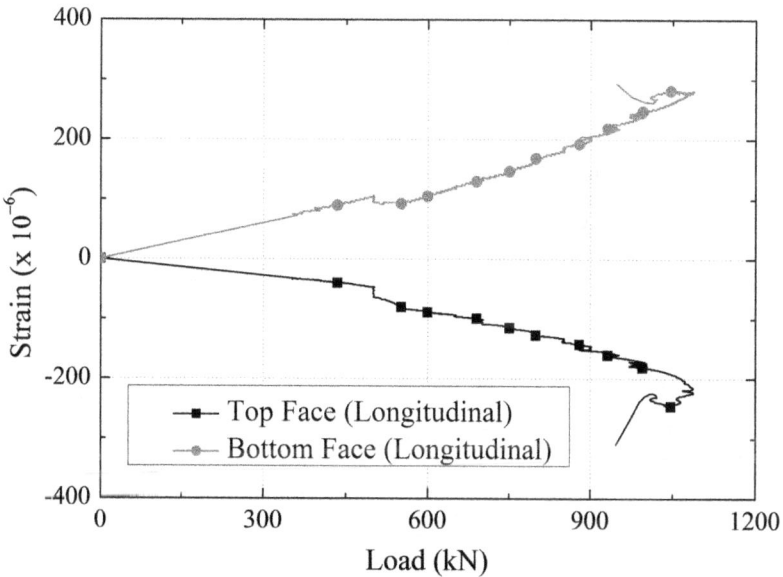

FIGURE 10.18   Load vs strain curve of a connector located at mid-height of a SWP.

10.6 SIMPLIFIED THEORETICAL INVESTIGATION

A simplified theoretical investigation has recently been conducted to predict the load capacity of SWPs subjected to concentric or eccentric axial load capacity (Kumar et al. 2021). In theoretical analysis, a sinusoidally deflected curve was assumed for SWPs, which provided a relationship between curvature at mid-height and deflection at mid-height (Bazant et al. 1991). A layered iterative sectional analysis considering non-linear material properties was conducted to obtain the moment–curvature curve corresponding to an axial load. An initial mid-span lateral deflection (δ) was assumed, and the exerting total moment ($M = P \times e + P \times \delta$) at the mid-height of the SWP was calculated. An iterative analysis was carried out to find the axial load that satisfied both the moment–curvature relationship and deflection–curvature relationship (obtained from the assumed sinusoidally deflected curve) at the mid-height of SWP. Following the steps mentioned above, an axial load–lateral deflection curve was obtained. The above procedure was used to obtain the axial load capacity of concentrically loaded SWPs and the theoretical N–M interaction curve of the eccentrically loaded SWPs. The section was treated as the three equivalent solid cases for sectional analysis, as shown in Figure 10.19.

A comparison of theoretical and experimental axial load capacities for different parameters is shown in Figure 10.20. For the ratio of insulation thickness to wythes' thickness (t_i/t_w) of 0.33, the experimental capacity was found near the theoretical capacity obtained following the Case 2 assumption for the studied range of slenderness ratio (i.e., 8–17). However, for $t_i/t_w = 0.66$, the experimental capacity was found to lie in between the theoretical capacities obtained following the Case 1 and Case 2 assumptions. Overall, the theoretical capacity obtained following the Case 2 assumption can be adopted as the most reasonable value for the studied range of parameters. However, the universality of the conclusion needs to be verified by doing a comparison study in the future, which includes sufficient test data and a wide range of parameters.

10.6.1 SWPs Subjected to Eccentric Axial Load

Axial load–moment interaction curves were obtained following the above methodology. Figure 10.21 shows the experimental and theoretical N–M curves for the tested squat and slender SWPs. The theoretical interaction curves obtained following the Case 1, Case 2, and Case 3 assumptions are referred hereafter as envelop-1, envelop-2,

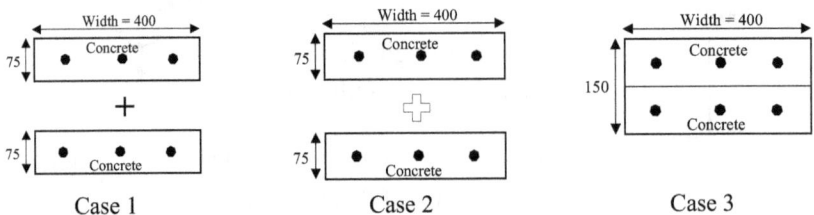

FIGURE 10.19 Equivalent solid wall panels. (Adapted from Kumar et al. (2021).)

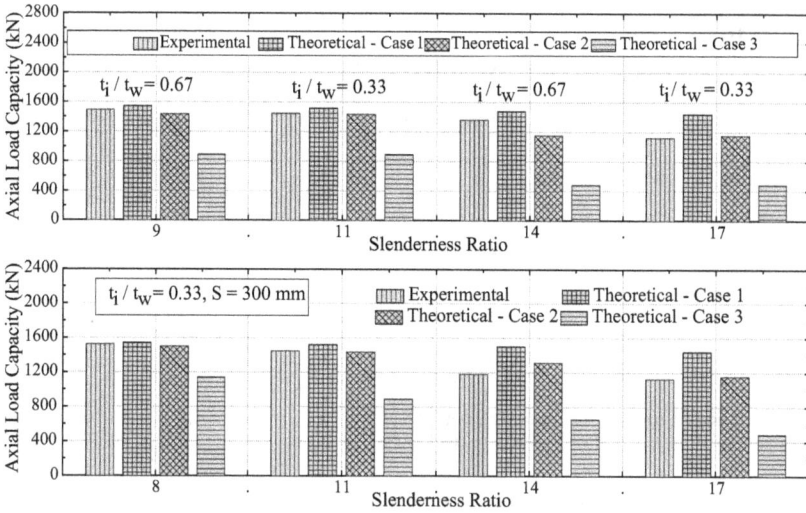

FIGURE 10.20 Comparison of theoretical and experimental axial load capacities. (Adapted from Kumar et al. (2021).)

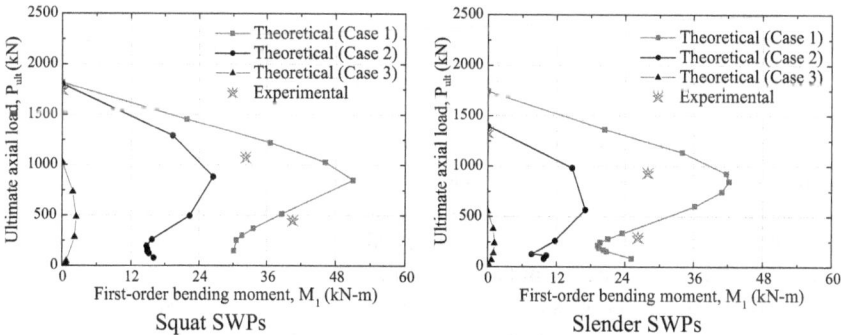

Squat SWPs

Slender SWPs

FIGURE 10.21 Experimental and theoretical N–M curves.

and envelop-3, respectively. It can be seen from Figure 10.21 that the experimental load capacity value of small-e-SWPs lies in between envelop-1 and envelop-2, while for large-e-SWPs, the experimental value falls beyond envelop-1. However, the universality of the conclusion needs to be verified by doing a comparison study in the future, which includes sufficient test data and a wide range of parameters.

10.7 PREDICTION OF LOAD CAPACITY FROM THE EXISTING EMPIRICAL DESIGN FORMULAE

The experimental load capacities of the eccentrically loaded ($e = t_t/6$) SWPs were compared with those obtained using the existing empirical design formulae, which are presented in Table 10.2. The slenderness of the SWP was calculated

TABLE 10.2

Available Empirical Design Formulae for Solid/Sandwich RC Wall Panels

	Design Formulae	H_{eff} / t_{total}	H_{eff} / L	e
Benayoune et al. (2007)	$P = 0.4 f_c' A_g \left[1 - \left(\dfrac{H}{40t} \right)^2 \right] + 0.67 f_y A_{sc}$	10.7–20	1.2–2	$\leq \dfrac{t}{6}$
ACI 318-19 (2019)	$P = 0.55 f_c' A_g \left[1 - \left(\dfrac{H}{32t} \right)^2 \right]$	0–25	All	$\leq \dfrac{t}{6}$
Pillai and Parthasarathy (1977)	$P = 0.57 f_c' A_g \left[1 - \left(\dfrac{H}{50t} \right)^2 \right]$	16–31.5	5–30	$\leq \dfrac{t}{6}$
Oberlender and Everard (1977)	$P = 0.60 f_c' A_g \left[1 - \left(\dfrac{H}{30t} \right)^2 \right]$	8–28	1–3.5	$\leq \dfrac{t}{6}$

Note: H_{eff} / t_{total}, slenderness ratio; H_{eff} / L, aspect ratio; e, load eccentricity.

following the Case 1 and 2 assumptions (Figure 10.19) for use in the empirical design formulae.

It may be noted that these empirical formulae are applicable for the design of solid wall panels except for the one proposed by Benayoune et al. (2007), which applies to SWPs. These formulae are applicable to wall panels subjected to load eccentricity up to one-sixth the total sectional thickness. Based on the comparison, it was found that the load capacities of SWPs obtained from the formula proposed by Benayoune et al. (2007) and those obtained following the assumption of Case 1 were closer to the experimental load capacities. The predicted load capacities (i.e., using the formula proposed by Benayoune et al. (2021) and those obtained following the Case 1 assumption were 9% and 12% smaller than the experimental values for squat and slender SWPs, respectively. The load capacities of squat specimens calculated using other design formulae, following the Case 1 or 2 assumption, were generally higher than the experimental values. For slender SWPs, experimental load capacity values were found between the predicted values while following the Case 1 and Case 2 assumptions. This comparison with these limited test data suggests that the existing design empirical formulae proposed for solid wall panels significantly overestimate the capacity of squat SWPs. However, the universality of the conclusion needs to be verified by doing a comparison study in the future, which includes sufficient test data and a wide range of parameters.

10.8   CONCLUDING REMARKS

This chapter presents the investigation findings of few available experimental studies that were conducted on axially loaded SWPs. A simplified numerical analysis is also presented, which predicts the load capacity of a SWP by treating it as an equivalent solid wall panel. Further, the applicability of the existing empirical design formulae was checked by carrying out a comparison study with the experimental results. However, the observations made from the simplified numerical analysis and comparison with empirical design formulae are based on limited test data. The universality of these observations needs to be verified by conducting future studies to gather sufficient test data with a wide range of parameters.

REFERENCES

ACI 318-14. 2014. *Building Code Requirements for Reinforced Concrete and Commentary*, American Concrete Institute, Farmington Hills, MI.

ACI 318-19. 2019. *Building Code Requirements for Reinforced Concrete and Commentary*, American Concrete Institute, Farmington Hills, MI.

ACI 440.1R-15. 2015. *Guide for the Design and Construction of Structural Concrete Reinforced with Fiber-Reinforced Polymer* (FRP) Bars (ACI 440. 1R-15). American Concrete Institute, Farmington Hills, MI.

Albitar M, Mohamed Ali MS, and Visintin P. 2017. Experimental study on fly ash and lead smelter slag-based geopolymer concrete columns. *Construction and Building Materials*, 141, 104–12.

Bazant ZP, Cedolin L, and Tabbara MR. 1991. New method of analysis for slender columns. *ACI Structural Journal*, 88, 391–401.

Benayoune A, Samad AAA, Trikha DN, Abang AAA, and Ashrabov AA. 2006. Structural behaviour of eccentrically loaded precast sandwich panels. *Construction and Building Materials*, 20(9), 713–24.

Benayoune A, Samad AAA, Abang AAA, and Trikha DN. 2007. Response of precast reinforced composite sandwich panels to axial loading. *Construction and Building Materials*, 21(3), 677–85.

Ding Y. 2017. *Experimental Study on Fracture Properties of Alkali-Activated Concrete*. PhD Thesis, The Hong Kong Polytechnic University, Hong Kong.

Ding Y, Dai JG, and Shi CJ. 2018. Fracture properties of alkali-activated slag and ordinary Portland cement concrete and mortar. *Construction and Building Materials*, 165, 310–20.

Einea A, Salmon DC, Fogarasi GJ., Culp T D, and Tadros MK. 1991. State- of the-art of precast concrete sandwich panels. *PCI Journal*, 36(6), 78–98.

Frankl BA, Lucier GW, Hassan TK, and Rizkalla SH. 2011. Behavior of precast, prestressed concrete sandwich wall panels reinforced with CFRP shear grid. *PCI Journal*, 56(2), 42–54.

Huang JQ, and Dai JG. 2019. Direct shear tests of glass fiber reinforced polymer connectors for use in precast concrete sandwich panels. *Composite Structures*, 207, 136–147.

Huang JQ, and Dai JG. 2020. Flexural performance of precast geopolymer concrete sandwich panel enabled by FRP connector. *Composite Structures*, 248, 112563.

Ibrahim AMA, Wu Z, Fahmy MFM, and Kamal D. 2016. Experimental study on cyclic response of concrete bridge columns reinforced by steel and basalt FRP reinforcements. *Journal of Composites for Construction*, 20 (3), 04015062.

Kumar S, Chen B, Xu Y, and Dai JG. 2021. Structural behavior of FRP grid reinforced geopolymer concrete sandwich wall panels subjected to concentric axial loading. *Composite Structures*, 270, 114117.

Lee NK, and Lee HK. 2013. Setting and mechanical properties of alkali-activated fly ash/slag concrete manufactured at room temperature. *Construction and Building Materials*, 47, 1201–1209.

Oberlender BGD, and Everard NJ. 1977. Investigation of reinforced concrete walls. *ACI Journal Proceedings*, 74(6), 256–63.

PCI Committee. 2011. State of the art of precast/prestressed concrete sandwich wall panels. *PCI Journal*, 56(2), 131–176.

Pillai SU, and Parthasarathy CV. 1977. Ultimate strength and design of concrete walls. *Building and Environment*, 12(1), 25–9.

Sarker PK. 2008. Analysis of geopolymer concrete columns. *Materials and Structures*, 42, 715–24.

Sim J, Park C, and Moon DY. 2005. Characteristics of basalt fiber as a strengthening material for concrete structures. *Composites Part B: Engineering*, 36(6–7), 504–12.

Tomlinson DG, and Fam A. 2015. Performance of concrete beams reinforced with basalt FRP for flexure and shear. *Journal of Composites for Construction*, 19(2), 04014036.

Tomlinson DG, Teixeira N, and Fam A. 2016. New shear connector design for insulated concrete sandwich panels using basalt fiber-reinforced polymer bars. *Journal of Composites for Construction*, 20(4), 04016003.

Tomlinson DG, and Fam A. 2018. Axial load-bending moment-interaction diagram of partially composite precast concrete sandwich panels. *ACI Structural Journal*, 115(6), 1515–1528.

Tran TT, Pham TM, and Hao H. 2019. Rectangular stress-block parameters for fly-ash and slag based geopolymer concrete. *Structures*, 19, 143–55.

Wang ZK, Zhao XL, Xian GJ et al. 2017. Durability study on interlaminar shear behaviour of basalt-, glass- and carbon-fibre reinforced polymer (B/G/CFRP) bars in seawater sea sand concrete environment. *Construction and Building Materials*, 156, 985–1004.

11 Enhanced Failure Behavior for Sandwiches with Hybrid Wire Mesh/ FRP Face Sheets

Çağrı Uzay
Kahramanmaraş Sütçü İmam University

Necdet Geren
Çukurova University

CONTENTS

11.1 Introduction 191
11.2 Significance of Wire Mesh Sheets for Hybrid Composites 192
11.3 Hybridization of Fiber/Polymer Matrix Face Sheets with
      Stainless Steel Wire Mesh Sheets..194
11.4 Evaluation of the Flexural Characteristics of the Developed
      Sandwiches with Carbon/Epoxy/Wire Mesh Face Sheets......................198
      11.4.1 Load–Deflection Curves..199
      11.4.2 Load-Carrying Capacities 205
11.5 Examination of Failure Behaviors..206
11.6 Conclusions..207
References..210

11.1 INTRODUCTION

The new engineering materials have arisen due to the need for better structural efficiency and mechanical performance compared to common types. However, it forces the boundaries of the materials universe with the incoming challenges and works that require additional efforts. The lightweight composite materials have emerged as one of the most important indicators in the currently growing technology (Vasiliev and Morozov 2007).

DOI: 10.1201/9781003143031-11

Sandwich structures are going to be more prevalent due to their attractive features, such as extremely high flexural stiffness and lightness. They also allow optimization of weight critical structural parts in the fields of aviation, marine, and automotive, space structures, naval structures, sporting goods, wind turbine blades, and infrastructures in civil engineering (Gibson 2012). Today, automotive manufacturers are in keen competition to reduce the emission rates to come across the requirements of EU legislation for the newly manufactured vehicles. Therefore, the use of sandwich panels in the body of vehicles does not only reduce the weight, but also provides eco-friendly vehicles with reduced fuel consumption and CO_2 emissions (Elmarakbi 2013).

Polymer foams, becoming more prevalent, have several advantages over metallic foams, honeycomb, truss, and web core materials, such as better resistance to water penetration (Wang et al. 2015). While the honeycomb core sandwiches have excellent structural performance, there is an increasing demand by airlines to find out alternatives due to high maintenance costs. The high maintenance cost is mostly associated with the possibility of water filling up, cell freezing, and expanding at low temperatures. Consequently, polymer foam cores, also in full contact with face sheets, can be better alternatives for sandwich structures and they can be used for both a cost reduction in production and structural applications instead of honeycombs. But the undesired failure behaviors of the sandwiches (i.e., indentation, and sudden and brittle fracture of both face sheet and core materials) restrict the use of extremely low-density polymer foam cores with fiber-reinforced polymer (FRP) composite face sheets, such as glass/epoxy and carbon/epoxy. In this case, hybrid face sheets can be a better alternative to overcome such failures. They not only eliminate undesired failures, but also provide multi-functionality such as combining high strength-to-weight ratio, stiffness-to-weight ratio, and better resistance to damage (Uzay et al. 2019). Hybridization can also be made for cost-effective sandwich designs.

11.2 SIGNIFICANCE OF WIRE MESH SHEETS FOR HYBRID COMPOSITES

The integrity of FRP composites to aerospace, aviation, and automotive industries has made the hybrid use of fiber metal laminates (FMLs) possible. However, the jointing mechanisms of fiber and metal alloys require the use of rivet elements, hole drilling, and/or strong adhesives to prevent failures such as delamination. Also, using holes and rivets can weaken the composite structure. Therefore, open metallic structures such as wire mesh sheets and perforated plates provide stabilization and additional reinforcement for FRP composites (Hasselbruch et al. 2015). The metallic wire mesh sheets such as aluminum and steel are in the form of porous structures and available in engineering fields from very small to large sizes with different wire thickness values.

The presence of wire mesh sheets in composite materials such as brittle glass- or carbon FRP composites, and polymer foam core sandwich structures with glass/epoxy and/or carbon/epoxy face sheets brings significant advantages. The plastic deformation behavior of the metallic wire mesh sheets provides the hybrid composites with resistance to more impact loads, prevents perforations, and helps to absorb more energy (Karunagaran and Rajadurai 2016). Due to the high ductility of the metallic wire mesh sheets, they can act as impact load absorbers and increase the impact load-bearing capacity of the polymer composite structures (Prakash and Jaisingh 2018). With increasing the reinforcement fraction of the wire mesh sheets within the brittle hybrid laminates, the failure mechanism of the structure can also change, and it becomes possible to obtain improved damage behavior at higher impact energies (Ahmed et al. 2007). The addition of wire mesh layers in the tensile side of the fiber/epoxy face sheet resulted in improved perforation energy and a smaller damage area compared to an equivalent thickness of pure fiber/epoxy laminate (Karunagaran et al. 2020). The hybridization of wire mesh sheets with FRP composites allows obtaining an improved crosslinking of the fiber and metal reinforcements together with the structural integrity (Sakthivel et al. 2018). Hasselbruch et al. (2015) revealed the favorable influence of wire mesh sheets for carbon fiber thermoplastic composites. Moreover, using different wire mesh sizes and examining the effects of various stacking sequences were also proposed to investigate the contribution to structural effectiveness. The researchers also stated that the polymer matrix surroundings provide better protection for metallic wire mesh sheets against corrosion and force transferring medium between the composite layers (Hasselbruch 2015, Sakthivel and Vijayakumar 2017, Qeshta et al. 2014).

On the other hand, an advanced hybrid material system, which is called FML, has mainly been used in aerospace applications due to its relatively lightweight and better impact and damage tolerance characteristics over both laminated FRP matrix composites and aluminum alloys (Sinmazçelik et al. 2011, Sadighi et al. 2012, Chai and Manikandan 2014). Most works related to FML applications are constrained to the combination of closed metal reinforcements such as conventional and thin sheets of metal foils in the thermoset FRPs. But some drawbacks existed in producing FMLs. The lap shear strength between fibers and metal alloys can be very weak unless some surface treatment processes (anodizing, mechanical grinding, acid etching, etc.) are applied to metal surfaces to improve adhesion bonding capability.

Integration of wire mesh sheets into FRP composites is a kind of strengthening mechanism and can be an alternative solution and substitute for FML applications. Due to the porous form of wire mesh sheets, it may not require additional surface treatments. Besides improved structural properties, this strengthening technique is also a cost-effective process (Meng et al. 2017).

The use of wire mesh sheets embedded in sandwich panels has recently been encountered in the field of civil engineering (Joseph et al. 2017, Matalkah et al. 2017, Shao et al. 2018). When compared to pure carbon/epoxy face sheets, hybrid face sheets including wire mesh sheets provided better flexural properties and high impact energy absorption capacity. Qeshta et al. (2014) reported that by increasing the number of wire mesh layers, significant enhancement can be obtained in the flexural strength, cracking behavior, and energy absorption capability of the concrete sandwich structures. However, the sizes of the used wire meshes are very coarse for civil engineering applications. Qi et al. (2019) used a very coarse wire mesh with 3 mm wire diameter to improve the interfacial integrity, which increased the flexural performance of the concrete beam by providing an additional fiber bridging mechanism. In mechanical engineering applications, Kurian et al. (2016, 2017) used wire mesh sheets to increase thermal and hydrodynamic performance by improving the heat transfer. They stacked the wire mesh blocks with various thickness values and utilized the porosity of the wire meshes for better cost-effective heat transfer capacity.

The advantages of hybrid FRP composites with stainless steel wire mesh sheets have been revealed according to a very limited number of researches. More attention should be given to investigating the effect of size, stacking sequences, number of layers within the face sheets, and surface treatment processes as well (Karunagaran et al. 2020).

11.3 HYBRIDIZATION OF FIBER/POLYMER MATRIX FACE SHEETS WITH STAINLESS STEEL WIRE MESH SHEETS

The hybridization of fiber/epoxy laminates with stainless steel (SS) wire mesh layers has been carried out by a limited number of researchers. In the previous section, the significance of using wire mesh sheets is declared, and now, the effect of embedding wire mesh sheets within the FRP composites for various purposes is scrutinized based on the existing literature.

Hasselbruch et al. (2015) hybridized the carbon fiber-reinforced thermoplastic (polyphenylene sulfide) composite with unalloyed steel wire mesh, where the wire mesh layer was placed between six plies of carbon laminates on each side, and the sandwiches were subjected to tensile tests. Compared to pure carbon FRP, the tensile strength of the hybrid sandwich was decreased. Similar findings were observed by Karunagaran and Rajadurai (2016). They sandwiched three plies of glass/epoxy with stainless steel wire mesh which was stacked in the middle as shown in Figure 11.1. This was explained by Hasselbruch et al. (2015) as the addition of wire mesh causes an increase in the amount of matrix material in the structure, thus lowering the tensile stiffness. Additionally, the effect of reinforcement orientation was revealed based on the obtained tensile stiffness values. The shear and bending tests were suggested for further investigations to determine the interface properties that affect the mechanical features.

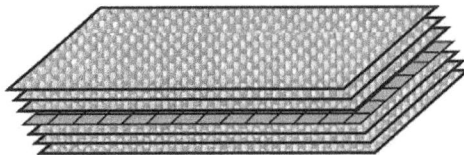

FIGURE 11.1   Schematic representation of glass/epoxy–wire mesh sheet sandwich.

Sakthivel and Vijayakumar (2017) hybridized the glass/epoxy polymer composites with SS wire mesh sheets and investigated the flexural and tensile strengths, low-velocity impact behavior, and interlaminar properties considering various stacking sequences. A single layer of wire mesh sheet was stacked at top, middle, and bottom to construct different sandwich specimens. While the fiber volume fraction over the epoxy was 50:40, the volume fraction of wire mesh was set to 10%. They found out that the mechanical properties and crack growth behavior of the hybrid laminates were improved and especially the hybrid laminate where the wire mesh sheets were stacked into middle of the structure (Figure 11.1) yielded the best results. Also, the other configurations considerably contributed to the tensile load-bearing capacity, percent elongation, flexural strength, and bending load capacity. The results are given in Table 11.1 for the corresponding configurations.

The researchers then investigated the effect of fiber volume fraction value with different fiber–matrix weights based on the position where the wire mesh sheet was sandwiched between the glass FRP laminates (Sakthivel et al. 2018). The wire mesh volume fraction was set to 10%, and the fiber volume fraction was arranged in between 50% and 60% with 2.5% increments. The best results were obtained for the fractions of 52.5% and 37.5% for glass fiber and epoxy matrix, respectively. This is because of the improved interfacial bonding between the layers. The addition of a wire mesh sheet in the FRP laminates increased the first crack load and improved the failure behavior, and it was declared that hybridizing with a wire mesh sheet is a new technique to obtain significant improvement in the mechanical properties. Wire mesh sheets also provided better thermal stability compared to pure FRP laminates.

Prakash and Jaisingh (2018) investigated the effect of the surface treatment process on the adhesion of wire mesh to glass fiber/epoxy laminates. Stainless steel and Aluminum 6061 wire mesh sheets were individually bonded to glass/epoxy laminates. The tensile, flexural, impact, and interlaminar shear strength results of hybrid composites based on both the as-received and surface-treated wire mesh sheets are given in Table 11.2. As seen, while the as-received wire mesh sheets adversely affected the mechanical properties, the surface treatment had a favorable influence on the results. Similar work was conducted by

TABLE 11.1

Tensile and Flexural Results for the Sandwich Specimens (Sakthivel and Vijayakumar 2017)

Stacking Sequences/ Mechanical Property	Pure GFRP	SSWM in Top Side	SSWM in Middle	SSWM in Bottom Side
Ultimate tensile load (N)	1,100	2,900	3,600	3,500
Tensile strength (MPa)	86	95	110	99
% Elongation	13.67	14.9	16.83	15.5
Bending load capacity (N)	310	480	520	430
Flexural strength (MPa)	75	80	95	85

TABLE 11.2

Mechanical Properties of the Composites Based on the As-Received and Surface-Treated Layers (Prakash and Jaisingh 2018)

	Mechanical Property (As-Received/Surface-Treated)			
The Specimen	Tensile Strength (MPa)	Flexural Strength (MPa)	Izod Impact (J)	I.L.Shear Strength (MPa)
Glass/epoxy	120/128	207/219	4.32/4.92	23/27
Glass/epoxy/SSWM	108/134	192/241	4.10/5.40	18/25
Glass/epoxy/Al-6061	99/127	180/226	4.22/5.04	20/23

Karunagaran and Rajadurai (2016) considering the glass/epoxy/wire mesh sheet sandwich specimens. Except for the tensile properties, the surface treatment of the composite layers improved the flexural, impact, and interlaminar strength values. The increment was due to the suppression of interfacial crack formation because surface modification provided an enhanced adhesion behavior between fibers, wire mesh sheets, and matrices. The stainlesssteel wire mesh sheet resulted in higher values than the Aluminum 6061 wire mesh sheet (Prakash and Jaisingh 2018).

Meng et al. (2018) hybridized the basalt fiber with wire mesh sheets and bonded it to polystyrene foam core to construct a structural insulated panel, which aimed to resist wind-borne debris. Placing the wire mesh layer in the outer surface was found more effective than placing it in between skin and core interface.

Matalkah et al. (2017) and Joseph et al. (2017) indicated that the existence of wire mesh sheets within the face sheets improved the bending rigidity of the fiber-reinforced concrete beams, and Shao et al. (2018) showed that the hybrid composite acted as a good absorber under impact loadings.

Uzay and Geren (2020) showed that a single layer of stainless steel wire mesh sheet within the top face sheet of PVC foam core sandwich enhanced the flexural behavior of the structure by preventing the sudden force drop after the flexural load reaches a maximum value, which is the common situation for pure carbon/epoxy polymer foam core sandwiches. A single layer of wire mesh was stacked either under the first carbon fiber fabric of the top face sheet or at the interface of top skin and foam core. The effects of two different mesh sizes, 100 mesh (0.2 mm wire diameter) and 200 mesh (0.1 mm wire diameter), were investigated for different staking configurations. The tensile strength of the face sheet was found the best for the configuration in which the 100-mesh sheet was embedded between the carbon plies compared to both the 200-mesh configuration and the configurations in which the wire mesh sheets were stacked to the outermost side of the face sheet. It was explained by the researchers as the bridging effect of hybridization. The wire mesh sheets embedded between the carbon plies provided an increased strain rate and retarded the crack growth rate. The face sheets were then bonded to

PVC foam core materials to investigate the flexural behavior of the hybrid face sheet sandwiches with wire mesh sheets. The results are evaluated in the next section.

When the literature is scrutinized, it can be noted that making a hybrid sandwich structure with wire mesh sheets is a new technique and a few researches have been made to investigate the mechanical behavior of the sandwich structure under particular loading conditions. Some of the studies are based on the examination of surface treatment effects such as mechanical, chemical, or electrical processes of wire mesh sheets (Karunagaran and Rajadurai 2016, Prakash and Jaisingh 2018, Sakthivel et al. 2018). Table 11.3 illustrates the literature studies describing the type of wire mesh, its size, and sandwich type. The table also indicates the applied manufacturing method. Hasselbruch et al. (2015) manufactured their sandwich specimens under the hydraulic press of 100t and 300t. They revealed that the 100t press provided better results than the heavier press. Because higher press loads resulted in lower matrix volume fraction; thus, sufficient resin impregnation could not be achieved. Qi et al. (2020) fabricated wire mesh–concrete sandwich structures with the mold casting method. However, a study carried out by Uzay and Geren (2020) investigated the effects of very fine wire mesh sheets embedded in polymer foam core sandwich structures on the flexural characteristics. In that study, all the constituents, carbon fiber fabrics, wire mesh sheets, and foam core materials, were co-cured simultaneously after stacking one on each other. They applied the vacuum bag method. For the rest of the studies in Table 11.3, the hand lay up method was applied.

TABLE 11.3
Previous Studies Related to the Use of Wire Mesh Sheets

The References	Wire Mesh Material	Mesh Size/Wire Diameter/Wire Thickness	Type of Sandwich	Mechanical Test	Manufacturing Method
Hasselbruch et al. (2015)	CS	1.2 mm thick	CF–WM–CF	Tensile	Hydraulic press
Karunagaran and Rajadurai (2016)	SS	0.151 mm wire dia.	GF–WM–GF	Tensile, flexural, interlaminar, izod impact	Hand lay-up
Sakthivel and Vijayakumar (2017)	SS	0.5 mm wire dia.	GF–WM–GF	Tensile, flexural	Hand lay-up and vacuum bag
Sakthivel et al. (2018)	SS	0.112 mm wire dia.	GF–WM–GF	Tensile, flexural, shear, Charpy impact	

(Continued)

TABLE 11.3 (*Continued*)
Previous Studies Related to the Use of Wire Mesh Sheets

The References	Wire Mesh Material	Mesh Size/Wire Diameter/Wire Thickness	Type of Sandwich	Mechanical Test	Manufacturing Method
Meng et al. (2018)	Gal.S	0.64 mm wire dia.	BF/ WM–EPS core	Wind-borne debris impact	Hand lay-up
Prakash and Jaisingh (2018)	SS		GF–WM– GF–WM	Tensile, flexural, interlaminar, drop weight impact	Hand lay-up and post-cure in a hot oven
Prakash and Jaisingh (2018)	Al-6061	0.25 wire dia.			
Truong et al. (2019)	SS	0.1 mm wire dia.	CF–WM– CF	Tensile	Hand lay-up
Karunagaran et al. (2020)	SS	0.151 mm wire dia.	GF–WM– GF	Drop weight impact	Hand lay-up
Qi et al. (2020)	CS	3 mm wire dia.	WM– concrete	Flexural	Mold casting
Uzay and Geren (2020)	SS	100 mesh (0.2 mm wire dia.) and 200 mesh (0.1 mm wire dia.)	CF–WM– PVC core	Tensile, flexural	Vacuum bag

Al-6061, Aluminum 6061; CF, carbon fiber; CS, carbon steel; Gal.S, galvanized steel; GF, glass fiber; SS, stainless steel; WM, wire mesh.

11.4 EVALUATION OF THE FLEXURAL CHARACTERISTICS OF THE DEVELOPED SANDWICHES WITH CARBON/EPOXY/WIRE MESH FACE SHEETS

In this section, the flexural characteristics of PVC foam core sandwich beams developed with carbon/epoxy/wire mesh hybrid face sheets are presented. The density of the foam core is 48 kg/m³, which is very low compared to the literature because the densities of the most common PVC foam core materials available in both applications and the literature are between 60 and 350 kg/m³ (Gupta 2019). As the technology related to materials science is continuously developing, and further lightness without sacrificing the stiffness is desired particularly in satellite and aviation industries, research studies should be performed to investigate the effect of extremely low-density core materials. But in this case, a precaution is required to prevent undesired failure modes because using lighter core materials may increase the possibility of indentation failure mode as in the case of using thinner face sheets. A practical solution was introduced for polymer foam core

TABLE 11.4
Sandwich Design Configurations Based on Face Sheet Hybridization

Sandwich Design	Face Sheet Code	Explanation of Face Sheet
	NMFS	Non-mesh face sheet
	MH1FS	Hybrid face sheet (H1) in which wire mesh is embedded under the first carbon fiber fabric of top face sheet
	MH2FS	Hybrid face sheet (H2) in which wire mesh is embedded at top face sheet–core interface
	MH3FS	Combination of H1 and H2 (H3)

sandwiches by hybridizing the brittle carbon/epoxy face sheets with a very fine size of wire mesh sheets (Uzay and Geren 2020). The carbon/epoxy face sheets of the sandwich structures were symmetrically manufactured, and the wire mesh sheets were stacked on top face sheets with different configurations. Table 11.4 shows the configurations developed by the researchers. In the table, the configurations are given based on the face sheet hybridization. The carbon/epoxy face sheets include four-ply woven plain fabrics, whereas the wire mesh sheet is used as either a single layer or two layers based on the configuration. In addition to this, two different wire mesh sizes (100 and 200 mesh) and various core thickness values (10, 20, and 30 mm) were taken into consideration in the study, so a sufficient number of configurations were considered for the evaluation as seen in Table 11.5.

11.4.1 LOAD–DEFLECTION CURVES

Quasi-static three-point flexural tests were applied to the sandwich configurations presented in Table 11.5. The crosshead speed was set to 2 mm/min. The span length was set to 150 mm. The flexural load–deflection data were plotted, as shown in Figure 11.2. Figure 11.2a–g provides the results according to the core thickness values. While Figure 11.2a belongs to control specimens (NMFS-CT10, NMFS-CT20, and NMFS-CT30) which have pure carbon/epoxy face sheets, the others are based on the type of hybrid face sheets.

When the non-mesh face sheet sandwiches are examined (Figure 11.2a), the following behaviors are obtained for the sandwiches of 10, 20, and 30 mm core thickness, respectively:

TABLE 11.5

Extended Sandwich Configurations Based on Hybrid Type, Mesh Size, and Core Thickness

Sandwich Code	Hybrid Type	Mesh Size (mesh)	Core Thickness (mm)
NMFS-CT10	Pure face sheet	Non-mesh	10
NMFS-CT20	Pure face sheet	Non-mesh	20
NMFS-CT30	Pure face sheet	Non-mesh	30
100MH1FS-CT10	H1	100	10
100MH1FS-CT20	H1	100	20
100MH1FS-CT30	H1	100	30
200MH1FS-CT10	H1	200	10
200MH1FS-CT20	H1	200	20
200MH1FS-CT30	H1	200	30
100MH2FS-CT10	H2	100	10
100MH2FS-CT20	H2	100	20
100MH2FS-CT30	H2	100	30
200MH2FS-CT10	H2	200	10
200MH2FS-CT20	H2	200	20
200MH2FS-CT30	H2	200	30
100MH3FS-CT10	H3	100	10
100MH3FS-CT20	H3	100	20
100MH3FS-CT30	H3	100	30
200MH3FS-CT10	H3	200	10
200MH3FS-CT20	H3	200	20
200MH3FS-CT30	H3	200	30

- NMFS-CT10 sandwich specimen firstly shows a linear increase of up to about 450 N, and then a gradual increase in slope continues non-linearly up to the maximum point of approximately 550 N. After the maximum force is achieved, the load is decreased slowly from a mid-span deflection of ~4 to ~10 mm. Fiber and matrix failures occur at that range. But the load is carried by the top face sheet, and core crushing can be observed. At the end of the curve, a sudden drop in bending load happens due to the compressive failure of the top face sheet at a deflection of approximately 10 mm.
- NMFS-CT20 sandwich specimen provides increased flexural stiffness compared to 10 mm thick core sandwich structure as expected. When the bending force reaches the maximum point at an approximate deflection of 6.29 mm, the specimen fails with a sudden drop of force after a very short plateau.
- NMFS-CT30 sandwich specimen behaves nearly similar to the 20 mm thick core pure face sheet sandwich structure. Although the flexural stiffness greatly increased with the increase in core thickness from 10 to 20 mm, 30 mm thick core sandwich specimen provided a very slight change because indentation failure mode controls the failure behavior of

the NMFS-CT30 sandwich. Therefore, the maximum flexural load of the sandwich is nearly the same as that of the NMFS-CT20 sandwich. The specimen suddenly fails whenever the maximum bending force is reached at a deflection of about 5.59 mm due to the indentation failure mode.

The flexural behaviors of the control sandwich specimens given in Figure 11.2a are summarized above. The common point of the NMFS-CT10, NMFS-CT20, and

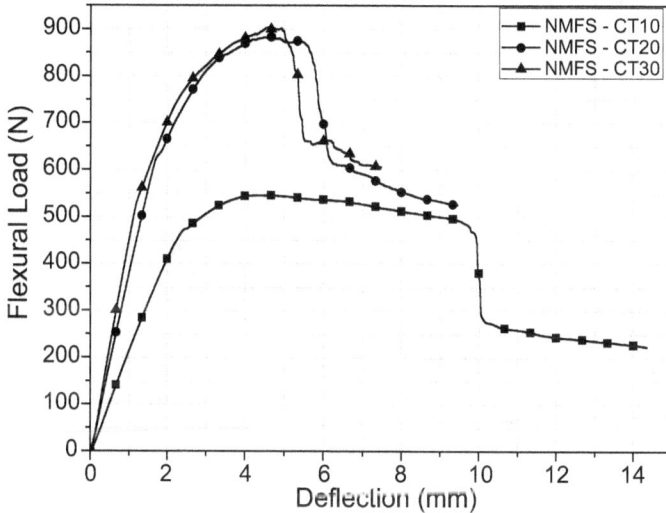

(a)

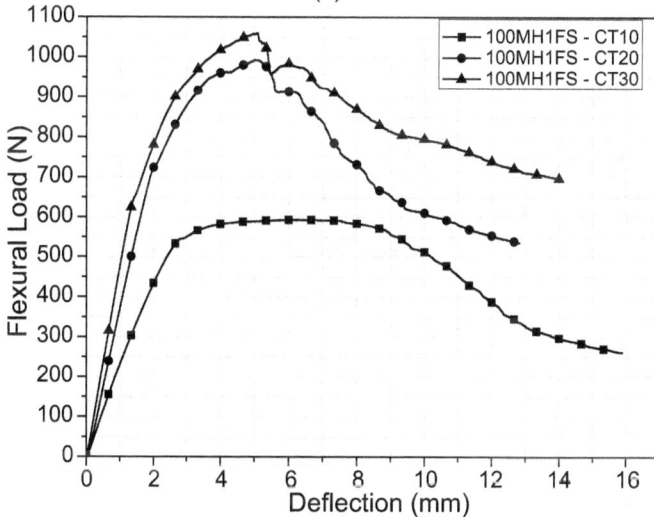

(b)

FIGURE 11.2   Load–deflection curves of the control (a) and developed sandwiches (b–g).

(Continued)

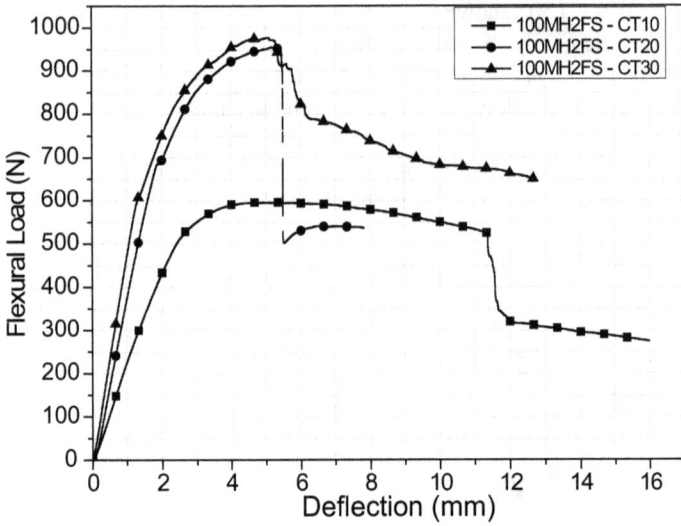

(c)

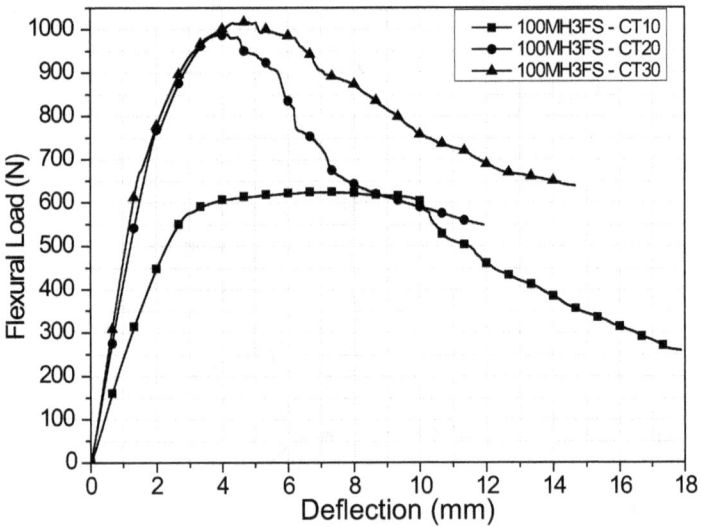

(d)

FIGURE 11.2 (*CONTINUED*)   Load–deflection curves of the control (a) and developed sandwiches (b–g).

(*Continued*)

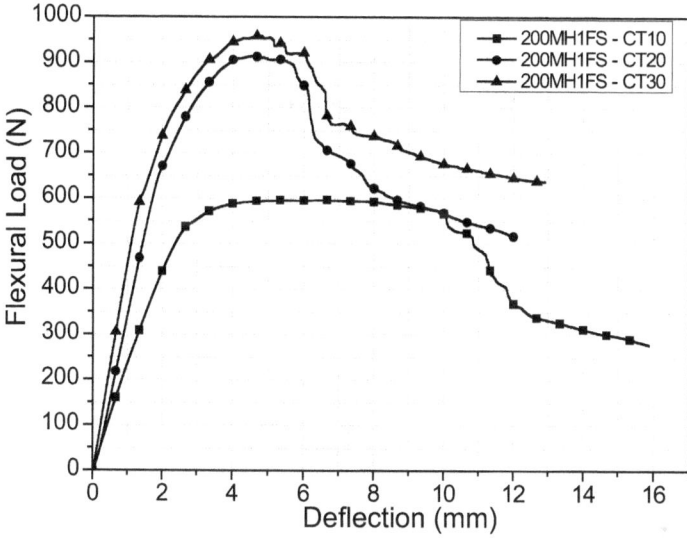

(e)

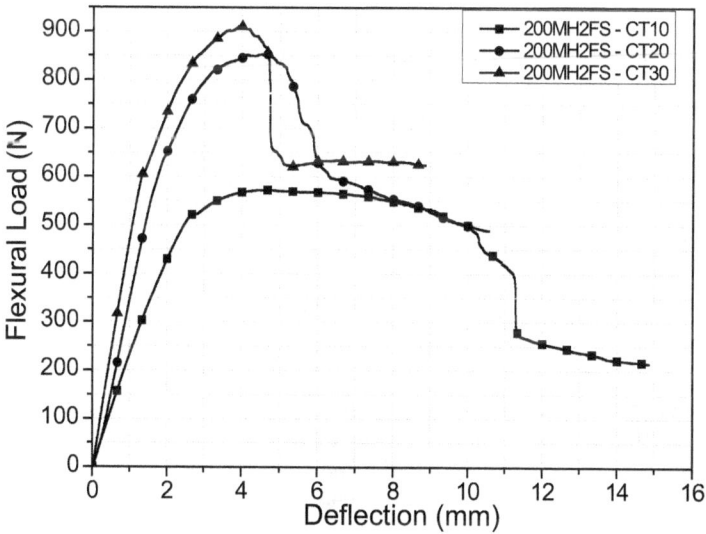

(f)

FIGURE 11.2 (*CONTINUED*) Load–deflection curves of the control (a) and developed sandwiches (b–g).

(*Continued*)

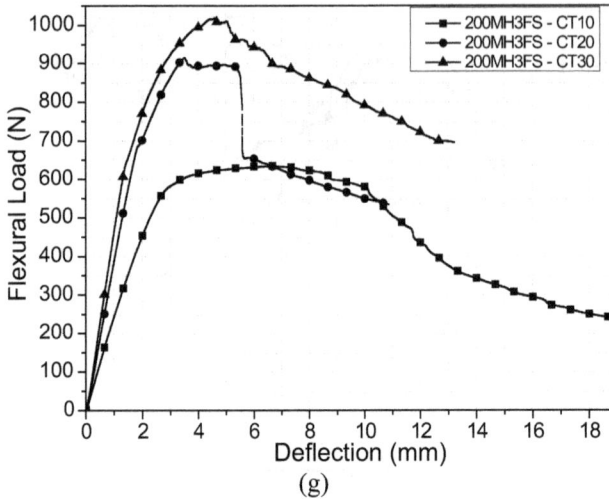

FIGURE 11.2 (*CONTINUED*) Load–deflection curves of the control (a) and developed sandwiches (b–g).

NMFS-CT30 sandwiches is the sudden force drop that causes a catastrophic failure at certain deflection values. Figure 11.2b–g presents the failure behaviors of the developed sandwich structures with wire mesh hybridization of the top face sheets. The hybrid face sheets increased the flexural load values of the sandwich specimens. Also, some configurations of the sandwiches with a 10 mm foam core provided a constant and larger plateau after the maximum flexural force was achieved (Figure 11.2b–g). Additionally, and most importantly, almost all the configurations prevented the sudden force drops in bending and provided a linear decrease in the slope of bending force. Therefore, the specimens with a particular type of hybrid wire mesh/carbon/epoxy face sheets eliminated the indentation failure mode. Although the carbon/epoxy face sheets resulted in a sudden catastrophic failure for low-density polymer foam core sandwiches due to their brittleness, the catastrophic failure behavior is also eliminated and the enhanced failure behaviors are obtained for the hybrid face sheet sandwich structures. Among all the configurations, the H2-type face sheet hybridization provided less or no contribution. Because the significant contribution of H1- and H3-type hybridization depends on the bridging effect, the bridging effect can be achieved with proper hybridization of the composite materials and provides to improve strain rates instead of stiffness or strength properties. Therefore, the sandwiches in which the wire mesh sheets were stacked between carbon fiber fabrics yielded enhanced flexural behavior compared to those with wire mesh sheets stacked at top face sheet–core interface. However, the 200 mesh size did not contribute to flexural behavior as high as the 100 mesh size due to the lower wire diameter. The study has contributed to the literature since the investigation of the effects of both stacking sequences of the wire mesh sheets and their sizes has been suggested in the literature (Hasselbruch et al. 2015 and Karunagaran et al. 2020).

11.4.2 LOAD-CARRYING CAPACITIES

To find out the significant effects of wire mesh hybridization on the flexural behavior of the developed sandwich structures, the flexural load-carrying capacities are compared in Figure 11.3. Furthermore, the load capacities at certain deflections where the sandwich beams with pure face sheets undergo sudden failure (10 mm for NMFS-CT10, 6.29 mm for NMFS-CT20, and 5.59 mm for NMFS-CT30) are also demonstrated in Figures 11.3–11.5.

As seen in Figures 11.3–11.5, pure face sheet sandwiches lose a larger amount of their load capacity after the sandwiches undergo a sudden failure at certain

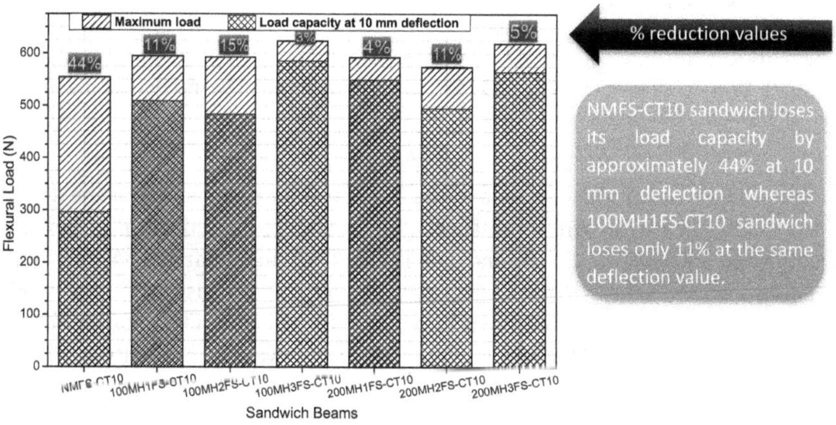

FIGURE 11.3 The load capacity of 10 mm thick core sandwich beams both at maximum and at 10 mm deflections.

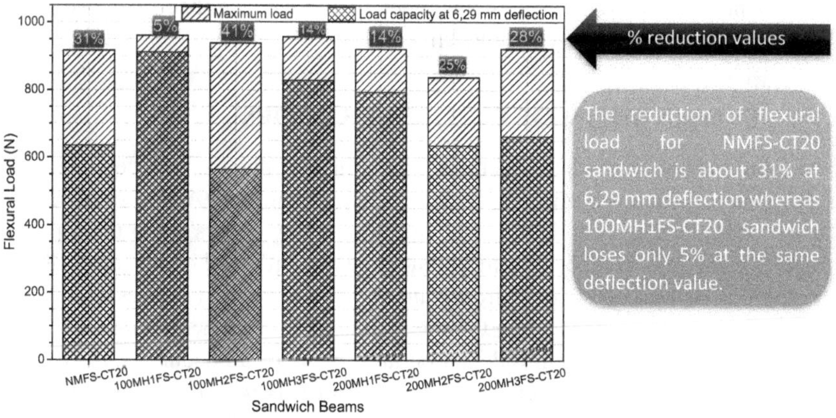

FIGURE 11.4 The load capacity of 20 mm thick core sandwich beams both at maximum and at 6.29 mm deflections.

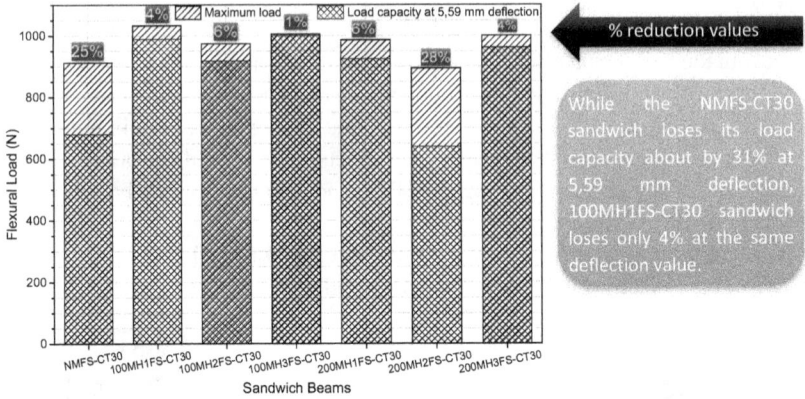

FIGURE 11.5   The load capacity of 30 mm thick core sandwich beams both at maximum and at 5.59 mm deflections.

deflections. However, the decrease in the load capacity is very low for the developed sandwiches with wire mesh sheets. For example, the reduction in flexural load for the 200MH1FS-CT10 sandwich is about 4% and it carried 85.75% more bending load than the NMFS-CT10 sandwich. This is because while the sandwiches with pure face sheets fail suddenly, the developed sandwiches still carry higher bending loads and the reduction in force occurs very slowly with the downward slope of the longer displacements. Therefore, the wire mesh hybridizations did not only prevent the sudden force drops in flexure, but also provide the sandwich structure with the ability to carry larger amounts of load with an improved failure behavior for a particular deflection value even after the failure. As the PVC foam core sandwiches with the carbon/epoxy face sheets show sudden and catastrophic failures without any warning due to the brittle nature of the composite constituents, the wire mesh sheets embedded within the face sheets can overcome the drawbacks of that kind of sandwich structures.

11.5   EXAMINATION OF FAILURE BEHAVIORS

The failure of the sandwich structures occurs mainly due to the face yield, face debonding, indentation, core shear, and facing wrinkling (Ashby et al. 2000, Daniel et al. 2002). Face yield or face micro-buckling occurs when the top face sheet's axial stress exceeds the micro-buckling strength of the face sheet. It can cause catastrophically sudden drops during load carrying (Steeves and Fleck 2004). Core shear happens when the shear stress of a sandwich attains the shear strength of the core material. Indentation failure mode can occur due to concentrated loads on a sandwich panel, which is a critical issue for sandwich designs. Daniel et al. (2002) described it as a local compressive failure of the core followed by local face sheet bending. Generally, the indentation failure mode is valid for thicker core sandwiches having thinner and more brittle face sheets. Face debonding can be described as the delamination of the face sheet from the core material or the delamination of the face sheet laminates. It can mostly be due to unsuccessful fabrication and happen in the case of insufficient resin impregnation, lack of bonding, and/or other defects made during manufacturing. Face wrinkling is a common failure mode for the sandwiches with honeycombs, metal foams, and web or truss core materials. The short wavelengths are observed on the top face sheet, which causes local elastic instability, which matches the findings of Steeves and Fleck (2004).

The polymer foam core sandwiches with fiber-reinforced composite face sheets generally undergo a failure mode of core shear and indentation when the laminates are assumed to be in perfect bonding. These failure modes can be controlled by arranging the density and thickness of the core materials or face sheet thickness in an appropriate manner. Ashby et al. (2000) showed the core shear failure mechanisms schematically as given in Table 11.6. While the plastic hinges are under the loading bar in Case I, the plastic hinges are both on the supporting rolls and under the loading bar in Case II. The sandwich of NMFS-CT10-S20 (S20 refers to the span length of 200 mm) is found to resemble more likely the core shear mechanism of Case I, whereas 100MH1FS-CT10-S20 sandwich fails based on the Case II mechanism. In addition, as shown in Figure 11.6, the wire mesh hybridization to the sandwich face sheet contributes to the core shear stress of the sandwich structure. It also contributes to the decrease in top face sheet stress as presented in Figure 11.6b.

11.6 CONCLUSIONS

Hybrids play an important role in achieving multi-functionality over conventional laminated composites by bringing the best characteristics of the constituents together. The wire mesh sheets, in the porous form, provide better hybridization and can contribute to the mechanical properties of the hybrid composite structures by improving the adhesion between laminates and conduct bridging effect. The investigation of the effects of wire mesh sheets on hybrid sandwich structures have been encountered in the recent literature, and its significance was revealed in this chapter by summarizing the current literature. In general, embedding the wire mesh sheets into composite materials has increased the tensile, impact, flexural, and interlaminar shear strength of the sandwich structures due to better adhesion between the laminates. In addition, although the polymer foam core sandwiches with FRP face sheets undergo a failure of sudden force drop under flexural loading, wire mesh hybridization within FRP face sheets improved flexural stiffness, provided higher load-carrying capacity, and enhanced failure behavior by both eliminating the sudden force drops and catastrophic failures. For instance, a 10 mm thick PVC foam core sandwich specimen with pure carbon/epoxy face sheets failed suddenly at an approximate deflection of 10 mm under three-point bending loading and lost the load capacity by 44%. On the other hand, the reduction was only 4% at the same deflection value for a specimen with the 200-mesh sheet stacked within the top face sheet, and the structure firstly provided a constant

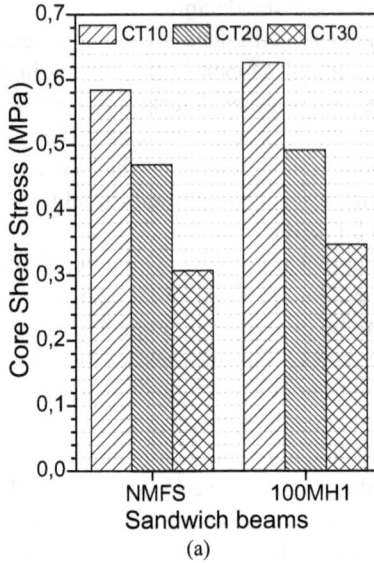

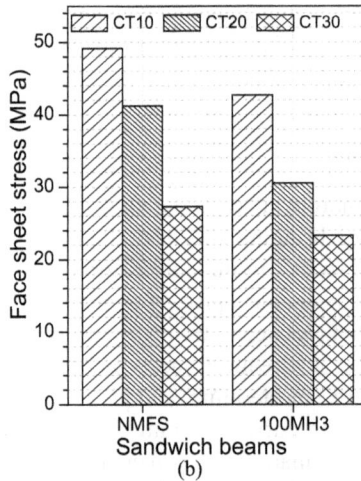

FIGURE 11.6 (a) Core shear stress of non-meshed and H1-type face sheet sandwich; (b) face sheet stress of non-meshed and H3-type face sheet sandwich.

and larger plateau after the maximum point of bending load, and then continued to carry the load with a downward slope instead of a sudden force drop.

FMLs have been developed to combine the best features of metal alloys and FRP composites. But establishing a strong adhesion between fibers and metals is difficult and requires additional surface treatments for metals. In the case of replacing porous wire mesh materials with metal sheets provide cost-effective solutions both with improved adhesion and eliminated additional precaution for metal surface.

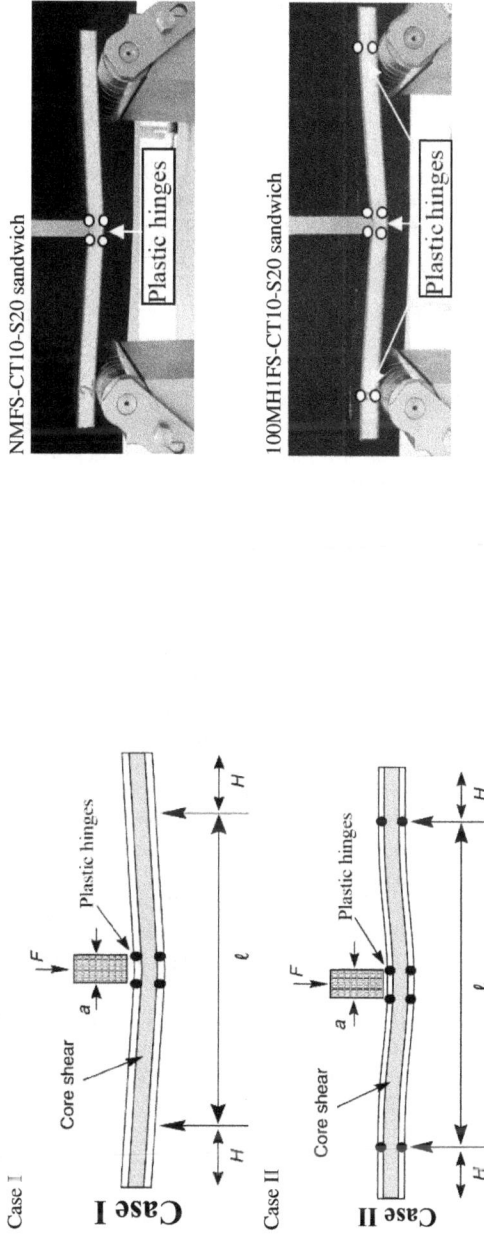

TABLE 11.6

Theoretical and Experimental Comparison of the Sandwich Core Shear Mechanisms

Failure Mode of Core Shear

Schematic Representation (Ashby et al. 2000)	Experimental Image

Case I

Case II

NMFS-CT10-S20 sandwich

100MH1FS-CT10-S20 sandwich

Although decreasing foam core density with thinner face sheets allows the sandwich structure to become more lightweight, hybridization of wire mesh sheets is one of the best candidates for protecting the sandwich structure from undesired failure behaviors and can be introduced for the cost-effective and weight critical applications in the fields of automotive, aviation, satellite, etc.

REFERENCES

Ahmed, T. J., H. E. N. Bersee, A. Beukers. 2007. Low velocity impact on woven glass compoistes reinforced with metal mesh layers. *16th International Conference on Composite Materials*, Kyoto, Japan, 1–6.
Ashby, M. F., A. G. Evans, N. A. Fleck, L. J. Gibson, J. W. Hutchinson and H. N. G. Wadley. 2000. *Metal Foams: A Design Guide*. London: Butterworth, Heinemann, 251.
Chai, G. B. and P. Manikandan. 2014. Low velocity impact response of fibre-metal laminates – a review. *Composite Structures* 107:363–381.
Daniel, I.M., E. E. Gdoutos, K. A. Wang and J. L. Abot. 2002. Failure modes of composite sandwich beams. *International Journal of Damage Mechanics* 11:309–334.
Elmarakbi, A. 2013. *Advanced Composite Materials for Automotive Applications Structural Integrity and Crashworthiness*. Automotive Series, John Wiley & Sons, Inc., 470.
Gibson, R.F. 2012. *Principles of Composite Material Mechanics*, 3rd Edition. Boca Raton, FL: CRC Press.
Gupta N, S. E. Zeltmann, D. D. Luong and M. Doddamani. 2019. Core materials for marine sandwich structures (Chapter 7). *Marine Composites: Design and Performance*, 187–224. Duxford: Woodhead Publishing.
Hasselbruch, H, H. A. Von and H. W. Zoch. 2015. Properties and failure behaviour of hybrid wire mesh/carbon fibre reinforced thermoplastic composites under quasi-static tensile load. *Mater Des* 66:429–436.
Joseph, J. D.R., J. Prabakar and P. Alagusundaramoorthy. 2017. Precast concrete sandwich one-way slabs under flexural loading. *Engineering Structures* 138:447–457.
Karunagaran, N. and A. Rajadurai. 2016. Effect of surface treatment on mechanical properties of glass fiber/stainless steel wire mesh reinforced epoxy hybrid composites. *Journal of Mechanical Science and Technology* 30(6):2475–2482.
Karunagaran, N., G. Bharathiraja, A. Muniappan and K. Yoganandam. 2020. Energy absorption and damage behaviour of surface treated glass fibre/stainless steel wire mesh reinforced hybrid composites. *Materials Today: Proceedings* 22: 1078–1084.
Kurian, R., C. Balaji and S. P. Venkateshan. 2016. Experimental investigation of convective heat transfer in a vertical channel with brass wire mesh blocks. *International Journal of Thermal Sciences* 99:170–179.
Kurian, R., C. Balaji and S. P. Venkateshan. 2017. An experimental study on hydrodynamic and thermal performance of stainless steel wire mesh blocks in a vertical channel. *Experimental Thermal and Fluid Science* 86:248–256.
Matalkah, F., H. Bharadwaj, P. Soroushian, W. Wu, A. Almalkawi, A. M. Balachandra and A. Peyvandi. 2017. Development of sandwich composites for building construction with locally available materials. *Construction and Building Materials* 147:380–387.
Meng, Q., W. Chen and H. Hao. 2018. Numerical and experimental study of steel wire mesh and basalt fibre mesh strengthened structural insulated panel against projectile impact. *Advances in Structural Engineering* 21(8):1183–1196.

Prakash, V. R. A. and S. J. Jaisingh. 2018. Mechanical strength behaviour of silane treated e-glass fibre/Al 6061 & SS-304 wire mesh reinforced epoxy resin hybrid composite. *Silicon* 10:2279–2286.

Qeshta, I. M. I., P. Shafigh, M. Z. Jumaat, A. I. Abdulla, Z. Ibrahim, U. J. Alengaram. 2014. The use of wire mesh–epoxy composite for enhancing the flexural performance of concrete beams. *Materials and Design* 60:250–259.

Qi, J., W. Jingquan, Z. Zhongwen, L. Wenchao and Y. Hu. 2020. Flexural behavior of an innovative dovetail ultra-high performance concrete joint using steel wire mesh interface treatment in composite bridges. *Advances in Structural Engineering* 23: 1142–1153.

Sakthivel, M and S. Vijayakumar. 2017. Influence of stainless-steel wire mesh on the mechanical behaviour in a glass-fibre-reinforced epoxy composite. *Mater Tehnol* 51:455–461.

Sakthivel, M., S. Vijayakumar and B. V. Ramnath. 2018. Investigation on mechanical and thermal properties of stainless steel wire mesh-glass fibre reinforced polymer composite. *Silicon* 10:2643–2651.

Sadighi, M., R. C. Alderliesten and R. Benedictus. 2012. Impact resistance of fiber-metal laminates: A review. *International Journal of Impact Engineering* 49:77–90.

Shao, R., C. Wu, Z. Liu, Y. Su, J. Liu, G. Chen and S. Xu. 2018. Penetration resistance of ultra-high-strength concrete protected with layers of high-toughness and lightweight energy absorption materials. *Composite Structures* 185:807–820.

Sinmazçelik, T., E. Avcu, M. Ö. Bora and O. Çoban. 2011. A review: Fibre metal laminates, background, bonding types and applied test methods. *Materials and Design* 32: 3671–3685.

Steeves, C.A. and N. A. Fleck. 2004. Collapse mechanisms of sandwich beams with composite faces and a foam core, loaded in three-point bending, part I: Analytical models and minimum weight design. *International Journal of Mechanical Sciences* 46:561–583.

Uzay, Ç. and N. Geren. 2020. Effect of stainless-steel wire mesh embedded into fibre-reinforced polymer facings on flexural characteristics of sandwich structures. *Journal of Reinforced Plastics and Composites* 39(15–16):613–633.

Uzay, Ç., D. Acer and N. Geren. 2019. Impact strength of interply and intraply hybrid laminates based on carbon-aramid/epoxy composites. *European Mechanical Science* 3(1):1–5.

Vasiliev, V. V. and E. Morozov. 2007. *Advanced Mechanics of Composite Materials*. 2nd, Elsevier, Edition, UK, 504.

Wang, J., B. Chen, H. Wang and A. M. Waas. 2015. Experimental study on the compression after- impact behavior of foam-core sandwich panels. *Journal of Sandwich Structures and Materials* 17(4):446–465.

12 Low-Velocity Impact Behaviour of Textile-Reinforced Composite Sandwich Panels

B.K. Behera, Manya Jain, Lekhani Tripathi, and Soumya Chowdhury
Indian Institute of Technology

CONTENTS

12.1 Introduction 214
    12.1.1 Mechanics of Impact..217
12.2 Concept of Low-Velocity Impact...218
12.3 Importance of Impact in Composites 220
12.4 Manufacturing of Sandwich Composites 220
    12.4.1 Face Material..221
        12.4.1.1 Manufacturing of Face Sheet.......................................222
    12.4.2 Core Material 222
        12.4.2.1 Balsa Wood..223
        12.4.2.2 Metallic..223
        12.4.2.3 Polymeric 223
    12.4.3 Honeycomb 225
        12.4.3.1 3D Woven Honeycomb Structures............................225
        12.4.3.2 3D Woven Hollow Spacer Structures 226
        12.4.3.3 Fold Cores..227
        12.4.3.4 Truss..227
        12.4.3.5 Integrated Woven Cores..228
        12.4.3.6 Manufacturing of Core 230
    12.4.4 Manufacturing of Sandwich Panel 231
        12.4.4.1 Skin–Core Joining Methods 231
    12.4.5 Different Core Orientations in Sandwich Panel.....................234
12.5 Characterization of Composite Sandwich Panel 234
    12.5.1 In-Plane Tensile Properties..235
    12.5.2 Flatwise Tensile Strength of Sandwich Panels.....................235
    12.5.3 Edgewise Compressive Strength of Sandwich Panels..............236
    12.5.4 Flatwise Compressive Properties of Cores in a Sandwich Panel.....236

DOI: 10.1201/9781003143031-12

12.5.5   Standard Test Method for Core Shear Properties of
         Sandwich Panel 237
12.5.6   Different Impact Tests 238
12.6   Impact Damage 241
12.6.1   Causes of Impact Damage 241
12.6.2   Mechanism of Impact Damage 241
12.6.3   Energy Absorption During Impact 242
12.7   Factors Affecting Impact in Composite Sandwich Panels 243
12.7.1   Effect of Impactor 244
12.7.2   Effect of the Target Structure 245
12.7.3   External Factors 246
12.8   Modes of Failure and Failure Load in Low-Velocity Impact 246
12.9   Structure and Properties of Textile Structure-Based Composite
       Sandwich Panels 247
12.9.1   Using Metal Face Sheets with Textile-Based Core 247
12.9.2   Auxetic Structure 247
12.9.3   Integrated Structure 248
12.9.4   Dual-Core-Based Structure 248
12.9.5   Incorporation of Nano-Structures 248
12.10   Modelling 248
12.10.1   Analytical Modelling 248
12.10.2   Numerical Modelling 249
12.11   Applications 249
12.11.1   Civil Infrastructure 249
12.11.2   Marine Structures 249
12.11.3   Aeronautics 249
12.11.4   Railway 250
12.12   Summary and Outlook 252
References 253

12.1   INTRODUCTION

A composite sandwich panel comprises of two components—a stiffened strong composite face and a lightweight core. A load transfer is obtained between them when they are bonded. These separating faces escalate the moment of inertia along with very little weight increase for effective bending and buckling resisting.

- A sandwich element efficiently utilizes the ultimate limit high specific stiffness as well as high specific bending strength [1].
- Sandwich composites overcome the problem of insufficient stiffness of composites due to the lower modulus despite equal or even higher strength than the metals such as aluminium or steel.
- The face sheet provides continuous support, unlike a stiffened structure, implying the flatness of surfaces despite high compressive stress without buckling.

- The cellular material in the core provides a low structural weight, and due to its low thermal conductivity, it shows thermal insulation.
- The modus operandi of the sandwich is like a beam of 'I' shape. The faces mimic the flanges to bear the more significant share of bending loads along with the loads in in-plane direction. The second component—core—mimics the web by sustaining transverse shear, which involves redistributing the normal towards the forces on the surface. Eventually, it maintains the structural integrity. While bearing the shear load, the distanced skins carry the stresses in the in-plane direction. During this, one skin behaves as under tension, while the other as under compression. The core steadies the faces against buckling or wrinkling. The outer face sheets carry the axial loads along with the in-plane shears and bending moments. The resistance to the shear and tensile stresses built is carried by the bond joining face sheets and core. The core's cellular material provides a low structural weight and thermal insulation due to its low thermal conductivity.

Sandwiches can be bifurcated into symmetrical sandwiches (Figure 12.1) and asymmetrical ones. The former resists the buckling as well as bending stiffness. From the application point of view, these are used where the structure demands being exposed to an aerodynamic load, for instance in a pressurized structure. Asymmetrical ones find applications in aircraft construction (Figure 12.2). Taking a specific example of classic fuselages, it was found that an asymmetrical sandwich consists of a first skin of laminate of carbon, which is called the '*working skin*'. The membrane stresses are taken by the working skin from the construction. The core provides buckling resistance. The second skin consists of single or dual plies of either Kevlar or carbon, which is called the '*stabilizing skin*' [2].

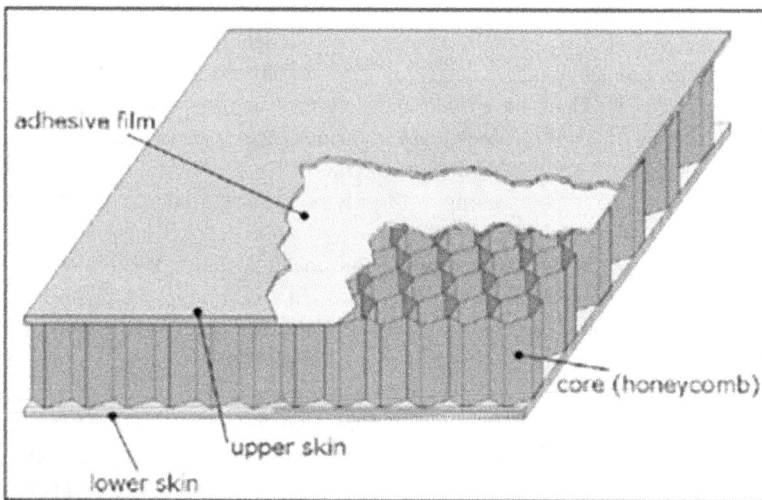

FIGURE 12.1 Symmetrical sandwich [2] (open access).

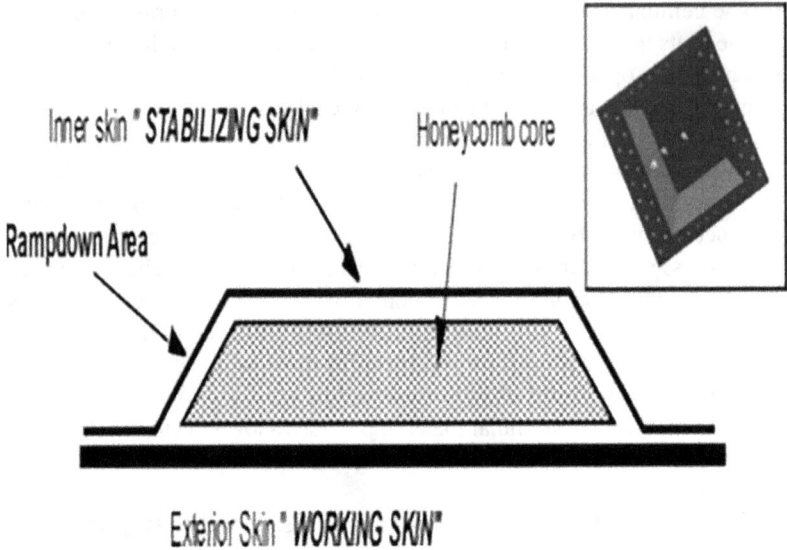

FIGURE 12.2 Asymmetrical sandwich [2] (open access).

Load-carrying sandwich structures are chosen to be lightweight composite materials and damage tolerant of being extensively used in numerous application fields. Broadly, these domains of application of sandwich structures could be in aerospace that requires high bending strength and stiffness strength along with lightweight structures. Another can be marine industry where strength and corrosion resistance are prime features. The automotive field and energy industries also incorporate sandwich structures. This is attributed to their excellent thermal insulation along with acoustic damping properties. The use of monolithic composites or the hybrid and sandwich composites gives enhanced properties in terms of weight and cost [3]. There have been many improvements in the sandwich structural composites in last three decades, and the advancements in applied industries and manufacturing technology enhance the facility of producing reinforced laminates of high-strength fibres along with polymers of cellular structure and also from monolithic materials [4].

During the service life of a sandwich composite structure used in any industry, they are vulnerable to impact loading, which could be in the form of a bird strike in air, hail stones or even the debris on runway [5]. The damage produced on being impacted can be core crushing and debonding between the face sheet and core [6]. The velocity of impact and its energy affect the damage produced. The shape, mass and diameter of the impactor are other factors contributing to the impact damage. The target parameters such as core material, thickness and face sheet type also significantly affect the impact characteristics. Therefore, the study of these factors in order to minimize impact damage can be the major area of research.

12.1.1 MECHANICS OF IMPACT

In mechanics, impact is a shock or high force that is applied over a period of short time of collision of two or more bodies. During an impact event, an object's energy is converted into work. For example, when an object falls from the resting position, its gravitational potential energy (PE) is converted to kinetic energy (KE). Therefore, the total initial potential energy (PE) converts into kinetic energy at the hitting surface. So, mathematically,

Initial PE = Final KE

$$PE = mgh \tag{12.1}$$

$$KE = \frac{1}{2}mv^2 \tag{12.2}$$

$$mgh = \frac{1}{2}mv^2 \tag{12.3}$$

where m = mass, g = gravity, v = velocity, and h = height.

Here, it can be observed that the impact velocity just before the impact is (Figure 12.3)

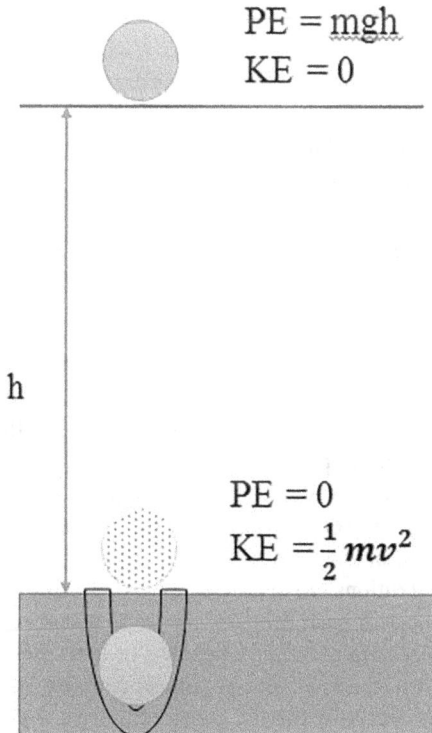

$$PE = mgh$$
$$KE = 0$$

$$h$$

$$PE = 0$$
$$KE = \frac{1}{2}mv^2$$

FIGURE 12.3 Free falling object (redrawn).

$$v = \sqrt{2gh} \qquad (12.4)$$

During the impact, the energy of the object changes. This change in the kinetic energy is basically the net work done. This is known as the work–energy principle.

$$W = \frac{1}{2}mv_f^2 - \frac{1}{2}mv_i^2 \qquad (12.5)$$

where W = work done, v_f = final velocity, v_i = initial velocity, and m = impactor mass.

The work done implies the amount of energy absorbed by the impacted structure. Moreover, the total work done can be simplified as the distance travelled during the impact multiplied by the average force of impact for collision in a straight line.

Therefore, the average impact force represented as 'F_{avg}' will be measured by the division of the change in kinetic energy to the distance travelled represented by 'd'.

Mathematically,

$$F_{\text{avg}} = \frac{\frac{1}{2}mv_f^2}{d} \qquad (12.6)$$

It concludes that the average impact force varies directly and is proportional to the energy which the target would absorb. During the event of impact, the distance that an impactor travels contributes an important role in determining the damage of impact that would cause potential target failure.

12.2   CONCEPT OF LOW-VELOCITY IMPACT

Generally, impacts are classified as low-velocity, high-velocity and, occasionally, hyper-velocity impacts. Cantwell and Morton [7] suggested 10 m/s as the highest threshold for low-velocity impacts considering the height limit of test facility, e.g. tup tower impact. Sjoblom et al. [8] and Shivakumar et al. [9] reported the velocity range from 1 to 10 m/s based on several conditions such as the mass and rigidity of impactors and the material properties of the target.

The stress wave propagation dominates the impact response through the material. High-velocity impact event causes much localized impact damage than the globalized one because of the lack of time of response and the effect of insignificant boundary conditions. In contrast, an overall impact response can be seen under a low-velocity impact event by the impacted structure, which absorbs more energy elastically than plastically because of long enough impact duration. Besides numerical limit, the low-velocity impact can also be recognized by analysing the material property, while the impacted target undergoes the through-thickness stress wave's negligible effect [10]. Types of impact damages can also be categorized to determine the type of impact. Matrix crack and delamination are the primary damages under a low-velocity impact event, whereas the high-velocity impact primarily causes fibre breakage and penetration [11].

In a low-velocity impact, there is more elastic energy absorption. This is because longer time is taken for responding to the impact. Therefore, the structural geometry substantially affects the impact response of the target. A popular approach, which is the energy balance method, can be considered for analysing the impact dynamics. This model is based on conserving total energy [12]. In this model, the kinetic energy variation for mass impactor is equated to the total internal energy (E_d) of deformation which is dissipated.

$$E_d = \frac{1}{2} mv - \frac{1}{2} mv_0^2 \qquad (12.7)$$

The total energy of deformation after impact in this structure can be considered as the total sum of few different forms of deformations. These deformations comprise of contact, bending and shear, along with deformations in the membrane [4]. Few other forms of deformation can be neglected for simple understanding. The neglected losses from material damping, higher mode vibrations and surface friction make the energy balance representation of plate–impactor system as:

$$E_T = E_{BS} + E + E \qquad (12.8)$$

Abrate [13] showed that

$$E_C = \frac{2}{5} Q_C \, w \qquad (12.9)$$

$$E_{BS} = \frac{1}{2} Q_{BS} w_{max}^2 \qquad (12.10)$$

$$E_{BS} = \frac{1}{4} Q_M w_{max}^4 \qquad (12.11)$$

where
  E_C is the contact energy,
  E_{BS} is the bending–shear energy,
  E_M is the membrane energy,
  E_T is the total energy of deformations,
  v is the velocity of the impactor after impact,
  v_0 is the velocity of the impactor before impact,
  W_{max} is the maximum displacement,
  m is the mass of the impactor,
  Q_{BS} is the bending–shear stiffness,
  Q_C is the contact stiffness,
  Q_M is the membrane stiffness, and
  w_{max} is the maximum displacement.

This model is result oriented rather than behaviour oriented as it only provides the maximum impact force. In order to completely understand the force time behaviour is significant which cannot be obtained by this method making it relevant for heavy impactor mass.

12.3   IMPORTANCE OF IMPACT IN COMPOSITES

Normally, in the case of metallic structures, damage due to impact is not treated as a threat because of the ductility of material wherein the energy absorption is immense. Energy absorption in composites is achieved through damage mechanisms and elastic deformation rather than plastic deformation owing to their non-resilient characteristics. The impact produced when a low-velocity impactor strikes a target can be fibre fracture, where the load is substantially borne by the fibre. It could even be delamination or the cracking of matrix. These types of damages affect the serviceability of the structure. Conversely, for high impact energies, there may be perforation, spalling and cracking. These types of damages degrade the ability of load bearing of the structures; that's why it is important to evaluate the behaviour of impact in composite materials so that the designing of these materials withstands impacts during service life, such as a flying debris or dropped tool. It has been found that transverse impact is more common than longitudinal one, and conventional composite structures exhibit poor transverse damage resistance because of the lack of through-thickness reinforcement, which has been overcome in 3D woven composites. Besides, ILSS (interlaminar shear stresses) is one of the vital reasons that cause primary failure due to the composite's low interlaminar strength [7,8].

12.4   MANUFACTURING OF SANDWICH COMPOSITES

The mechanical properties of a composite sandwich panel depend on the core and face materials as well as the geometrical design. There are three or more constituents that form the sandwich structure, such as faces, core and adhesive joints. The utilization of the best materials for a specific application and overcoming drawbacks is possible by performing geometrical sizing. Few reinforced plastics are not as stiff as metals, but it can be overcome by increased core thickness. Materials are chosen on the basis of mechanics as well as factors such as surface finish, environmental resistance, a specific manufacturing method, resistance to wear and cost obstruction. The choice of materials depends on the application, which in turn determines lifetime loading. The availability and cost of material are also the determining features of materials to be used. It is possible to achieve the desired performance and various properties by changing the thickness, the core and the material of face sheet of sandwich structure. For instance, a nominal reduction in the thickness of core leads to a remarkable decrease in flexural rigidity.

12.4.1   FACE MATERIAL

The face sheets can consist of either woven composite materials or a single metal-lic layer, laminated. Broadly, the face can be metallic (steel, stainless steel and aluminium alloys) or non-metallic (plywood, veneer, cement, fibre composites and reinforced plastic). Fibre composites are some of the important non-metallic materials because

- Properties of composites are akin to metals, although their stiffness is lower. To achieve high rigidity, sandwiching with a light core is done.
- They are easier to manufacture.
- They have an anisotropic behaviour.

The constituents of fibre composites consist of the matrix, which can be either thermoplastic or thermoset (i.e. polyesters, vinyl esters, epoxies, phenolics, poly-urethane), and the reinforcement (glass, carbon, or polyamide such as Kevlar or Nomex). Construction of sandwich is a particular kind of laminate consisting of weak material, lightweight thickened core sandwiched between face sheets of strong material as shown in Figure 12.4.

Conventionally, synthetic fibres such as glass and carbon fibres were used to reinforce composite sandwich structures, as they provided higher mechanical properties with high cost [15]. Environmental conservation has made efforts to produce new sustainable resources that include natural fibre-based and wood-based composites as lightweight engineering materials [16–18] (Figure 12.5).

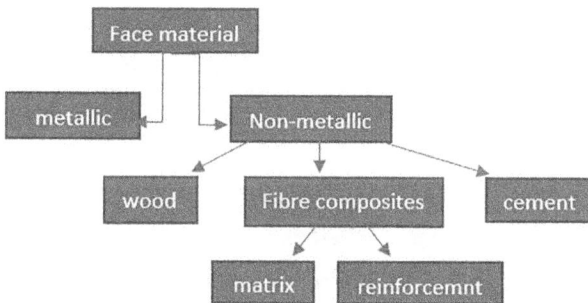

FIGURE 12.4   Classification of types of face material—unpublished.

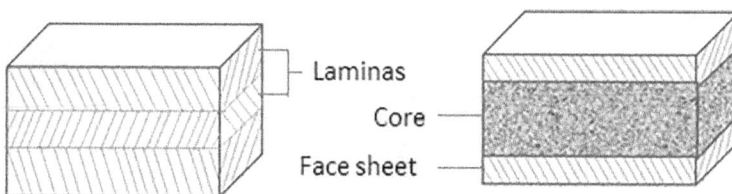

FIGURE 12.5   Laminate composite and sandwich composite [14]—redrawn.

12.4.1.1 Manufacturing of Face Sheet

Non-composite face materials such as wood veneer and sheet are always produced in one step by bonding with the core, while polymer composite face sheets are manufactured by pre-impregnating fabrics. The methods can be bulk moulding compound (BMC), sheet moulding compound (SMC) and glass mat-reinforced thermoplastic (GMT).

12.4.2 CORE MATERIAL

The core is often less stiff than face, with poor strength. On the basis of materials, the core materials are typically divided into the following groups:

- Balsa woods.
- Corrugated and fold cores or truss and metallic cores.
- Honeycombs—There are different varieties of cell shapes, for instance rectangular, square, corrugated or triangular [19], but the hexagonal shape is universally the most used shape made of metal (aluminium) or composite materials (glass, Nomex, thermoplastic or phenolic glass).
- Polymer cellular and foams—The foams that are often used are phenolic foams, polyurethane foams, extruded and expanded polystyrene (EPS and XPS) foams and polymethacrylimide (PMI) foams (Figures 12.6 and 12.7).

Homogeneous core materials:

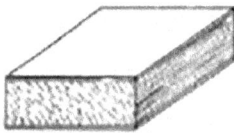   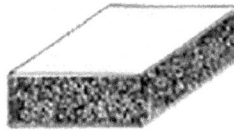

Wood cores   Foam cores

Structured core materials:

   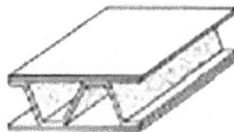   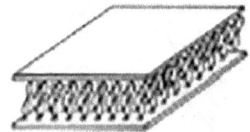

Honeycomb cores   Corrugated cores   Textile cores

FIGURE 12.6 Types of core [19].

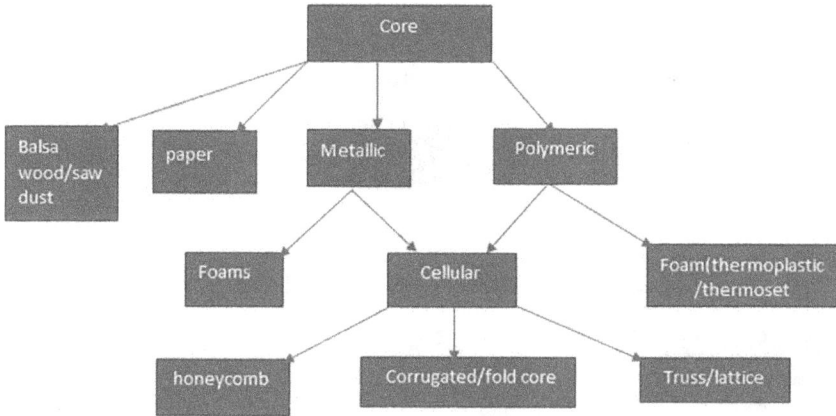

FIGURE 12.7 Classification of types of core materials—unpublished.

12.4.2.1 Balsa Wood

It is a primitive core material utilized perpendicular to the faces of the sandwich. The inconstant density, but high stiffness and transverse strength with moderate shear stiffness contribute to its application.

12.4.2.2 Metallic

Sandwich panels made of metals with open cells and periodic cores are the new important structures. The peak strength of textile and truss cores is superior to the low relative densities of honeycomb, because of their higher buckling resistance. Core of Particulate composite materials (PCMs) in sandwich structure gives the benefit of higher bending and compression [20]. The cores can be prismatic elements, such as textile, honeycomb, and corrugated or assembly of struts, such as pyramidal, Kagome, tetrahedral or shell elements (egg-box). The textile and truss topologies (Figure 12.8) can have either hollow or solid ligaments [21].

12.4.2.3 Polymeric

The cores in a sandwich panel range from stochastic cellular materials, for instance, metallic and polymeric foams, to periodic lattice materials. The latter ones are stretch-dominated and efficient structures, which are appropriate for multifunctional applications [22]. Metallic structures were produced with different varieties of topologies such as pyramidal [23,24], tetrahedral [25,26], hollow truss lattice [27–29] and Kagome [30]. The recent attention incorporates hybrid of composite materials with optimal lattice topology [31].

Multifunctional sandwich panels consisting of cores with prismatic diamond and corrugated cores (Figure 12.9) were analysed, and it was concluded that diamond prismatic topology is more weight-effective among the designs for cooling as well as for load bearing [32].

(a) tetrahedral   (b) pyramidal   (c) 3-D Kagome

(d) diamond weave   (e) hollow truss   (f) egg-box

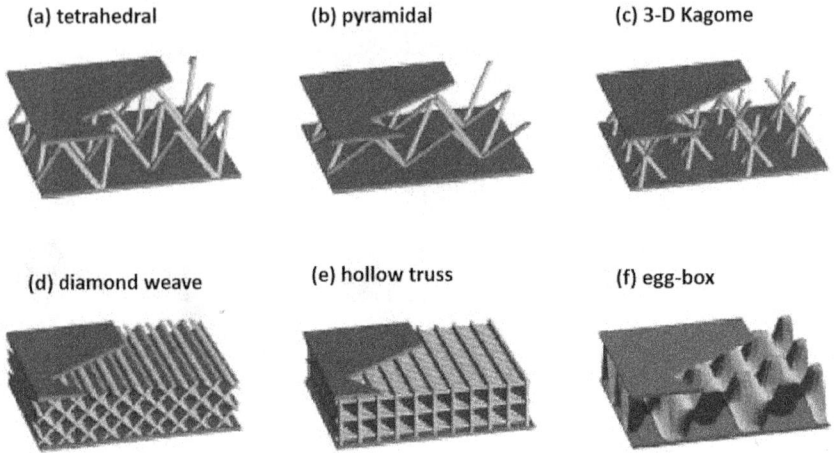

FIGURE 12.8   Examples of sandwich panels with periodic cellular metal cores [21].

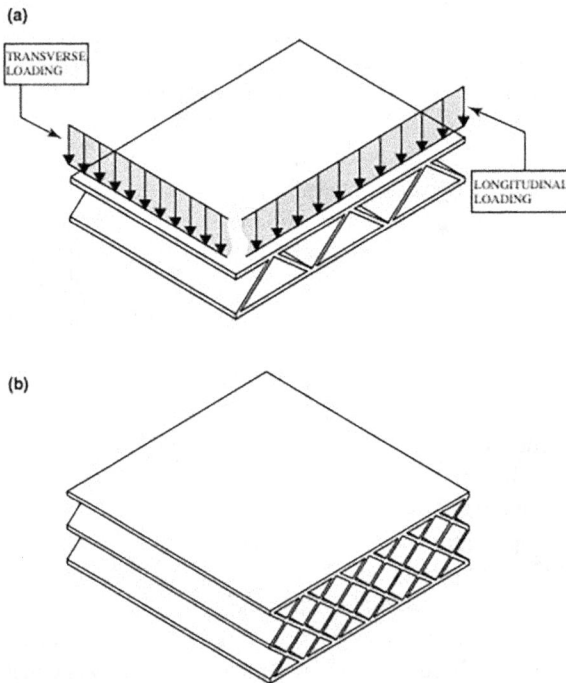

FIGURE 12.9   Schematic of a prismatic panel: (a) corrugated core ($n = 1$) and (b) diamond core ($n = 4$) [32].

12.4.3 HONEYCOMB

Honeycomb material being a unique core material offers many advantages, for example good mechanical properties, low dielectric properties, low thermal conductivity coefficients, excellent crushing properties and fluid control. With small cross-sectional area, good acoustic properties and a large exposed area within the cells are obtained. Honeycomb composites give excellent energy absorption [33].

Honeycomb has series of hexagonal cells that are arranged together and form cells which has similar appearance as the cross-sectional slice of beehive as shown in Figure 12.10 [34].

Honeycomb structure exists also in nature as in balsa wood, in which unit cells nucleate and grow, as it develops rearrangement. In man-made honeycombs, the structure of honeycomb is of different sizes and shapes such as square, triangle and hexagon, and they have regular patterns as shown in Figure 12.11 [34]. Materials having structured cores are corrugated cores, textile cores and honeycombs. The honeycomb that is often used is made of thermoplastic honeycombs, impregnated glass or aluminium or aramid fibre mats such as Nomex [14].

12.4.3.1 3D Woven Honeycomb Structures

To form honeycomb-like structures with multiplicity of cells, adhesive-coated thin sheets are laminated and bonded. Plane woven fabric as a sheet material with impregnation of a thermoset resin such as epoxy can be used to create honeycomb, but has less sufficient peel strength, tensile strength and shear strength, and delamination occurs easily. A multilayer structure by interlacement of warp and wefts of adjacent layers of woven fabric forms the textile honeycomb structure. This does not give the problem of delamination between the layers of honeycomb structure [35]. Xiaozhou Gong [33] investigated the mechanical performance

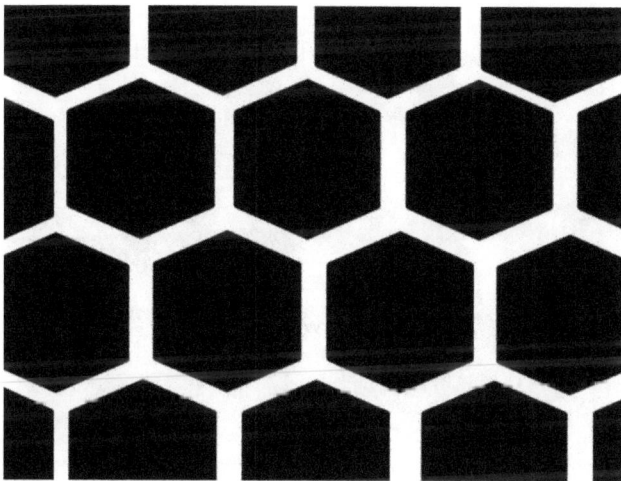

FIGURE 12.10 Two-dimensional honeycomb [34]—redrawn.

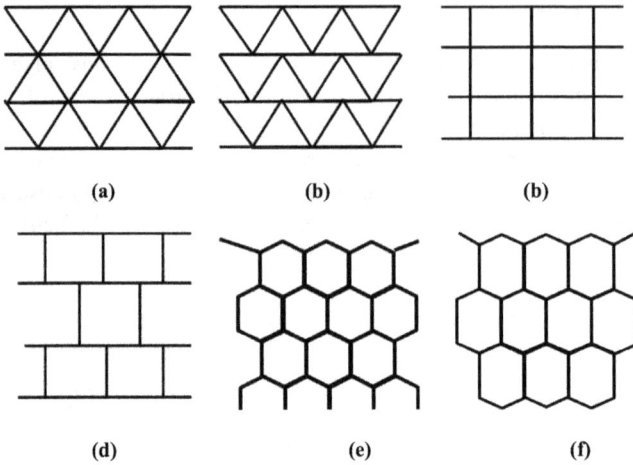

FIGURE 12.11 Schematic illustration of honeycombs structure with different cell shapes: (a, b) two packings of equilateral triangles; (c, d) two packings of squares; (e) a packing of regular hexagon; (f) a packing of irregular hexagon [33]—redrawn.

(low-velocity impact test) of honeycomb textile composites with different geometric structure parameters such as opening angle, cell size, ratio of wall lengths and composite of same thickness with different cell sizes.

12.4.3.2 3D Woven Hollow Spacer Structures

This fabric can be called hollow, spacer and distant fabric. This incorporates hollow spaces between two different fabric surfaces, providing thickness to the fabric structure. There can be variation of connecting spacer yarns [36]. Sandwich structure consisting of three-dimensional spacer structure increases the core–skin bond strength. These structures have high delamination resistance and high specific stiffness as compared to a 2D laminated structure of same thickness (Figure 12.12).

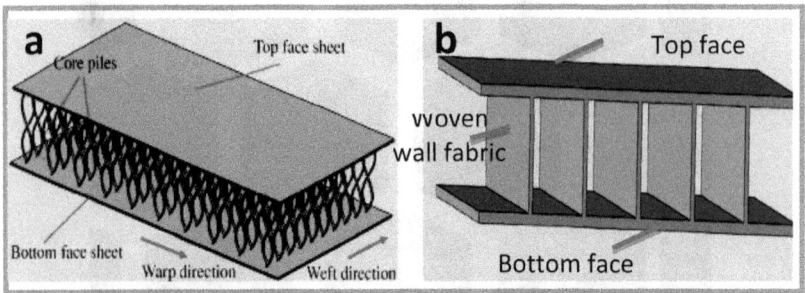

FIGURE 12.12 Woven spacer fabric integrally woven with (a) vertical pile yarn, (b) vertical woven fabric between the outer fabric layers [37]—open access.

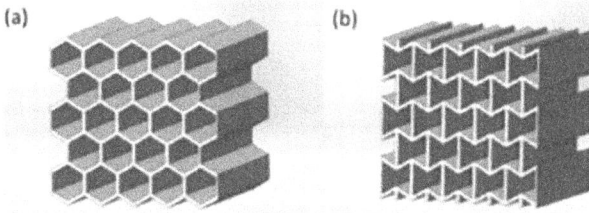

FIGURE 12.13   Hexagonal and re-entrant structures [40].

Earlier spacer woven fabric made with pile yarns connecting the hollow spaces between the two face sheets. The pile yarns do not maintain the structural stability and pose serious problems during composite processing, and resin is not distributed evenly throughout the structure. There is angle distortion in pile threads that consist of thicker fabrics, and they have limited capacity to endure high shear stress and pressure. The researchers experimented that woven cross-links (Figure 12.11) give higher tensile strength, compressive strength and also higher bending forces [36,38]. 3D shapes are formed in weaving stage itself, which reduces the additional labour cost in making the composite as compared to manufacturing using assembly technology.

Hexagonal honeycomb possesses a positive Poisson's ratio, which leads to anticlastic behaviour. An alternate to the core of hexagonal shape is represented as auxetic structure (Figure 12.13). These structures are characterized by the stiffening geometric effects, transverse shear modulus, enhanced in-plane indentation resistance and impact energy absorption. The topologies of auxetic involve double arrow head, chiral, re-entrant and rotating rigid units [39].

Packaging industry utilizes paper honeycomb-based sandwich structure and others like furniture, aviation, and building industry because of their light weight, favourable cushioning properties, and furniture. The products of paper honeycomb are important in design of packaging because of the amount of energy absorbed by it [41].

12.4.3.3   Fold Cores

The closed hexagonal cells may cause trapping of condensing water, leading to increased weight and decreased mechanical properties in aeronautical structures [42]. To overcome this, a new generation of core materials came into existence called fold cores (Figure 12.14) [43]. These fold cores are three-dimensional structures produced by simple folding process in an origami-like manner from a flat sheet material. Diverse fold core geometries can be formed in a large number. These can be simple zigzag structures or complex, curved and unsymmetrical with even tapered geometries [44].

12.4.3.4   Truss

Lattice truss cores have been investigated as a new sandwich core material as it has a higher specific stiffness and strength and the interconnected void space is

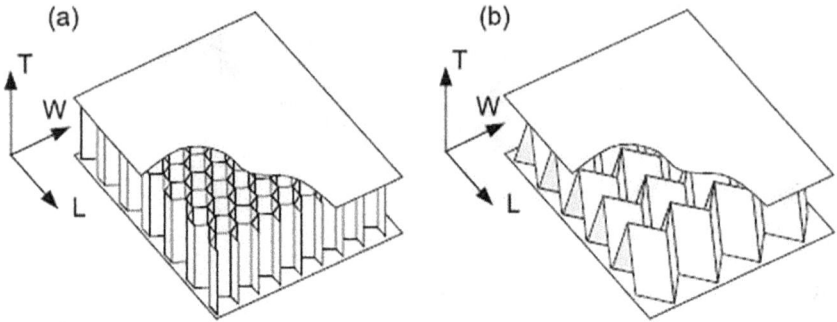

FIGURE 12.14 Illustration of honeycomb and fold core sandwich structures [43].

also large. The measurements show that lattice sandwich structures have necessary mechanical properties to find applications in lightweight and compact structural heat exchangers [45]. The structural behaviour of sandwich panel depends on the parent material and topology.

Pyramidal lattice cores are of low relative density, and their sandwich panels are used to study the fracture response and large-scale bending of the model of system of sandwich panel, and little stretch resistance is shown by the core. The sandwich panels were made of ductile stainless steel [46]. Sandwich panel structures of square and triangular honeycomb cores made from ductile metals show better of mitigating the effects of localized shock loading in water and air [47,48].

Sandwich panels consolidated with cores of low density find extensive applications in multifunctional structures. Latest techniques are developed to fabricate composite truss cores to optimize multifunctional structures. In a research work [49], the authors evaluated the failure mechanisms and mechanical response of two-layer carbon fibre composite sandwich panels having core of pyramidal truss (Figure 12.14) under low-speed concentrated impact and uniform quasi-static compressive loading.

Polymeric materials including Nomex, aluminium honeycomb or other polymers and metal foams are typically utilized to build pyramidal cores [25,50–52]. Simple methods serve as an alternative to solid cores for the construction of pyramidal core-based metal sandwich panel [53,54]. When the sandwiches are subjected to bending load, the truss is exposed to non-axial tension or pressure, while foams with the similar density are deformed, if curved [55].

CFRP (carbon fibre-reinforced polymer) composite sandwich panels with 2D lattice truss core (Figure 12.15) are produced by the hot pressing method. Different tests were conducted in this material (three-point bending test, out-of-plane compression and shear tests) [56].

12.4.3.5 Integrated Woven Cores

The high debonding led to a new type of structure known as integrally woven sandwich structure that provides effective resistance to delamination. These can be used in aircraft or civil infrastructures [57,58]. For skin enhancement,

pyramidal configuration

tetrahedral configuration

X-type configuration

Kagome configuration

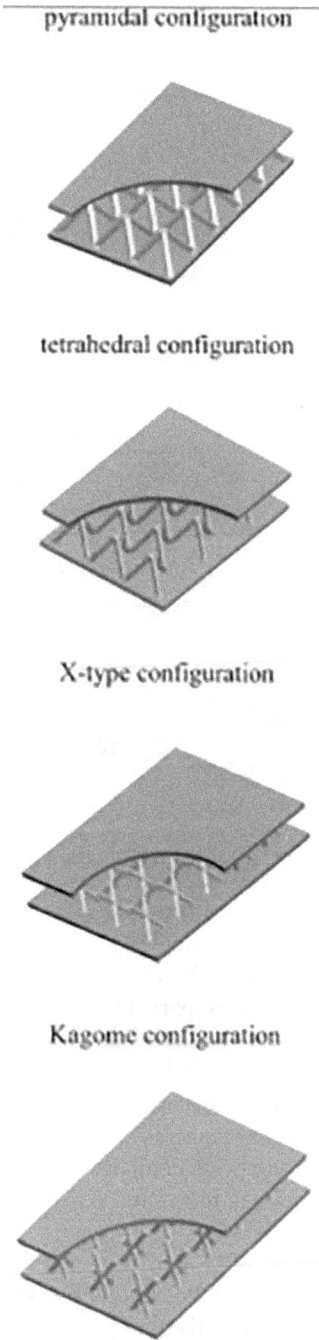

FIGURE 12.15 Schematic illustrations of lattice structures with tetrahedral, pyramidal, Kagome and woven textile truss topologies [40].

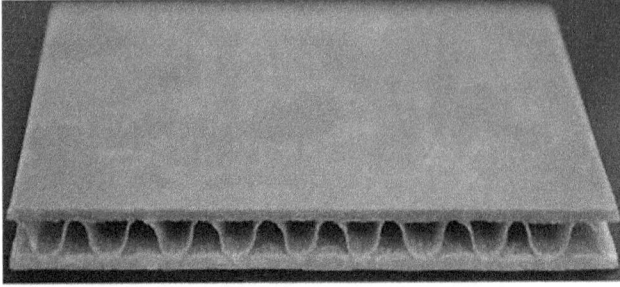

FIGURE 12.16   Sandwich panel with strengthened skin and integrated corrugated woven cores [58].

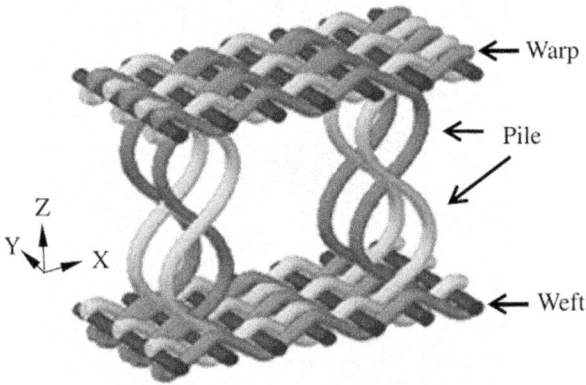

FIGURE 12.17   Typical IWTSC weave structure [38].

Shiah et al. [59] incorporated woven textile fabrics into core which was co-cured with laminated plastic layers to form integrated woven composites.

Johnson et al. [60] developed a cellular structure like a box section consisting of a pair of woven fibre/epoxy skins, parted by braided glass fibre/epoxy cores of square sections. These woven materials are appropriate for automated production and low cost. The mechanical characterization of a woven textile material with orthotropic integrated woven corrugated cores is shown in Figure 12.16, which was evaluated by quasi-static experiments [58].

Integrated woven textile sandwich composite (IWTSC) is a lightweight sandwich structure with core piles and integrated skins as shown in Figure 12.17.

12.4.3.6   Manufacturing of Core

The most primitive balsa wood can be cut into blocks to form any curved shape of the sandwich composite. Commonly used polymer foams include polyurethane (PUR) and polystyrene (PS). PUR may be produced in situ between the faces without pre-foaming into blocks, thus eliminating the need to machine complex core geometries.

12.4.4 Manufacturing of Sandwich Panel

The three methods of manufacturing sandwich structures specifically of aeronautical quality in an autoclave are as follows:

- Co-curing: During curing, the two skins which can be with or without adhesive are bonded to the core. This is called one-curing.
- Co-bonding process: One skin is separately cured, and the other is freshly bonded to the core while curing. This way can be named as two-curing.
- Secondary bonding: The two skins, cured separately, are bonded to the core with an adhesive film. This manufacturing consists of three curing steps.

12.4.4.1 Skin–Core Joining Methods

For manufacturing a durable sandwich structure, an efficacious bond has to be obtained between the skins and core. Figure 12.18 provides an insight into various joining techniques for the same. The methods for these can broadly be either adhesive bonding or fusion bonding.

12.4.4.1.1 Adhesive Bonding

The adhesives used to bond sandwich structures are narrowly different from the normal ones as they require bonding of two very dissimilar constituents (a solid compact to a softer cellular). Its selection is primarily based on meeting the mechanical requirements of the structure in order to provide a strong adhesion between material components along with consideration of essential factors

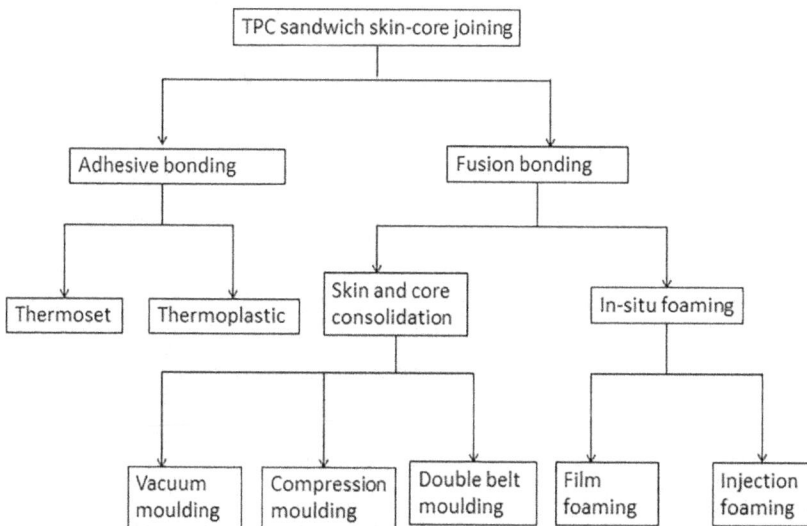

FIGURE 12.18 Skin–core joining methods [61]—redrawn.

such as heat resistance, fatigue, strength, creep and ageing. Providentially, there is a range of adhesives such as epoxy resins, phenolics, polyurethane and polyester [61]. The choice of adhesives can be either thermoplastic based or thermoset based.

12.4.4.1.2 Fusion Bonding Processes

Fusion bonding is exclusively applicable for thermoplastic materials. The intermolecular diffusion of polymers forms the joint which is called adherents in contrast to adhesive joining which creates mechanical interlocking of dual different substrates on the surface [62].

Fusion bonding creates a weak bond at the interface at lower temperatures and collapsing core and delamination of skin at higher temperatures along with the complexity to control pressure and temperature distribution.

One method is to produce either knitted or woven 3D integrated preforms called spacers or distance fabrics [63,64] with very high delamination resistance. Woven spacer connected with core piles demonstrated excellent energy absorption and ductile failure under dynamic compression and quasi-static loading conditions. Higher energy per unit volume and compressive strength in compression were found in thinner panels, while there is a decrease in the damage threshold and compressive load with an increase in pile height under low-velocity impact. Despite the enhancement of several properties, the flexural performance was not significant because of the weaker face sheets.

The mechanisms for bonding skins and core are akin to 2D sandwich structures except that for lightly curved panels, a shaped mould is used (Figure 12.19). Thermoforming is one such method of the sandwich panel construction (Figure 12.19).

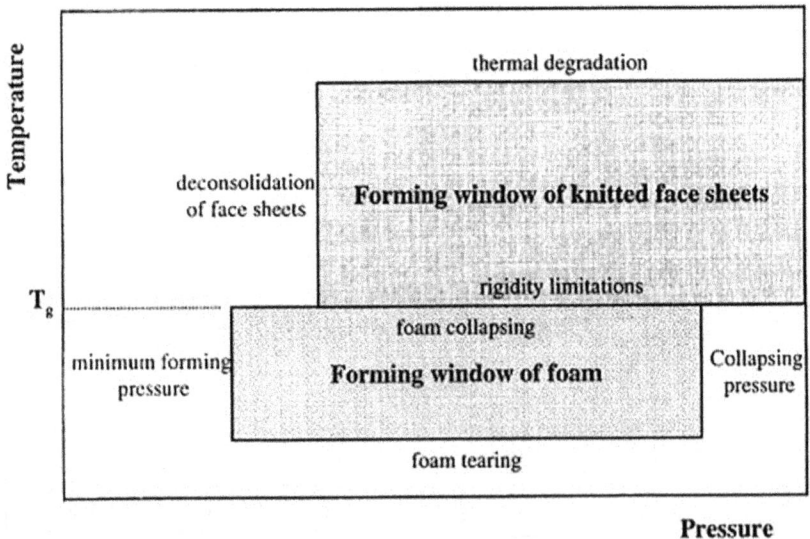

FIGURE 12.19 Thermoforming [65].

In a particular case, polypropylene composites and PP honeycomb core were compressed and heated in an IR field and then fashioned and fused during moulding in a solo step [63,65]. Initially, assembled flat sandwich structures were thermoformed, which can then be employed to produce parts of 3D sandwich. This is an effectual method to obtain complex sandwich structures (Figure 12.20).

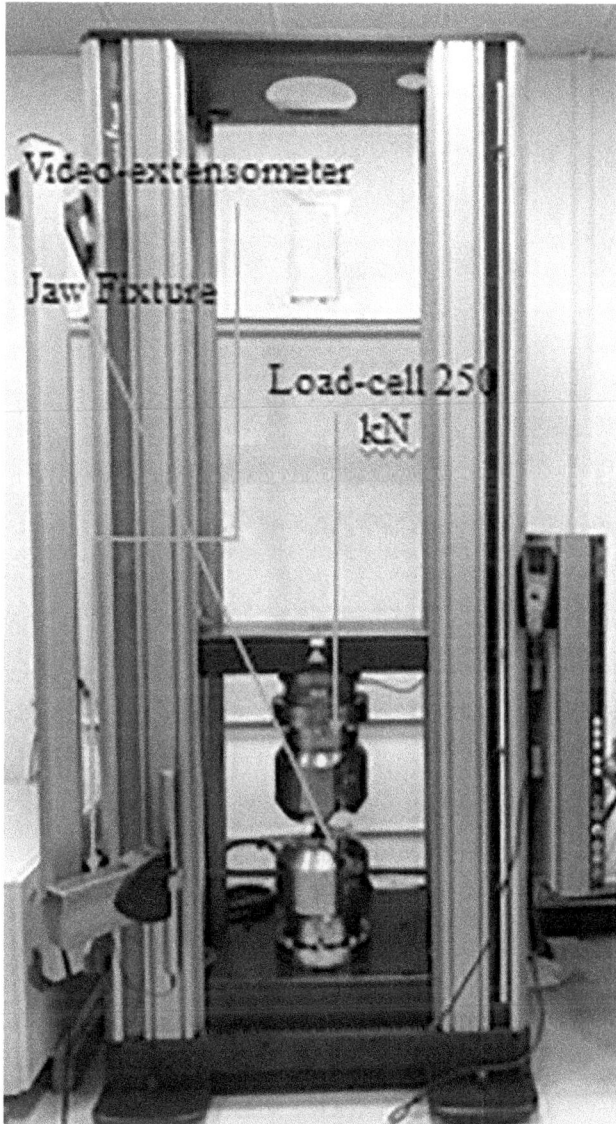

FIGURE 12.20 In-plane tensile testing—unpublished.

FIGURE 12.21   Edgewise compressive test—unpublished.

12.4.5   DIFFERENT CORE ORIENTATIONS IN SANDWICH PANEL

The common feature to characterize sandwich structures of two flat face sheets and core which takes many generic forms such as the number of discrete or continuous corrugated sheets but aligned zed, top-hat, or channel sections (Figure 12.21a–e). The facings and core plates can be joined by self-trapping screws, rivets or spot welds [1].

12.5   CHARACTERIZATION OF COMPOSITE SANDWICH PANEL

The failure of the sandwich structures is caused due to large normal local stress concentrations because of the heterogeneous nature of assembly of the face sheet and core. In general, the sandwich panels are characterized for their tensile loading (in plane and flatwise), failure behaviour and residual strength subjected to compression after-impact testing, low- and high-velocity impact, blast loading and fatigue performance.

12.5.1 IN-PLANE TENSILE PROPERTIES

The in-plane tensile properties were determined by the test procedure given in the ASTM D3039 standard of the laminated composite materials such as polymer matrix composite reinforced with fibres of high modulus.

Properties which are usually measured by this method are:

- ultimate tensile strength,
- ultimate tensile strain,
- tensile chord modulus of elasticity,
- Poisson's ratio, and
- transition strain.

Tensile stress/tensile strength—The ultimate tensile strength can be calculated by the following equation:

$$F_{tu} = P_{max}/A$$

$$\sigma_i = P_i/A$$

where:

F_{tu} = ultimate tensile strength, MPa [psi];
P_{max} = maximum force before failure, N [lbf];
σ_i = tensile stress at ith data point, MPa [psi];
P_i = force at ith data point, N [lbf]; and
A = average cross-sectional area [in^2].

12.5.2 FLATWISE TENSILE STRENGTH OF SANDWICH PANELS

ASTM D297M-16 test method consists of subjecting a sandwich construction to a uniaxial tensile force normal to the plane of the sandwich. This test method is useful to determine the strength and quality of facing-to-core bonding and also used to analyse flatwise tensile strength data of core material.

Ultimate strength—This can be calculated by the following equation

$$F_{zftu} = P_{max}/A$$

where:

F_{zftu} = ultimate flatwise tensile strength, MPa [psi],
P_{max} = ultimate force prior to failure, N [lbf], and
A = cross-sectional area, mm^2 [in^2].

12.5.3 EDGEWISE COMPRESSIVE STRENGTH OF SANDWICH PANELS

ASTM 364M test method consists of subjecting the specimen to monotonically augmenting compressive force which is to be kept parallel to the plane of its faces. The edgewise compressive strength assists us to judge the load-carrying capacity of short sandwich construction specimens in terms of the developed facing stress.

Ultimate strength—This can be calculated from the following equation:

$$\sigma = P_{max} / [w(2t_f)]$$

where:

σ = ultimate edgewise compressive strength, MPa [psi],
P_{max} = ultimate force prior to failure, N [lbf],
w = width of specimen, mm [in.], and
t_f = thickness of a single face sheet, mm [in.].

12.5.4 FLATWISE COMPRESSIVE PROPERTIES OF CORES IN A SANDWICH PANEL

ASTM D365M test method consists in subjecting a sandwich core to a uniaxial compressive force normal to the plane of the facings of the core. Flatwise compressive strength and modulus can be obtained by this method. From a complete force versus deformation curve, deformation and compressive stress at any applied force and the effective modulus of the core can be obtained.

Ultimate strength—The ultimate flatwise compressive strength can be calculated using the below equation:

$$F_{zfcu} = P_{max} / A$$

where:

F_{zfcu} = ultimate flatwise compressive strength, MPa [psi],
P_{max} = ultimate force prior to failure, N [lbf], and
A = cross-sectional area, mm^2 [in.^{2}].
Compressive modulus—This can be calculated by the following equation:

$$E_{zfc} = ((P0.003 - P0.001) * t) / ((\delta 0.003 - \delta 0.001) * A)$$

where:

E_{zfc} = core flatwise compressive chord modulus, MPa[psi],
$P0.003$ = applied force corresponding to $\delta 0.003$, N [lbf],
$P0.001$ = applied force corresponding to $\delta 0.001$, N [lbf],
$\delta 0.003$ = recorded deflection value such that δ/t is closest to 0.003, and
$\delta 0.001$ = recorded deflection value such that δ/t is closest to 0.001 (Figure 12.22).

FIGURE 12.22   Flatwise compressive test—unpublished.

12.5.5   STANDARD TEST METHOD FOR CORE SHEAR
PROPERTIES OF SANDWICH PANEL

ASTM D393M test method determines the core shear properties of sandwich structures subjected to flexure such that the applied moments produce curvature of the sandwich facing planes. It determines the sandwich flexural stiffness, the core shear modulus and shear strength, or the facing tensile and compressive strengths.

This method can calculate the following parameters:

• Force–displacement behaviour.
• Core shear ultimate stress—This can be calculated by the below equation:

$$F_{\text{sult}} = P_{\max} / \left((d + c) \ {}^{*} b \right)$$

where:
F_{sult} = core shear ultimate strength, MPa [psi],
$P_{\max}$ = maximum force prior to failure, N [lb],

FIGURE 12.23 Core shear test—unpublished.

t = nominal facing thickness, mm [in.],
d = sandwich thickness, mm [in.],
c = core thickness, mm [in.] ($c = d - 2t$), and
b = sandwich width, mm [in.] (Figure 12.23).

12.5.6 DIFFERENT IMPACT TESTS

The actual impact that a sandwich panel will be subjected to during its service life can be simulated by using specific impactors for testing. Accordingly, several test procedures are proposed by many researchers. Mainly, drop weight and pendulum impact tests are performed (Figures 12.24 and 12.25).

I. Drop weight testers (Figure 12.26) are extensively used. A rail guides the heavy impactors during free fall from a certain height. Generally, a mechanical device activated by the sensor prevents from multiple impacts due to the bouncing back impactor. Quantitatively, the relative values were compared of damage resistance parameters with different constituents of composite materials. The parameters for damage response may include damage dimensions, dent depth, through-thickness locations, $F1$, $Fmax$, $E1$ and $Emax$, as well as the force versus time curve.

Impact velocity—This can be calculated by the following equation:

$$v_i = \left(W12 / (t2 - t1)\right) + g\left(t_i - ((t1 + t2) / 2)\right)$$

where:

v_i = impact velocity, m/s [in./s],

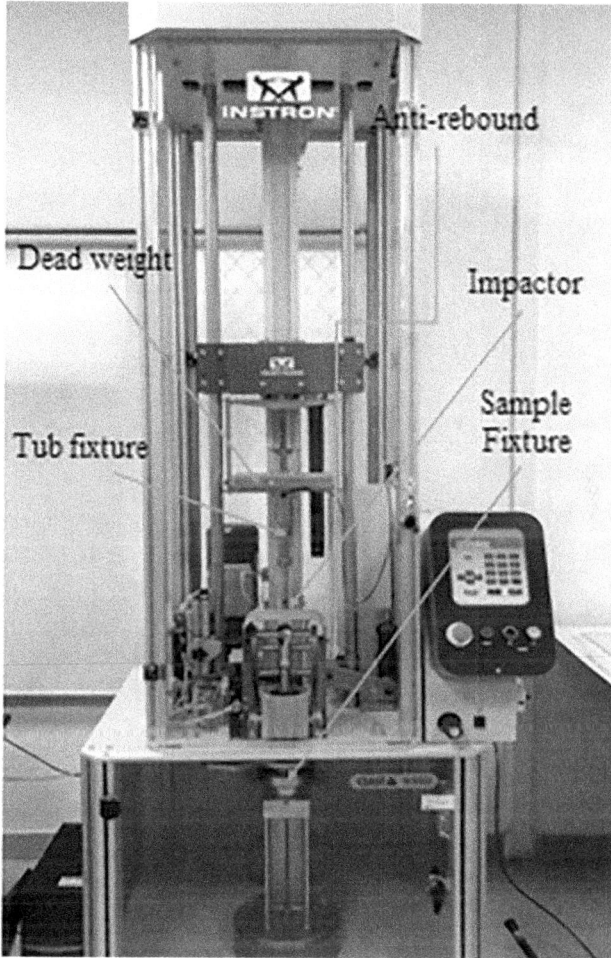

FIGURE 12.24 Drop weight impact test—unpublished.

$W12$ = distance between leading edges of the first (lower) and second (upper) flag prongs, m [in.],

$t1$ = time first (lower) flag prong passes detector,

$t2$ = time second (upper) flag prong passes detector, and

t_i = time of initial contact obtained from force versus time curve, s.

Measured impact energy—This can be calculated by the below equation:

$$Ei = \left(m * v_i^2 \right) / 2$$

where:

Ei = measured impact energy, J [in.-lbf], and

m = mass of impactor, kg [lbm].

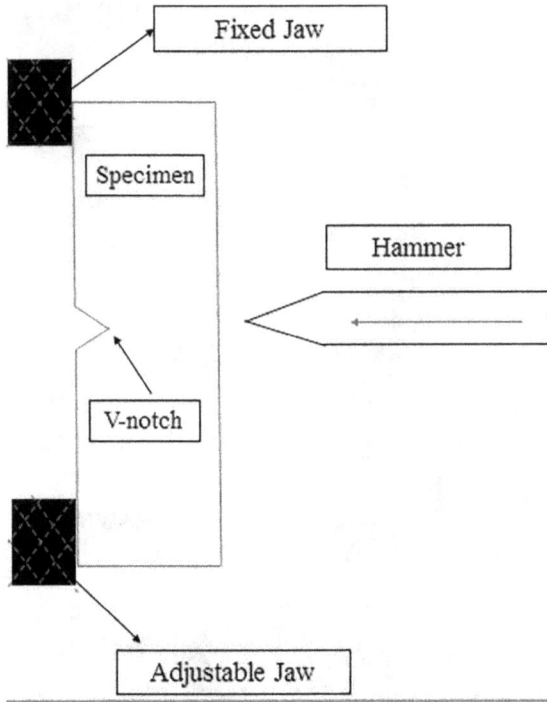

FIGURE 12.25 Schematic of Charpy impact test with vertical jaws—redrawn.

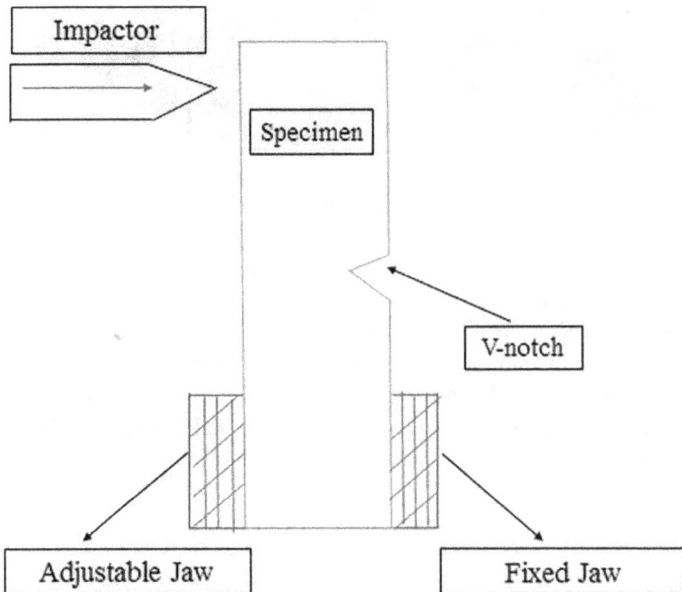

FIGURE 12.26 Schematic of Charpy impact test with horizontal jaws—redrawn.

II. Pendulum-type systems: The principle of performing impact test by using the principle of pendulum is used in Charpy impact tests which generate low-velocity impacts. In this test method, a steel ball is hung by a string. Also, for heavier projectile, it can be equipped with velocity sensors or force transducers. One of the techniques that can be incorporated is the Hopkinson-type pressure bar. A Charpy impact device is shown in Figure 12.25. The dial shown in the figure is used to provide the energy absorbed by the material in joules.

12.6   IMPACT DAMAGE

12.6.1   Causes of Impact Damage

There are many causes for the impact damage of composite materials. Some of them are given below [66].

1. Tools dropped during the repair or maintenance.
2. Hailstone, meteorite, lightning or bird strike.
3. Debris from burst tyres, blade loss, rotor burst, and engine debris or foreign object impact.
4. Walking on a structure.
5. Wheel threats, accidental collision with other handling equipment or ditching and crash of the structure.
6. Ballistic impact by bullets.

12.6.2   Mechanism of Impact Damage

When a blunt object strikes laminated composite materials under the low-velocity impact threshold, it causes internal damage. It is also known as barely visible impact damage (BVID), and often it is not noticeable the naked eye, but creates deep within the structure in the back face [64,65]. This type of damage causes a serious threat to overall structural integrity and residual strength [64,66,67] (Figure 12.27).

Most of the composites are brittle and absorb energy through damage mechanism in elastic deformation and not via plastic deformation [7,68–70]. The low-velocity impact is the primary cause of the onset of delamination rather than laminate's catastrophic failure [66], and it may increase during different times of service life of the laminates. The nature of low-velocity impact can be considered as quasi-static [10,11]. It has already been mentioned earlier that matrix cracking and delamination are the modes for primary damage [71] in the low-velocity impact, and this impact is the most occurring one on composite structures in real time from manufacturing to the end of service life. Damage mechanisms of sandwich structures are the combination of damage propagation of face sheet, failure modes and energy absorption of core structures.

Significant internal damage dependent on impact tip radius and face sheet layup was found at energy levels of 2 J [72,73]. Damaged panels with specific core

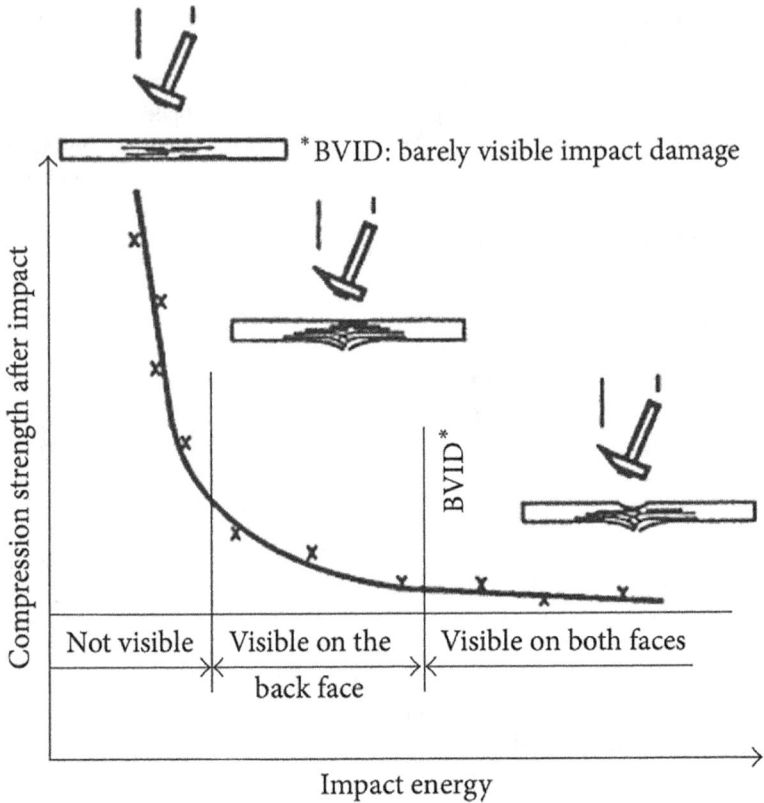

FIGURE 12.27 Effect of BVID on compression after-impact strength [70].

crushing undergo sudden fracture in the core, leading to complete loss in load-bearing capacity and catastrophic failure [74,75]. The effect of face sheet material on damage generation till catastrophic failure portrayed that thinner panels undergo greater overall deformation [76]; woven glass fibre skins were prone to puncture and core crushing, while cross-plied composite skins spread impact load and minimize core crushing. Low-velocity impact damage is dependent on loading rate, and thick or damaged sandwiches are not quasi-static [77,78]. It was found that that impact damage is a function of impact energy, but not the mass of the impactor or its velocity alone [79].

12.6.3 ENERGY ABSORPTION DURING IMPACT

Energy absorption during impact by sandwich structures is a complex and sequential process, where face sheets and core both have individual roles to play [71,80–82].

During impact, both the target and the impactor are responsible for the energy absorption behaviour. Here, the target is the sandwich panel which absorbs the

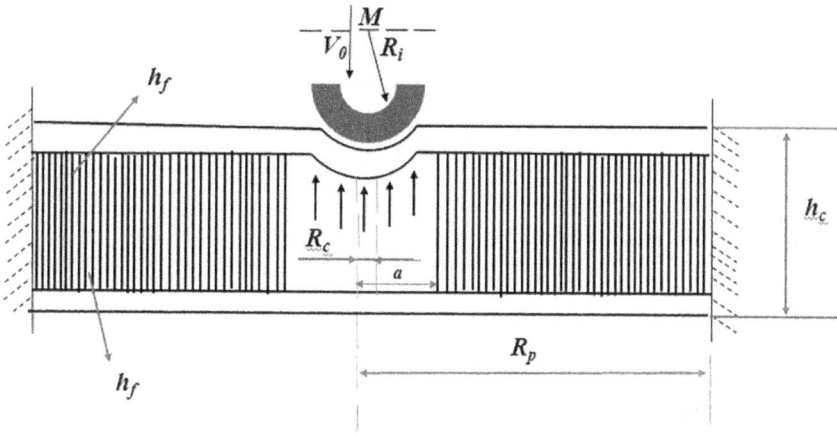

FIGURE 12.28 Basic model of a sandwich panel subjected to low-velocity impact [6]—redrawn.

strain energy. The impactor can be treated as rigid. The face sheet of the sandwich panel absorbs the energy caused by fracture in both high- and low-velocity impacts. Compression after-impact test is used for determining the residual strength of the impact-damaged structures, which provides data on structures' stiffness degradation [75,83] In the subsequent stage, the core and the face sheet get debonded because of the external energy caused by the impact event. Debonding between face and core usually happens due to the lower bond strength, and the wave reflection from boundary causes intense load levels. Debonding as a part of total energy is of trivial amount [73]. Friction, sound, heat, etc., are also negligible, accounting for 3% of the total impact energy.

A great number of factors [73] were formulated to analyse the energy absorption competence of structural materials. The capabilities of specific energy absorption of composite sandwich panels exhibit superiority to metals [84]. From various research studies, it is reported that with the increment of impact velocity, the absorption capability increases [80,85,86] (Figure 12.28).

12.7 FACTORS AFFECTING IMPACT IN COMPOSITE SANDWICH PANELS

The following parameters affect the impact response of composite sandwich panels (Figure 12.29) [14,76,79,80,84,87,88].

1. Geometrical parameters including skin and core thickness, support span and the core type.
2. Material properties such as Young's modulus of the face sheet, shear modulus of the core and Young's modulus of the impactor.

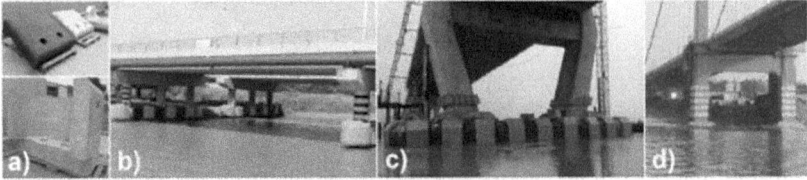

FIGURE 12.29 (a) Components of a fixed composite bumper system and applications in (b) Xinmengge Bridge, (c) Oubei Bridge and (d) side tower of Wuhan Yingwuzhou Yangtze River Bridge [119].

3. Impact velocity.
4. Mass and radius of the impactor, radius of contact and mass of the sandwich structure.

The skins of sandwich panel which are high in failure stress and modulus are required for stiffness optimization. The mechanical parameters of the skins include Poisson's ratio, elastic modulus and the skin thickness. The lightweight core has low strength and low modulus. Gluing the core to skin during deformation produces complex interaction. The weak and soft core causes damage initiation in the sandwich panel [6,79,87,89–91], but debonding of the interface may also cause initial damage.

Honeycomb sandwich structures have multicellular core with effective engineering properties. The finite element simulation models geometrically correct core using shell elements for obtaining more realistic distribution of strains and stresses. The mechanical parameters of the core comprise of core height, Poisson's ratio and shear modulus.

12.7.1 Effect of Impactor

1. Impactor mass: Researchers have used different impactor masses, but keeping other factors constant to see the influence of impactor mass on energy absorption. Comprehensively, it was found that a larger deflection in the target was produced when high impactor mass was used. This could be attributed to the longer duration of impact [6].
2. Impactor velocity: The velocity with which the impactor hits the target is of prime value. The speed substantially affects the contact force and deflection response. A high impact velocity causes a greater deflection with large contact force. This, however, did not affect the impact duration. Thus, higher impact velocity was found to complement the capacity of the energy absorption of the target [6].
3. Impactor geometry: The type of surface dimension of the impactor that is directly going to have contact with the target significantly determines the damage. Spherical-ended impactor was found to cause lower

ultimate load due to greater stress concentration than the flat-ended one around the edge of impactor. Also, spherical impactor is most common in research than other shapes [6]. In a particular case, the impact response of carbon fibre and PEEK sandwich structure was dependent on the impactor diameter, but the number of unit cells of the core did not affect the impact response [92].

12.7.2 Effect of the Target Structure

1. Thickness of the face sheet: For engineering of the structure to be used in load bearing, the thickness of face sheet in a sandwich panel finds extensive research to attain optimum value for minimum weight with maximum properties. The thicker sheet provided larger contact force at constant impact energy, but smaller central displacement. The duration of the impact decreases with thicker sheets. A change in the mechanism of the load transfer was caused by the increased thickness between the core and the top face sheet [6].

2. Core thickness: Another structural parameter for consideration during manufacturing of sandwich panels is the core thickness. For instance, a Kevlar honeycomb core of 3 and 5 mm thickness with basalt and carbon fibre face sheets-based sandwich panels resulted in similar impact strength, and it was inferred that the thickness of the core has a lower impact on the strength value of impact [93]. When using polyethylene terephthalate (PET) and polyvinyl chloride (PVC) foams as the core material, with an increase in core thickness, the elastic behaviour, contact time, deflection value and energy absorption were found to increase. Area of the impact damage increases with impact energy for each type of core materials and their thicknesses [94]. The perforation achieved under impact energy of 10 J was 38 and 98 for cores of 15 and 25 mm thickness, respectively. The material was balsa core and glass fibre thermoplastic sheets [95].

3. Face thickness ratio: It is the ratio between the thicknesses of outer and inner sheets. In boat construction, thicker outer sheets provide enhanced face impact penetration strength. A ratio of 70:30 increased the impact of outer and inner faces by 65% and 24%, respectively. The same can also be achieved by increasing the core thickness [96].

4. Reinforcement material: While constructing a surfboard, E-glass on being replaced by S-glass displayed enhanced impact resistance [97]. The highest absorbed energy was seen for glass fibre even with just 50% hybridization over aramid and polyethylene fibres, though better damage propagation can be attained by alternative fibres. Keeping other factors constant, glass fibre outperformed biaxial carbon fibre in terms of impact resistance, but carbon epoxy foam gave equivalent impact performance to glass monolithic sandwich at 50% weight [98,99].

5. Matrix material: Epoxy provided better impact resistance than polyester in a surfboard [97,99]. It was seen that for impact events, global bending and shearing were trivial and local indentation had the major influence, and therefore, the resin type, be it polyester or phenolic, did not contribute significantly to the performance. Epoxy being expensive and time-consuming was replaced by vinyl ester in a yacht hull [100]. Using flexible resin which has 15% elongation at break in aluminium honeycomb with glass sheet, the impact strength was improved with an 80% increase in penetration strength of inner face without affecting the outer face [96].

6. Core density: Out of all other material parameters, core density prominently affects the impact strength [96]. At high energies, denser PVC panels provided greater damage development as well as outer skin penetration strength [98]. The peak load was directly proportional, while the duration of impact was inversely proportional to the core density at constant impact energy [101]. The denser foams resulted in greater energy absorption along with greater skin failure resistance.

12.7.3 EXTERNAL FACTORS

Temperature—Different sandwich panels find applications in different range of temperatures during their service life. In order to determine their effectiveness and durability, monitoring of the impact of temperature on the structure's mechanical properties is inevitable. From the studies, it was confirmed that the lower temperature reduces the resistance of composite to delamination [102]. It was also found that with increasing temperature from 25°C to 75°C, the impact force decreases [6].

12.8 MODES OF FAILURE AND FAILURE LOAD IN LOW-VELOCITY IMPACT

The failure process involves both interlaminar damage mechanisms, such as penetration and delamination, and intralaminar damage mechanisms, including fibre fracture and matrix crack.

Broadly, the anisotropic nature of laminates of fibre-reinforced plastic (FRP) may result in four major modes of failure.

1. Matrix mode: Cracking occurs parallel to the fibres due to compression, tension or shear.
2. Delamination mode: Produced by interlaminar stresses.
3. Fibre mode: In tension, fibre breakage occurs, and in compression, buckling of fibres occur.
4. Penetration: The impacted surface is completely perforated by the impactor.

The mode of failure provides information about the event of impact as well as the residual strength of structure. Interaction between failure modes is crucial to understanding damage mode initiation and propagation [14,103].

12.9 STRUCTURE AND PROPERTIES OF TEXTILE STRUCTURE-BASED COMPOSITE SANDWICH PANELS

The impact resistance and impact tolerance of sandwich structures were found to be enhanced in different ways.

12.9.1 Using Metal Face Sheets with Textile-Based Core

3D fibreglass fabric sandwiched between thin sheets of magnesium alloy exhibits outstanding capacity of impact absorption, although the resistance of impact energy is lower when the core is formed by woven fabrics [104]. Thick skins of steel with various foams as core indicate that to minimize the weight and maximize the energy absorption of the crash component at the same time, a low-density foam can be used as core material [105]. Fabrication and impact testing of fibre composite axial and circular corrugated sandwich cylindrical panels (ACSCPs and CCSCPs) observed that the peak forces and absorbed energies ascend with relative density increase. Impact position affects impact response, but not impact energy. Also, the ACSCPs excel in impact resistance and energy absorption over CCSCPs [106]. Aluminium face sheets with core of composite laminates (glass and kenaf hybrids) with different fibre orientations demonstrated higher impact strength than pure kenaf. Also, the strength in edgewise orientation surpassed the flatwise one [107]. In a particular case, magnesium-based specimens were found to withstand more impact energy without delamination than steel which completely delaminated and failed under the same energy. This was because of the relatively lesser stiff and thicker magnesium sheets which developed higher stresses [108]. Moreover, enhanced impact performance can be achieved using 3D fibre metal laminates with 3D fibreglass as core and thicker metal sheets, but the resistance would be lower [104,108].

12.9.2 Auxetic Structure

The auxetic structures are the re-entrant honeycomb with negative Poisson's ratio. These can be used as core material in sandwich panels. The 3D auxetic textile composite was found to possess stable impact behaviour under repeated impact testing. These structures also had better impact protective performance than the non-auxetic textile composite because of the higher energy absorption capability and lower transmitted peak force [109]. In a specific study of creating three structures of Kevlar fibre and epoxy resin: regular honeycomb, re-entrant auxetic, auxetic–strut structures—the last one being a hybrid of regular and strut—it was concluded that highest impact energy was absorbed by the hybrid structure [39].

12.9.3   INTEGRATED STRUCTURE

The integrated structures perform efficiently under mechanical loadings due to the lack of adhesive joints between skin and core [110]. Polyester and glass fibres used as sheath and core were co-wrapped to produce a hybrid yarn which was used as core in 3D integrated woven sandwich composite panels. The contact time, deflection and residual strength were greater than those of homogeneous glass/epoxy composites. However, the impact load and index were reported to decrease [39,111].

12.9.4   DUAL-CORE-BASED STRUCTURE

Conventionally, sandwich structures were made by using a single core sandwiched between two skins. Using two fold cores exhibits two-phase energy absorption performance [43]. A novel method to improve energy absorption was attained by using dual cores. In a particular case, glass fibre-reinforced epoxy face sheets with a dual core composed of extruded polystyrene foam and alumina trihydrate-filled epoxy confirmed that the top face sheet, i.e. the latter one, governs the initial impact behaviour. In contrast, the recoverable compression deformation of core with the former provided the mechanism of primary energy absorption [112].

12.9.5   INCORPORATION OF NANO-STRUCTURES

Many researches in different domains have confirmed the increase in energy absorption using nano-materials which have very high surface area-to-volume ratio. Recently, composite sandwich panels made of 3D woven glass fibre and reinforcement was nano-structured zeolite or polyurethane foam being the core material. The study shows that the prevention of crack propagation and connection rupture is because of the creation of connective bridges of fibril in vertical direction with respect to loading [113].

12.10   MODELLING

12.10.1   ANALYTICAL MODELLING

The three approaches used by researchers for analytical modelling are Hertzian contact law, energy balance and spring–mass models [114]. Limited attempts have been made to develop analytical models for predicting the low-velocity impact response of sandwich composite structures. An analytical model was developed by Olsson [115] to predict impact damage caused by large mass of the impactor in quasi-static conditions, which was only applicable for significant mass impacts excluding fibre fracture. Using a two -degrees-of-freedom mass--spring model, M. K. Chan et al. [116] developed seven dimensionless parameters to represent factors affecting sandwich beam behaviour. These non-dimensional equations of motions reproduced the dynamic response subjected to low-velocity impacts of composite sandwich beams.

12.10.2 Numerical Modelling

Numerous computational analysis models were developed to predict the damage behaviour before and after the initial damage. Menna [117] developed a model using finite element (FE) code LS DYNA to reproduce and predict the overall force–displacement curves along with the extent and shape of impact damage, but is inaccurate for high velocities and thick face sheets.

12.11 APPLICATIONS

Sandwich systems consisting of FRPs have been economically and effectively utilized in railways, aerospace, civil infrastructure and marine industry due to their efficient load-carrying capacity and light weight.

12.11.1 Civil Infrastructure

The roof structure in Switzerland was constructed using a multifunctional glass fibre-reinforced sandwich structure. This could also provide energy supply functions by encapsulating photovoltaic cells into almost transparent GFRP skins. These sandwich systems with ribs with thermal insulation were also utilized as the structural cladding for buildings [118]. A hybrid fibre-reinforced deck was made with tension skin of glass and compression skin of ultrahigh-performance reinforced concrete. The core was of lightweight concrete. Another material used was of balsa core bridge deck made in Louisiana with skins made of high-strength steel reinforced with biaxial glass composite. This provided the required flexural stiffness [118].

12.11.2 Marine Structures

Floating structure with impact resistance and energy absorption characteristics was made from fibre-reinforced sandwich structure of glass skin with fibre-reinforced foam core in China for collision avoidance structure due to its low weight and high strength [118] (Figures 12.30 and 12.31).

12.11.3 Aeronautics

Initially, non-structural parts such as bag racks, sidewalls, interior parts and galleys, or flooring were made from sandwich structures using composites. Later, secondary structures such as rudders, ailerons, spoilers and flaps found the application of sandwich structures. The most famous Beechcraft Starship aircraft that was made using sandwich structures made its first flight on 15 February 1986 [2] (Figures 12.32 and 12.33).

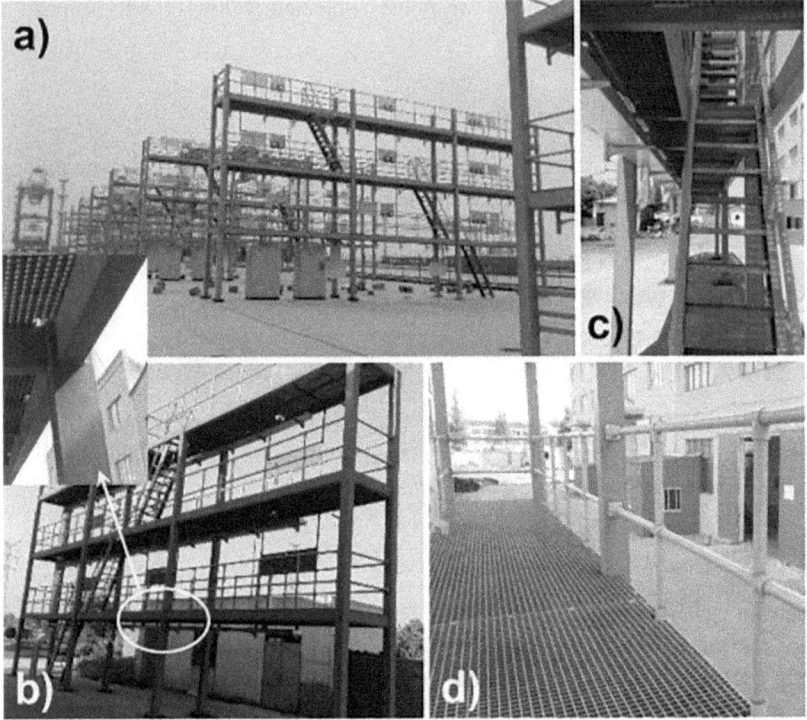

FIGURE 12.30 Supporting frames for cooling containers at Yangshan Port: (a) and (b) overall view, (c) FRP stair and (d) FRP grid deck and handrail [119].

FIGURE 12.31 An application of floating composite bumper system in Fuzhou Wulong River Bridge with (a) segments and completed installation and (b) an overview with FRP guardrails [119].

12.11.4 Railway

Composite sandwich structures made by either pultrusion or filament winding are widely used in the railway rolling stock or freight wagons. The interior panel of the vehicle or the luggage compartments can be made from these structures. The flooring of rail cars or the partitions in sleeper use the sandwich structures. The structures with thermal insulation characteristics can be used in diesel engine or the cable supports [120] (Figures 12.34 and 12.35).

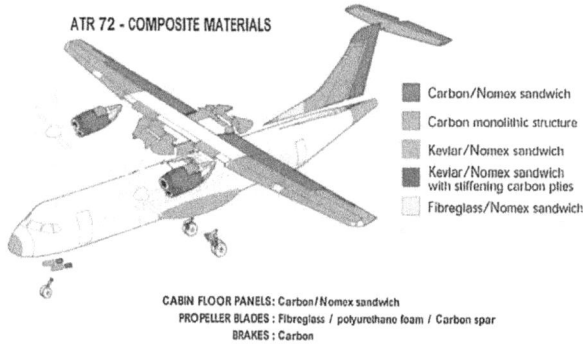

Fig. 19. ATR 72 composite materials.

ATR 72 - COMPOSITE MATERIALS

Carbon/Nomex sandwich
Carbon monolithic structure
Kevlar/Nomex sandwich
Kevlar/Nomex sandwich with stiffening carbon plies
Fibreglass/Nomex sandwich

CABIN FLOOR PANELS: Carbon/Nomex sandwich
PROPELLER BLADES : Fibreglass / polyurethane foam / Carbon spar
BRAKES : Carbon

FIGURE 12.32   Sandwich structures in aerospace (open access).

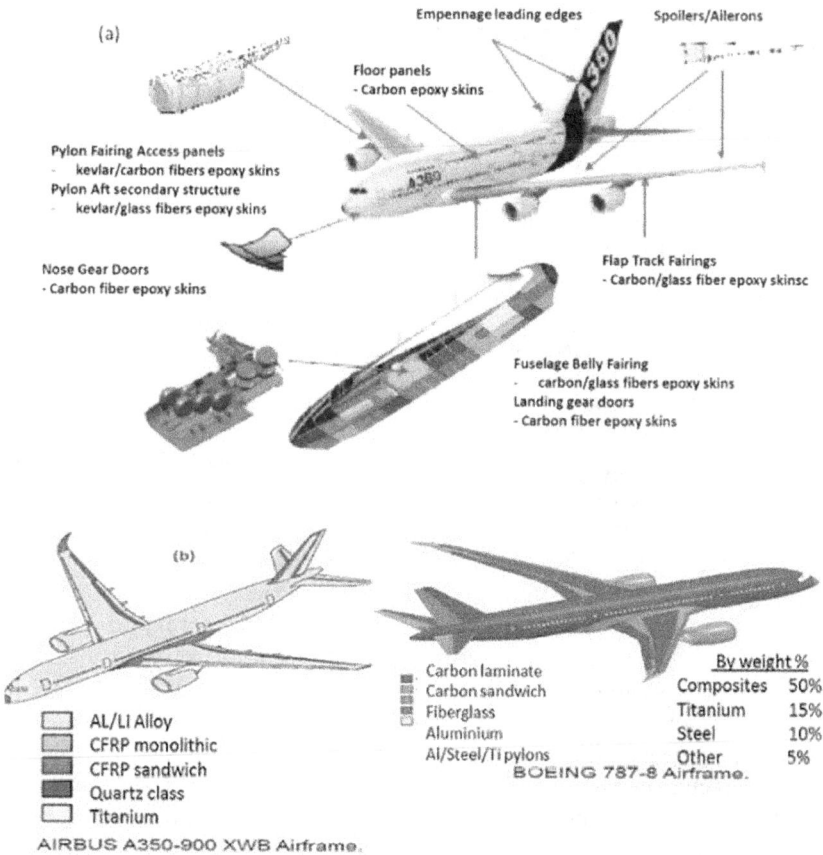

(a)

Empennage leading edges   Spoilers/Ailerons

Floor panels
- Carbon epoxy skins

Pylon Fairing Access panels
-   kevlar/carbon fibers epoxy skins
Pylon Aft secondary structure
-   kevlar/glass fibers epoxy skins

Nose Gear Doors
- Carbon fiber epoxy skins

Flap Track Fairings
- Carbon/glass fiber epoxy skinsc

Fuselage Belly Fairing
-   carbon/glass fibers epoxy skins
Landing gear doors
- Carbon fiber epoxy skins

(b)

Carbon laminate
Carbon sandwich
Fiberglass
Aluminium
Al/Steel/Ti pylons
BOEING 787-8 Airframe.

By weight %
Composites   50%
Titanium   15%
Steel   10%
Other   5%

AL/Li Alloy
CFRP monolithic
CFRP sandwich
Quartz class
Titanium

AIRBUS A350-900 XWB Airframe.

FIGURE 12.33   Sandwich and composites in aircraft (open access).

FIGURE 12.34 Latest examples from Japan Railways showing the complex 3D designs and curves of front nose section of high-speed trains: Alfa-X (left) and the newest series N700 [121].

(a)   (b)

FIGURE 12.35 Sandwich application in railway side body (a) vehicle body and (b) roof made of composite sandwich panels [122].

12.12 SUMMARY AND OUTLOOK

Composite sandwich panels with the textile-reinforced structure are lightweight materials with high specific stiffness and strength, finding direct applications in structural components of various automobiles and civil constructions. For a specific purpose, these can be manufactured by optimizing processing and changing materials such as fibre, resin and core material. The structural parameters of sandwich panel, such as the thickness of the face sheet and core or their ratio, have been reported to affect the structure's mechanical properties. The latest research on sandwich structures uses 3D woven structures in the core with enhanced properties. A proposed research aims to produce a smart honeycomb sandwich structure using small-diameter optical fibre sensors to repair typical damage in aerospace applications [123]. Further research can be conducted in this domain with varying geometrical parameters and the innovative materials. In addition, auxetic structure or nano-fillers can be incorporated in the research for enhanced results. Eventually, the modelling of impact properties has limited work and can prove to be a promising field for future research.

REFERENCES

1. T. S. Lok and Q. H. Cheng, "Free vibration of clamped orthotropic sandwich panel," *J. Sound Vib.*, 2000, doi: 10.1006/jsvi.1999.2485.
2. B. Castanie, C. Bouvet, and M. Ginot, "Review of composite sandwich structure in aeronautic applications," *Compos. Part C Open Access*, vol. 1, no. June, p. 100004, 2020, doi: 10.1016/j.jcomc.2020.100004.
3. A. Uzal, F. O. Sonmez, F. E. Oz, K. Cinar, and N. Ersoy, "A composite sandwich plate with a novel core design," *Compos. Struct.*, vol. 193, no. March, pp. 198–211, 2018, doi: 10.1016/j.compstruct.2018.03.047.
4. W. X. Yuan and D. J. Dawe, "Free vibration of sandwich plates with laminated faces," *Int. J. Numer. Methods Eng.*, vol. 54, no. 2, pp. 195–217, 2002, doi: 10.1002/nme.411.
5. K. R. Ramakrishnan, S. Guérard, P. Viot, and K. Shankar, "Effect of block copolymer nano-reinforcements on the low velocity impact response of sandwich structures," *Compos. Struct.*, vol. 110, no. 1, pp. 174–182, 2014, doi: 10.1016/j.compstruct.2013.12.001.
6. G. B. Chai and S. Zhu, "A review of low-velocity impact on sandwich structures," *Proc. Inst. Mech. Eng. Part L J. Mater. Des. Appl.*, vol. 225, no. 4, pp. 207–230, 2011, doi: 10.1177/1464420711409985.
7. S. Agrawal, K. K. Singh, and P. Sarkar, "Impact damage on fibre-reinforced polymer matrix composite – A review," *J. Compos. Mater.*, vol. 48, no. 3, pp. 317–332, Feb. 2014, doi: 10.1177/0021998312472217.
8. P. O. Sjoblom, J. T. Hartness, and T. M. Cordell, "On low-velocity impact testing of composite materials," *J. Compos. Mater.*, vol. 22, no. 1, pp. 30–52, 1988, doi: 10.1177/002199838802200103.
9. K. N. Shivakumar, W. Elber, and W. Illg, "Prediction of low-velocity impact damage in thin circular laminates," *AIAA J.*, vol. 23, no. 3, pp. 442–449, 1985, doi: 10.2514/3.8933.
10. P. Robinson, and G. A.-O. Davies, "Impactor mass and specimen geometry effects in low velocity impact of laminated composites," *Int. J. impact Eng.*, vol. 12, no. 2, pp.189–207, 1992.
11. S. P. Joshi and C. T. Sun, "Impact induced fracture in a laminated composite," *J. Compos. Mater.*, vol. 19, no. 1, pp. 51–66, 1985, doi: 10.1177/002199838501900104.
12. K. N. Shivakumar, W. Elber, and W. Illg, "Prediction of impact force and duration due to low-velocity impact on circular composite laminates," *J. Appl. Mech. Trans. ASME*, vol. 52, no. 3, pp. 674–680, Sep. 1985, doi: 10.1115/1.3169120.
13. S. Abrate, Impact on composite structures. Cambridge University Press, 2005.
14. A. K. Jha, "Free vibration analysis of sandwich panel," 2007, [Online]. Available: http://ethesis.nitrkl.ac.in/4320/.
15. M. Hao, Y. Hu, B. Wang, and S. Liu, "Mechanical behavior of natural fiber-based isogrid lattice cylinder," *Compos. Struct.*, vol. 176, pp. 117–123, 2017, doi: 10.1016/j.compstruct.2017.05.028.
16. M. Jin, Y. Hu, and B. Wang, "Compressive and bending behaviours of wood-based two-dimensional lattice truss core sandwich structures," *Compos. Struct.*, vol. 124, pp. 337–344, 2015, doi: 10.1016/j.compstruct.2015.01.033.
17. E. Labans, K. Kalniņs, and C. Bisagni, "Flexural behavior of sandwich panels with cellular wood, plywood stiffener/foam and thermoplastic composite core," *J. Sandw. Struct. Mater.*, vol. 21, no. 2, pp. 784–805, 2019, doi: 10.1177/1099636217699587.

18. J. Susainathan, F. Eyma, E. De Luycker, A. Cantarel, and B. Castanie, "Experimental investigation of impact behavior of wood-based sandwich structures," *Compos. Part A Appl. Sci. Manuf.*, vol. 109, no. January, pp. 10–19, 2018, doi: 10.1016/j. compositesa.2018.02.029.

19. A. S. Herrmann, P. C. Zahlen, and I. Zuardy, "Sandwich structures technology in commercail aviation: present applications and future trends," *Sandw. Struct. Adv. with Sandw. Struct. Mater.*, vol. 7, pp. 13–26, 2005, doi: 10.1007/1-4020-3848-8_2.

20. T. Bitzer, *Honeycomb Technology: Materials, Design, Manufacturing, Applications And Testing.* 1997.

21. H. N. G. Wadley, N. A. Fleck, and A. G. Evans, "Fabrication and structural performance of periodic cellular metal sandwich structures," *Compos. Sci. Technol.*, 2003, doi: 10.1016/S0266–3538(03)00266-5.

22. A. G. Evans, J. W. Hutchinson, N. A. Fleck, M. F. Ashby, and H. N. G. Wadley, "The topological design of multifunctional cellular metals," *Progress in Materials Science.* 2001, doi: 10.1016/S0079-6425(00)00016-5.

23. F. W. Zok, S. A. Waltner, Z. Wei, H. J. Rathbun, R. M. McMeeking, and A. G. Evans, "A protocol for characterizing the structural performance of metallic sandwich panels: Application to pyramidal truss cores," *Int. J. Solids Struct.*, 2004, doi: 10.1016/j.ijsolstr.2004.05.045.

24. R. Biagi and H. Bart-Smith, "Imperfection sensitivity of pyramidal core sandwich structures," *Int. J. Solids Struct.*, 2007, doi: 10.1016/j.ijsolstr.2006.11.049.

25. G. W. Kooistra, V. S. Deshpande, and H. N. G. Wadley, "Compressive behavior of age hardenable tetrahedral lattice truss structures made from aluminium," *Acta Mater.*, 2004, doi: 10.1016/j.actamat.2004.05.039.

26. V. S. Deshpande, N. A. Fleck, and M. F. Ashby, "Effective properties of the octet-truss lattice material," *J. Mech. Phys. Solids*, 2001, doi: 10.1016/S0022-5096(01)00010-2.

27. H. J. Rathbun et al., "Structural performance of metallic sandwich beams with hollow truss cores," *Acta Mater.*, 2006, doi: 10.1016/j.actamat.2006.07.016.

28. D. T. Queheillalt and H. N. G. Wadley, "Pyramidal lattice truss structures with hollow trusses," *Mater. Sci. Eng. A*, 2005, doi: 10.1016/j.msea.2005.02.048.

29. D. T. Queheillalt and H. N. G. Wadley, "Cellular metal lattices with hollow trusses," *Acta Mater.*, 2005, doi: 10.1016/j.actamat.2004.09.024.

30. J. Wang, A. G. Evans, K. Dharmasena, and H. N. G. Wadley, "On the performance of truss panels with Kagomé cores," *Int. J. Solids Struct.*, 2003, doi: 10.1016/ S0020-7683(03)00349-4.

31. S. Yin, L. Wu, L. Ma, and S. Nutt, "Pyramidal lattice sandwich structures with hollow composite trusses," *Compos. Struct.*, 2011, doi: 10.1016/j.compstruct.2011.06.025.

32. L. Valdevit, J. W. Hutchinson, and A. G. Evans, "Structurally optimized sandwich panels with prismatic cores," *Int. J. Solids Struct.*, 2004, doi: 10.1016/j.ijsolstr.2004.04.027.

33. X. Gong, Investigation of different geometric structure parameter for honeycomb textile composites on their mechanical performance. Dissertation, The University of Manchester, United Kingdom, 2011.

34. L. J. Gibson and M. F. Ashby, *Cellular Solids: Structure and Properties*, 2d ed., pp. 175–231, Cambridge: Press Syndicate of the University of Cambridge, 1997.

35. X. Chen, Y. Sun, and X. Gong, "Design, Manufacture, and Experimental Analysis of 3D Honeycomb Textile Composites Part I: Design and Manufacture," *Text. Res. J.*, vol. 78, no. 9, pp. 771–781, 2008, doi: 10.1177/0040517507087855.

36. D. sen Li, C. qi Zhao, L. Jiang, and N. Jiang, "Experimental study on the bending properties and failure mechanism of 3D integrated woven spacer composites at room and cryogenic temperature," *Compos. Struct.*, 2014, doi: 10.1016/j. compstruct.2013.12.026.

37. R. N. Manjunath, V. Khatkar, and B. K. Behera, "Influence of augmented tuning of core architecture in 3D woven sandwich structures on flexural and compression properties of their composites," *Adv. Compos. Mater.*, vol. 29, no. 4, pp. 317–333, 2020, doi: 10.1080/09243046.2019.1680925.

38. M. Li, S. Wang, Z. Zhang, and B. Wu, "Effect of structure on the mechanical behaviors of three-dimensional spacer fabric composites," *Appl. Compos. Mater.*, vol. 16, no. 1, pp. 1–14, 2009, doi: 10.1007/s10443-008-9072-4.

39. Y. Hou, R. Neville, F. Scarpa, C. Remillat, B. Gu, and M. Ruzzene, "Graded conventional-auxetic Kirigami sandwich structures: Flatwise compression and edgewise loading," *Compos. Part B Eng.*, vol. 59, pp. 33–42, 2014, doi: 10.1016/j.compositesb.2013.10.084.

40. V. Birman and G. A. Kardomateas, "Review of current trends in research and applications of sandwich structures," *Compos. Part B Eng.*, vol. 142, no. November 2017, pp. 221–240, 2018, doi: 10.1016/j.compositesb.2018.01.027.

41. D. Wang, "Impact behavior and energy absorption of paper honeycomb sandwich panels," *Int. J. Impact Eng.*, vol. 36, no. 1, pp. 110–114, 2009, doi: 10.1016/j.ijimpeng.2008.03.002.

42. J. E. Shafizadeh, J. C. Seferis, E. F. Chesmar, B. A. Frye, and R. Geyer, "Evaluation of the mechanisms of water migration through honeycomb core," *J. Mater. Sci.*, 2003, doi: 10.1023/A:1023985925293.

43. S. Heimbs, J. Cichosz, M. Klaus, S. Kilchert, and A. F. Johnson, "Sandwich structures with textile-reinforced composite foldcores under impact loads," *Compos. Struct.*, 2010, doi: 10.1016/j.compstruct.2009.11.001.

44. I. M. Zakirov and K. A. Alexeev, "New folded structures for sandwich panels," In *Proceedings of SAMPE Conference*, 1–11, 2006 Apr 30.

45. T. J. Lu, L. Valdevit, and A. G. Evans, "Active cooling by metallic sandwich structures with periodic cores," *Progress in Materials Science*. 2005, doi: 10.1016/j.pmatsci.2005.03.001.

46. K. P. Dharmasena, H. N. G. Wadley, K. Williams, Z. Xue, and J. W. Hutchinson, "Response of metallic pyramidal lattice core sandwich panels to high intensity impulsive loading in air," *Int. J. Impact Eng.*, 2011, doi: 10.1016/j.ijimpeng.2010.10.002.

47. K. P. Dharmasena, H. N. G. Wadley, Z. Xue, and J. W. Hutchinson, "Mechanical response of metallic honeycomb sandwich panel structures to high-intensity dynamic loading," *Int. J. Impact Eng.*, 2008, doi: 10.1016/j.ijimpeng.2007.06.008.

48. Z. Wei et al., "The resistance of metallic plates to localized impulse," *J. Mech. Phys. Solids*, 2008, doi: 10.1016/j.jmps.2007.10.010.

49. J. Xiong, A. Vaziri, L. Ma, J. Papadopoulos, and L. Wu, "Compression and impact testing of two-layer composite pyramidal-core sandwich panels," *Compos. Struct.*, 2012, doi: 10.1016/j.compstruct.2011.09.018.

50. W. Lestari and P. Qiao, "Damage detection of fiber-reinforced polymer honeycomb sandwich beams," *Compos. Struct.*, 2005, doi: 10.1016/j.compstruct.2004.01.023.

51. M. Ashby, et al., "Metal Foams: A Design Guide," *Appl. Mech. Rev.*, 2001, doi: 10.1115/1.1421119.

52. N. J. Mills, *Polymer Foams Handbook: Engineering and Biomechanics Applications and Design Guide*. Elsevier, 2007.

53. H. N. G. Wadley, "Multifunctional periodic cellular metals," *Philos. Trans. R. Soc. A Math. Phys. Eng. Sci.*, 2006, doi: 10.1098/rsta.2005.1697.

54. N. a Fleck, "An overview of the mechanical properties of foams and periodic lattice materials," *Cell. Met. Polym.*, vol. 2004, pp. 3–7, 2004.

55. V. S. Deshpande and N. A. Fleck, "Collapse of truss core sandwich beams in 3-point bending," *Int. J. Solids Struct.*, 2001, doi: 10.1016/S0020-7683(01)00103-2.

56. B. Wang, G. Zhang, Q. He, L. Ma, L. Wu, and J. Feng, "Mechanical behavior of carbon fiber reinforced polymer composite sandwich panels with 2-D lattice truss cores," *Mater. Des.*, 2014, doi: 10.1016/j.matdes.2013.10.025.

57. J. Romanoff and P. Varsta, "Bending response of web-core sandwich beams," *Compos. Struct.*, 2006, doi: 10.1016/j.compstruct.2005.02.018.

58. F. Jin, H. Chen, L. Zhao, H. Fan, C. Cai, and N. Kuang, "Failure mechanisms of sandwich composites with orthotropic integrated woven corrugated cores: Experiments," *Compos. Struct.*, 2013, doi: 10.1016/j.compstruct.2012.09.056.

59. Y. C. Shiah, L. Tseng, J. C. Hsu, and J. H. Huang, "Experimental characterization of an integrated sandwich composite using 3D woven fabrics as the core material," *J. Thermoplast. Compos. Mater.*, vol. 17, no. 3, pp. 229–243, 2004, doi: 10.1177/0892705704035410.

60. A. F. Johnson, G. D. Sims, and F. Ajibade, "Performance analysis of web-core composite sandwich panels," *Composites*, 1990, doi: 10.1016/0010-4361(90)90346-X.

61. J. Grünewald, P. Parlevliet, and V. Altstädt, "Manufacturing of thermoplastic composite sandwich structures: A review of literature," *J. Thermoplast. Compos. Mater.*, vol. 30, no. 4, pp. 437–464, 2017, doi: 10.1177/0892705715604681.

62. D. Grewell and A. Benatar, "Welding of plastics: Fundamentals and new developments," 2007, doi: 10.3139/217.0051.

63. T. Renault, "Sandwiform®: Thermoplastic composite sandwich structures for automotive applications," Tech. Pap. - Renault, T. (*2002*). Sandwiform® Thermoplast. Compos. Sandw. Struct. Automot. Appl. Tech. Pap. - Soc. Manuf. Eng. EM.Society Manuf. Eng. EM, 2002.

64. U. Breuer, M. Ostgathe, and M. Neitzel, "Manufacturing of all-thermoplastic sandwich systems by a one-step forming technique," *Polym. Compos.*, vol. 19, no. 3, pp. 275–279, 1998, doi: 10.1002/pc.10100.

65. O. Rozant, P. E. Bourban, and J. A. E. Månson, "Manufacturing of three dimensional sandwich parts by direct thermoforming," *Compos. - Part A Appl. Sci. Manuf.*, 2001, doi: 10.1016/S1359-835X(00)00184-6.

66. M. Ali, S. Joshi, and M.T.H. Sultan, undefined, "Palliatives for low velocity impact damage in composite laminates," hindawi.com, 2017.

67. P. Kumar, Rai, B., "Delaminations of barely visible impact damage in CFRP laminates," *Compos. Struct.*, vol. 23, no. 4, pp. 313–318, 1993.

68. P.W. Manders, "A parametric study of composite performance in compression-after-impact testing," *SAMPE J*, vol. 22, pp. 47–51, 1986.

69. A. Rotem, "Residual flexural strength of FRP composite specimens subjected to transverse impact loading," *SAMPE J*, 1988, ci.nii.ac.jp.

70. S. Petit, C. Bouvet, A. Bergerot, and J. J. Barrau, "Impact and compression after impact experimental study of a composite laminate with a cork thermal shield," *Compos. Sci. Technol.*, vol. 67, no. 15–16, pp. 3286–3299, Dec. 2007, doi: 10.1016/j.compscitech.2007.03.032.

71. D. Liu and L. E. Malvern, "Matrix cracking in impacted glass/epoxy plates," *J. Compos. Mater.*, vol. 21, no. 7, pp. 594–609, 1987, doi: 10.1177/002199838702100701.

72. "Impact and after-impact properties of carbon fibre reinforced composites enhanced with multi-wall carbon nanotubes," *Elsevier*.

73. N. Jones, undefined. "Energy-absorbing effectiveness factor," *Int J Impact Eng*, vol. 37, no. 6, pp. 754–765, 2010 *Elsevier*.

74. P. Feraboli and K. T. Kedward, "A new composite structure impact performance assessment program," *Compos. Sci. Technol.*, vol. 66, no. 10, pp. 1336–1347, Aug. 2006, doi: 10.1016/j.compscitech.2005.09.009.

75. C. C. Foo, G. B. Chai, and L. K. Seah, "A model to predict low-velocity impact response and damage in sandwich composites," *Compos. Sci. Technol.*, vol. 68, no. 6, pp. 1348–1356, May 2008, doi: 10.1016/j.compscitech.2007.12.007.
76. M. Elamin, B. Li, and K. T. Tan, "Compression after impact performance of carbon-fiber foam-core sandwich composites in low temperature arctic conditions," *Compos. Struct.*, vol. 261, p. 113568, Apr. 2021, doi: 10.1016/j.compstruct.2021.113568.
77. D. W. Zhou and W. J. Stronge, "Low velocity impact denting of HSSA lightweight sandwich panel," *Int. J. Mech. Sci.*, vol. 48, no. 10, pp. 1031–1045, Oct. 2006, doi: 10.1016/j.ijmecsci.2006.05.011.
78. R. Olsson, "Mass criterion for wave controlled impact response of composite plates," *Compos. Part A Appl. Sci. Manuf.*, vol. 31, no. 8, pp. 879–887, Aug. 2000, doi: 10.1016/S1359–835X(00)00020-8.
79. J. A. Artero-Guerrero, J. Pernas-Sánchez, J. López-Puente, and D. Varas, "Experimental study of the impactor mass effect on the low velocity impact of carbon/epoxy woven laminates," *Compos. Struct.*, vol. 133, pp. 774–781, Dec. 2015, doi: 10.1016/j.compstruct.2015.08.027.
80. G. L. Farley and R. M. Jones, "Crushing characteristics of continuous fiber-reinforced composite tubes," *J. Compos. Mater.*, vol. 26, no. 1, pp. 37–50, 1992, doi: 10.1177/002199839202600103.
81. A. H. Fairfull, undefined, "Energy absorption of polymer matrix composite structures: frictional effects," *Struct. Failure* ci.nii.ac.jp, 1988.
82. M. Meo, R. Vignjevic, and G. Marengo, "The response of honeycomb sandwich panels under low-velocity impact loading," *Int. J. Mech. Sci.*, vol. 47, no. 9, pp. 1301–1325, Sep. 2005, doi: 10.1016/j.ijmecsci.2005.05.006.
83. R. Olsson, "Engineering method for prediction of impact response and damage in sandwich panels," *J. Sandw. Struct. Mater.*, vol. 4, no. 1, pp. 3–29, 2002, doi: 10.1177/1099636202004001192.
84. G. Zhou, M. Hill, and N. Hookham, *Investigation of Parameters Governing the Damage and Energy Absorption Characteristics of Honeycomb Sandwich Panels*, vol. 9, no. 4. SAGE Publications Ltd, 2007.
85. S. Sayahlatifi, G. H. Rahimi, and A. Bokaei, "The quasi-static behavior of hybrid corrugated composite/balsa core sandwich structures in four-point bending: Experimental study and numerical simulation," *Eng. Struct.*, vol. 210, p. 110361, May 2020, doi: 10.1016/j.engstruct.2020.110361.
86. "Review of Damage Tolerance for Composite Sandwich Airframe Structures," Wichita State University, Kansas, 1999.
87. S. Zhu and G. B. Chai, "Damage and failure mode maps of composite sandwich panel subjected to quasi-static indentation and low velocity impact," *Compos. Struct.*, vol. 101, pp. 204–214, Jul. 2013, doi: 10.1016/j.compstruct.2013.02.010.
88. E. J. Herup and A. N. Palazotto, "Low-velocity impact damage initiation in graphite/epoxy/Nomex honeycomb-sandwich plates," *Compos. Sci. Technol.*, vol. 57, no. 12, pp. 1581–1598, Jan. 1998, doi: 10.1016/S0266–3538(97)00089-4.
89. J. Dean, C. Dunleavy, P.M. Brown, and T.W. Clyne, undefined, "Energy absorption during projectile perforation of thin steel plates and the kinetic energy of ejected fragments," *Inter. J. Impact Eng.*, vol. 36, no. 10–11, pp. 1250–1258, 2009, Elsevier.
90. V. Skvortsov, J. Kepler, and, E. Bozhevolnaya, undefined, "Energy partition for ballistic penetration of sandwich panels," *Int. J. Impact Eng.*, vol. 28, no. 7, pp. 697–716, 2003, Elsevier.
91. L. Sun, R.F. Gibson, F. Gordaninejad, and, J. Suhr, undefined, "Energy absorption capability of nanocomposites: a review,", *Compos. Sci. Technol.*, Vol. 69, no. 14, pp. 2392–2409, 2009, Elsevier.

92. J. Hu et al., "Novel panel-core connection process and impact behaviors of CF/ PEEK thermoplastic composite sandwich structures with truss cores," *Compos. Struct.*, vol. 251, no. June, p. 112659, 2020, doi: 10.1016/j.compstruct.2020.112659.

93. A. Kumar, S. Angra, and A. K. Chanda, "Analysis of the effects of varying core thicknesses of Kevlar Honeycomb sandwich structures under different regimes of testing," *Mater. Today Proc.*, vol. 21, pp. 1615–1623, 2020, doi: 10.1016/j. matpr.2019.11.242.

94. O. Ozdemir, R. Karakuzu, and A. K. J. Al-Shamary, "Core-thickness effect on the impact response of sandwich composites with poly(vinyl chloride) and poly(ethylene terephthalate) foam cores," *J. Compos. Mater.*, vol. 49, no. 11, pp. 1315–1329, 2015, doi: 10.1177/0021998314533597.

95. O. Ozdemir, N. Oztoprak, and H. Kandas, "Single and repeated impact behaviors of bio-sandwich structures consisting of thermoplastic face sheets and different balsa core thicknesses," *Compos. Part B Eng.*, vol. 149, no. May, pp. 49–57, 2018, doi: 10.1016/j.compositesb.2018.05.016.

96. M. Hildebrand, "Improving the impact strength of FRP-sandwich panels for ship applications," 1996, Accessed: Jun. 04, 2021. [Online]. Available: https://cris.vtt.fi/en/publications/improving-the-impact-strength-of-frp-sandwich-panels-for-ship-app-2.

97. J. Manning, A. Crosky, and S. Bandyopadhyay, "Flexural and impact properties of sandwich panels used in surfboard construction," 1993, Accessed: Jun. 04, 2021. [Online]. Available: https://www.osti.gov/biblio/143174.

98. "Impact properties of different shell structures in... - Google Scholar." https://scholar.google.com/scholar?hl=en&as_sdt=0%2C5&q=Impact+properties+of+different+shell+structures+in+relation+to+rule+requirement&btnG= (accessed Jun. 04, 2021).

99. "Impact testing of high technology laminates and structure - Google Scholar." https://scholar.google.com/scholar?hl=en&as_sdt=0%2C5&q=Impact+testing+of+high+technology+laminates+and+structure&oq=Impact+testing+of+high+technology+laminates+and+stru (accessed Jun. 04, 2021).

100. M. H. Arvidson and P. H. Miller, "Hull material evaluation for Navy 44 sail training craft," *Nav. Eng. J.*, vol. 113, no. 2, pp. 71–78, 2001, doi: 10.1111/j.1559–3584.2001. tb00036.x.

101. I. M. Daniel, J. L. Abot, P. M. Schubel, and J. J. Luo, "Response and Damage Tolerance of Composite Sandwich Structures under Low Velocity Impact," *Exp. Mech.*, vol. 52, no. 1, pp. 37–47, Jan. 2012, doi: 10.1007/s11340-011-9479-y.

102. K. S. Raju and J. S. Tomblin, "Energy absorption characteristics of stitched composite sandwich panels," *J. Compos. Mater.*, vol. 33, no. 8, pp. 712–728, 1999, doi: 10.1177/002199839903300804.

103. F. Choon Chiang, "Energy absorption characteristics of sandwich structures subjected to low-velocity impact foo Choon Chiang school of mechanical and aerospace engineering 2009 energy absorption characteristics of sandwich structures subjected to low-velocity impact."

104. Z. Asaee, S. Shadlou, and F. Taheri, "Low-velocity impact response of fiberglass/ magnesium FMLs with a new 3D fiberglass fabric," *Compos. Struct.*, vol. 122, pp. 155–165, 2015, doi: 10.1016/j.compstruct.2014.11.038.

105. A. Reyes and T. Børvik, "Low velocity impact on crash components with steel skins and polymer foam cores," *Int. J. Impact Eng.*, vol. 132, no. 7491, p. 103297, 2019, doi: 10.1016/j.ijimpeng.2019.05.011.

106. J. S. Yang et al., "Low velocity impact behavior of carbon fibre composite curved corrugated sandwich shells," *Compos. Struct.*, vol. 238, no. January, pp. 1–16, 2020, doi: 10.1016/j.compstruct.2020.112027.

107. S. Dhar Malingam, F. A. Jumaat, L. F. Ng, K. Subramaniam, and A. F. Ab Ghani, "Tensile and impact properties of cost-effective hybrid fiber metal laminate sandwich structures," *Adv. Polym. Technol.*, vol. 37, no. 7, pp. 2385–2393, 2018, doi: 10.1002/adv.21913.

108. D. De Cicco and F. Taheri, "Performances of magnesium- and steel-based 3D fiber-metal laminates under various loading conditions," *Compos. Struct.*, vol. 229, no. May, p. 111390, 2019, doi: 10.1016/j.compstruct.2019.111390.

109. L. Zhou, J. Zeng, L. Jiang, and H. Hu, "Low-velocity impact properties of 3D auxetic textile composite," *J. Mater. Sci.*, vol. 53, no. 5, pp. 3899–3914, 2018, doi: 10.1007/s10853-017-1789-8.

110. T. Dastan, A. Safian, and M. Sheikhzadeh, "The crashworthiness performance of integrally woven sandwich composite panels made using natural and glass fibers," *J. Compos. Mater.*, 2020, doi: 10.1177/0021998320964847.

111. A. Mirdehghan, H. Nosraty, M. M. Shokrieh, and M. Akhbari, "Manufacturing and drop-weight impact properties of three-dimensional integrated-woven sandwich composite panels with hybrid core," *J. Ind. Text.*, no. 424, pp. 1–29, 2020, doi: 10.1177/1528083719896764.

112. R. Ouadday, A. Marouene, G. Morada, A. Kaabi, R. Boukhili, and A. Vadean, "Experimental and numerical investigation on the impact behavior of dual-core composite sandwich panels designed for hydraulic turbine applications," *Compos. Struct.*, vol. 185, no. December 2016, pp. 254–263, 2018, doi: 10.1016/j.compstruct.2017.11.007.

113. H. Safari, M. Karevan, and H. Nahvi, "Mechanical characterization of natural nano-structured zeolite/polyurethane filled 3D woven glass fiber composite sandwich panels," *Polym. Test.*, vol. 67, no. March, pp. 284–294, 2018, doi: 10.1016/j.polymertesting.2018.03.018.

114. S. Abrate, "Modeling of impacts on composite structures," *Compos. Struct.*, vol. 51, no. 2, pp. 129–138, Feb. 2001, doi: 10.1016/S0263-8223(00)00138-0.

115. R. Olsson, "Analytical prediction of large mass impact damage in composite laminates," *Compos. - Part A Appl. Sci. Manuf.*, vol. 32, no. 9, pp. 1207–1215, Sep. 2001, doi: 10.1016/S1359-835X(01)00073-2.

116. M. K. Chan, *UC San Diego UC San Diego Electronic Theses and Dissertations Title Low Velocity Blunt Impact on Lightweight Composite Sandwich Panels*. San Diego: University of California, 2014.

117. C. Menna, A. Zinno, D. Asprone, and A. Prota, "Numerical assessment of the impact behavior of honeycomb sandwich structures," *Compos. Struct.*, vol. 106, pp. 326–339, Dec. 2013, doi: 10.1016/j.compstruct.2013.06.010.

118. A. Manalo, T. Aravinthan, A. Fam, M. Asce, and B. Benmokrane, "State-of-the-Art Review on FRP Sandwich Systems for Lightweight Civil Infrastructure," vol. 21, no. 1, pp. 1–16, 2017, doi: 10.1061/(ASCE)CC.1943-5614.0000729.

119. H. Fang, Y. Bai, W. Liu, Y. Qi, and J. Wang, "Connections and structural applications of fibre reinforced polymer composites for civil infrastructure in aggressive environments," *Composites Part B: Engineering*, vol. 164. Elsevier Ltd, pp. 129–143, May 01, 2019, doi: 10.1016/j.compositesb.2018.11.047.

120. M. Robinson, E. Matsika, and Q. Peng, "Application of composites in rail vehicles," *ICCM Int. Conf. Compos. Mater.*, vol. 2017-August, no. August, pp. 20–25, 2017, doi: 10.1016/b978-0-12-803581-8.03965-5.

121. A. Önder and M. Robinson, "Investigating the feasibility of a new testing method for GFRP/polymer foam sandwich composites used in railway passenger vehicles," *Compos. Struct.*, vol. 233, no. August 2019, 2020, doi: 10.1016/j.compstruct.2019.111576.

122. A. Zinno, E. Fusco, A. Prota, and G. Manfredi, "Multiscale approach for the design of composite sandwich structures for train application," *Compos. Struct.*, vol. 92, no. 9, pp. 2208–2219, Aug. 2010, doi: 10.1016/j.compstruct.2009.08.044.

123. N. Takeda, S. Minakuchi, and Y. Okabe, "Smart Composite Sandwich Structures for Future Aerospace Application -Damage Detection and Suppression-: a Review," *J. Solid Mech. Mater. Eng.*, vol. 1, no. 1, pp. 3–17, 2007, doi: 10.1299/jmmp.1.3.

13 Drilling and Repair of the Composite Sandwich Panels

Shashikant Verma
Bundelkhand University

Lalit Ranakoti
National Institute of Technology

Brijesh Gangil and Manoj Kumar Gupta
H.N.B. Garhwal University

CONTENTS

13.1 Introduction 262
13.2 Input Parameters 263
    13.2.1 Operating Variables 263
    13.2.2 Drill Point Geometry 264
    13.2.3 Drill Material 264
13.3 Output Characteristics 265
    13.3.1 Drilling Force 265
    13.3.2 Drilling-Induced Damage 265
13.4 Damage 265
    13.4.1 Damage Sources 266
        13.4.1.1 Processing Irregularities 266
        13.4.1.2 Environmental Damages 266
        13.4.1.3 In-Service Damages 266
    13.4.2 Damage Types 267
        13.4.2.1 Combinations of Damages 267
        13.4.2.2 Fiber Breakage 267
        13.4.2.3 Cracks 267
        13.4.2.4 Dents 267
        13.4.2.5 Punctures 267
    13.4.3 Impact Damage 267
    13.4.4 Sandwich Damage 268

DOI: 10.1201/9781003143031-13

13.5 Repair..268
    13.5.1 Resin Infusion or Injection Repair 269
    13.5.2 Chopped Fiber 269
    13.5.3 Plug Repair 269
    13.5.4 Scarf Repair Patch 270
    13.5.5 External Patch Repair...270
    13.5.6 Structural Mechanically Fastened Repair 271
    13.5.7 Typical Repair Procedure 271
13.6 Self-Healing of Polymer Materials..272
13.7 Conclusions...272
References..273

13.1 INTRODUCTION

Sandwich panels are examples of sandwich structure composites, consisting of two thin laminates, lighter outer skin, and thick core structure. These materials having less weight prove to have excellent mechanical properties than traditional materials. Sandwich panels are suitable for light and substantial work that eventually increases material application area [1–3]. Their design is so flexible that facilitates them to achieve a complicated structure. Achievement of lightweight materials is a crucial aspect of modern engineering. Such a property of the material directly enhances the efficiency of carrying out operations. To obtain significant stiffness in sandwich panels at a lower density is the major purpose of their manufacturing. Sandwich structures are light materials in weight and possess a high strength-to-weight ratio, which is essential in considering the material's stiffness. These sandwich panels are designed so that the outer surfaces withstand bending and compressive loads, while the core withstands shearing loads. On that basis, the macroscopic structure of sandwich panels can be compared to that of an I-beam [4–6]. The sandwich composite panels are anisotropic, which means the change in strength depends on the loads applied. Anisotropy literature survey enables the development of composite materials that exhibit specific properties, according to requirements, in the desired direction. In materials, stiffness, strength, and specific volume, along with their thermal insulating power, acoustic resistance, ability to absorb energy, and hydrostatic weighing, are having major significance [7–10]. The sandwich panels are widely used in applications in which materials having stiffness properties are required. F. C. Campbell (2010) observed that doubling the core thickness increased the material's stiffness more than seven times with a rise in product weight by 4%, and increased the flexural strength [7].

In the later stage of parts production of any application, machine operations play a vital role. Drilling is the most widely used machining operation in composite structures to join parts by bolts, rivets, and screws. After machining machined surface due to reinforcement tool wear, surface appearance and delamination are major constituents [11]. Lystrup and Vorm [12] stated that the cutting process reinforced plastics in directions perpendicular and parallel to the fiber orientation, which produced a series of fractures and created chips. Delamination is one of the most common errors in the drilling operation. Several studies have been carried out to minimize the risk of delamination.

Davim and Reis [13] examined the impact of drilling operational parameters on surface roughness, delamination, and cutting forces in CFRP materials and reported that feed rate is a major factor affecting thrust force, which increases the damage. Hocheng and Tsao [14] concluded that the drill geometry also affects the delamination and found that the thrust force differs with the geometry of the drill and the feed rate, enabling the option of drill geometry to achieve a higher feed rate. Hocheng et al. [15] studied the effects on the delamination of various drill geometries (core drill and stage drill, saw drill, and candlestick drill). Bosco et al. [16] investigated the impact of cutting parameters on delamination of glass fiber-reinforced sandwich composites in arm or steel drilling. Zitoune et al. [17] investigated and found that the thrust force can be reduced during drilling by double cone drill geometry that influences the machine performance during drilling of the sandwich composite. The different types of input parameters that influence the life of the drilled surface are discussed below.

13.2 INPUT PARAMETERS

13.2.1 OPERATING VARIABLES

There are several operating variables in the drilling method, among which the cutting speed and feed rate are the two most relevant operating variables. Higher cutting speeds result in lesser wear and lower drilling damage, while higher feed rates lead to increased drilling forces and higher risk of damage caused by drilling. Thrust forces and torque are directly influenced by the cutting forces and feed rate [18]. The chances of thermal damage increase when high-speed cutting is performed, which effectively softens the matrix material, but reduces the damage at the matrix's entry and exit at the other end [19]. As cutting speed and feed rate are essential, so the drill material and drill point geometry also play a very important role. The operating parameter, drill parameter, and material parameter with output characteristics are shown in Figure 13.1.

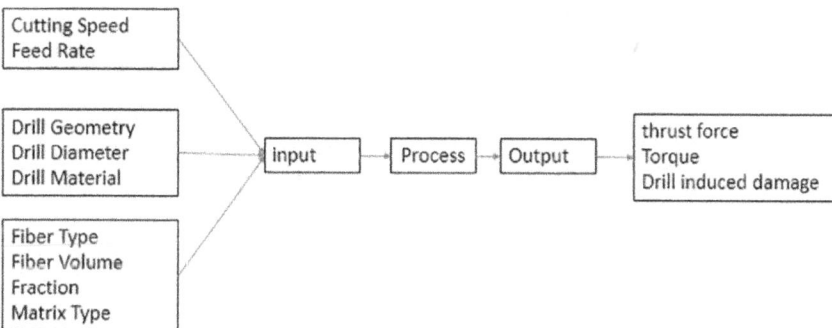

FIGURE 13.1 Element of the drilling system in the context of PMCs [20].

13.2.2 Drill Point Geometry

Drill point geometry plays a significant role in producing damage-free holes. It optimizes the drilled hole quality. Drill point geometry mainly focuses on the minimization of induced damage and the drilling forces. Different types of drill point geometry are shown in Figure 13.2. Among all the types of drills, 8-facet drills provide the best result by combining the free cutting and the other qualities of a split point twist drill with a long taper angle at the shoulder, which tends to minimize fiber break down, whereas the Jo drill results in minimum torque as compared to 4-facet and 8-facet drills [21]. The drilling-induced damage depends on drill point angle, drill material, and other drill parameters [22].

13.2.3 Drill Material

In the drilling of Polymer matrix composites (PMCs), the selection of the drill material plays a major role. As the reinforcing material is tougher in the PMCs, it is difficult to penetrate such robust materials that don't wear out quickly.

Basically, three different types of materials that are used for drilling purpose are as follows:

- Carbide drill
- High-speed steel (HSS)
- Cobalt high-speed steel (HSS-Co).

As the cutting speed increases, drill wear increases; there are more area damage, larger entry and exit burrs, and a decreased number of holes. Carbide drills are the best of all the types of drill materials because of the supreme hardness, which produces the smallest burrs, lower thrust force, the highest quality holes,

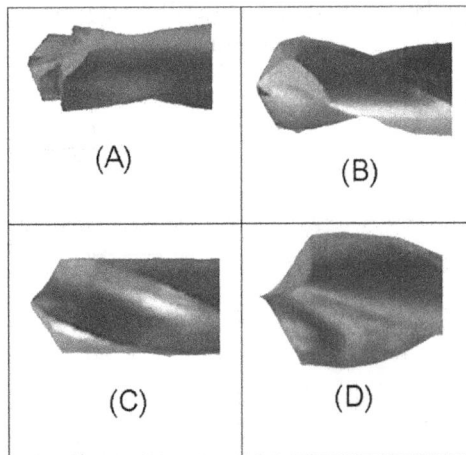

(A)

(B)

(C)

(D)

FIGURE 13.2   Different types of drill geometry.

and a negligible damaged area [23]. Feed rate increases during drilling with HSS and HSS-Co, thus increasing thrust power, torque, surface roughness, and drill damage caused along the composite thickness. Shyha et al. [24] investigated and reported that the helical flute cemented carbide drill is better than the HSS drill. Murphy et al. [25] found in studies that comparison with HSS drill and carbide drill give better wear resistance with or without coating. Cemented diamond drills show better surface quality than plated diamond and plated solid drills while drilling CFRP composites [26].

13.3   OUTPUT CHARACTERISTICS

13.3.1   DRILLING FORCE

The thrust force depends on the feed rate and produces a lower thrust force if the feed rate is lower. Thus, thrust force is a highly effective output that affects the drilling surface. Abrao et al. [27] investigated and found that the thrust force increases with the feed rate with an increased area of the drill. Similarly, Ei-Sonbaty et al. [28] investigated that the drill size influences thrust force and torque and noted that thrust force and toque increase with drill diameter due to an increase in shear area. The trepanning tool for GFRP has reduced the thrust force and torque as compared to the twist drill. Torque behavior is also similar, influenced by the tool geometry and feed rate. With the decrease in feed rate, the thrust force also decreases, which can reduce the delamination [29].

13.3.2   DRILLING-INDUCED DAMAGE

Delamination is one of the most common forms of damage that occur in PMCs. There are various types of other damages such as fiber pullout, spalling, chipping, circularity defects, debonding, matrix cracking, fuzzing, and matrix burning. There are numerous recent methods by which we can analyze the damage around the hole and quantify the drilling-induced damage. The techniques include image analysis, acoustic emissions, radiography, ultrasonic C-scan, computerized tomography (CT), damage quantification using microscope, and laser-based imaging. Drilling-induced damage causes inferior surface and affects the long-term performance of the composite product [30].

13.4   DAMAGE

In the failure mechanism of conventional metals, mainly fatigue grows under dynamic cyclic loading. The failure mode of composite structures under tension and compression is entirely different, and the sites repaired are highly intricate than in metals [31]. We can't deny that composite structures have sufficient mechanical strength such that even after the rough use of the material, it can also be remarkable. Basically, composite structures can fail in three modes – shear, compression, and tension. The composite's damage accumulation is a progressive event, so it cannot fail for a moment, but eventually [32–33]. The weakened composite's residual strength can be restored by laminate repair or laminate replacement [34].

Professional operators with inspection techniques are responsible for assessing the amount of injury. Restoring or improving the strength, protection, reliability, longevity, and efficiency of the material is our primary perspective on a repair.

The repair of composites is a massive issue for several reasons. At present, most maintenance organizations have data and knowledge deficient in repairing composites. That's why the repair of composites is intricate.

13.4.1 DAMAGE SOURCES

- Processing irregularities.
- Environmental damages.
- In-service damages.

13.4.1.1 Processing Irregularities

There are lots of defects in composite structures while processing, and there are some anomalies that occur from these types of processing and can be identified on a rigorous examination [35].

- Delamination, porosity, and voids are the processing irregularities and are generated in the curing phase. An improperly cured component has lower thermal and mechanical stability.
- While performing ply layup, inadequate care leads to the development of inclusion.

13.4.1.2 Environmental Damages

High-intensity radiated field (HIRF), ultraviolet radiation, pressure, and moisture induce damage to composite structures.

- UV radiation: It has a moderate effect on carbon fibers; UV radiation can produce embrittlement on the surface of unprotected composite materials.
- Fluid ingression: Absorption of moisture takes place in damaged composites from the surroundings; in the case of metal bond components, the fluid or moisture causes corrosion in composite materials.

13.4.1.3 In-Service Damages

The damages in aircraft components occur during the operation on the ground or flight. These damages occur due to the dropped parts, bird strikes, aircraft handling accidents, and incorrectly installed removable fasteners.

- Solvent absorption: The reduction in mechanical properties is due to the absorption of solvents. It includes all kinds of fluid: grease, oils, hydraulic fluids, and water.

- High-heat source: High-heat sources may impact composite substances. They are power plants, auxiliary power units (APUs), hot air feed ducts, thermal dicing ducts, and air-conditioning units.

13.4.2   Damage Types

- Combinations of damages.
- Delamination and fiber breakage.
- Cracks.
- Dents.
- Punctures.

13.4.2.1   Combinations of Damages

When a large mass of object strikes with high velocity on the composite, it can result in broken elements. The damage may involve substantial dissolved elements, broken clamps, core damage, and fiber failure.

13.4.2.2   Fiber Breakage

Composites are typically engineered and manufactured here to be fiber-intensive, high-speed blunt bodies that split internal structural components such as spars or ribs and stiffeners, but keep the laminate's exterior parts intact.

13.4.2.3   Cracks

The cracks are composite laminate fractures of various thicknesses that involve both the cracking of the matrix and the breakage of the fabric. Damages associated with cracks that involve matrix damage or delamination may have less critical stress concentration and react more like a soft inclusion in the structure.

13.4.2.4   Dents

The low-velocity impact phenomenon causes the dents; the concrete that is bound in the substructure can damage the dent substructure in laminate areas. The back and substructure behind and around the part should be inspected.

13.4.2.5   Punctures

Because impact damage penetration on the laminate sheet is produced, puncture boundaries must be cleaned based on the type of energy and impact.

13.4.3   Impact Damage

Aerospace structures are compositely vulnerable to low-velocity impact damage. In metallic structures, the damage is not essential, although impact damage in the form of an indent will obviously be found in traditional metallic structures. Due to the low-velocity effect, interlaminar delamination greatly influences the residual compression strength when a fastening hole is present. That is why more consideration is required to analyzing the open-hole compression of laminated

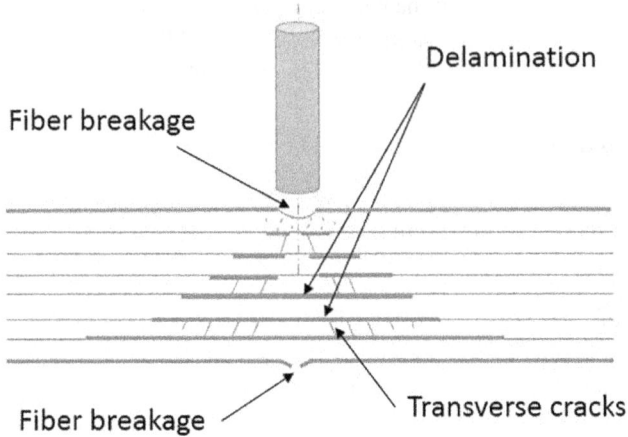

FIGURE 13.3   Impact damage configuration.

composites after impact strength. Different failure mechanisms occur in composite laminates and are as follows:

- Fiber breakage
- Rear-side
- Crater
- Transverse cracks
- Interlaminar delamination.

Damage also depends on the impactor shape [36]. Figure 13.3 shows the different failures that occur during impact damage.

13.4.4   SANDWICH DAMAGE

As shown in the composite sandwich, the high-strength fiber-reinforced plastic is bonded to a comparatively dense, lighter core. Carbon fiber-reinforced plastic (CFRP) structures are particularly susceptible to impact damage relative to glass fiber-reinforced plastic (GFRP) sandwich structures, and the primary mode of failure (GFRP) sandwich and CFRP are different from the crushing core. Different types of modes of sandwich damage are shown in Figure 13.4.

13.5   REPAIR

There are some techniques, such as cosmetic repair, used to repair the damaged area. For repairing the damaged holes, the hole is occupied with epoxy resin. The damages which occur due to the fastening holes or the face damage are cured by filling the hole. The degradation of resin occurs due to moisture. When the damaged hole is filled with the filler material, it is further sealed from the outside.

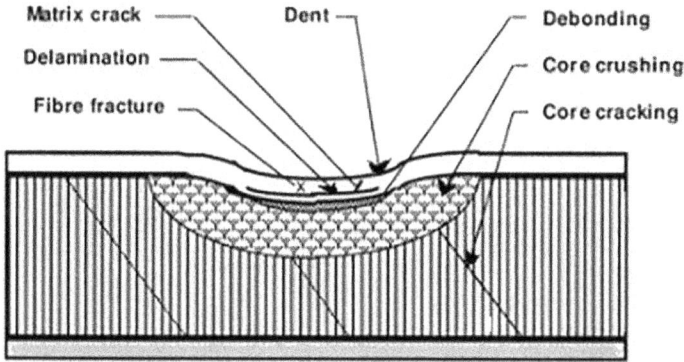

FIGURE 13.4   Sandwich damage configuration.

The material which is filled in aerospace is permeated according to the weight and requirement necessity.

The repair methods used for repairing are as follows:

1. Cosmetic repair (filling and sealing)
2. Semistructural repair (filling and employing a patch)
3. Scarf repair
4. Bolted repair.

13.5.1   Resin Infusion or Injection Repair

In this repair technique, the damages that occur are stored within the system through compressive and shear strength produced by the injection or infusion of low viscosity. In both marine and aircraft repair applications, this approach is employed. It is a cost-effective and straightforward approach that removes the need to prepare surfaces and be done in situ. There is a chance of risk of surface degradation. Resin injection repair is basically used to prevent the short-term damages of composites [37–38].

13.5.2   Chopped Fiber

When the aircraft compilation of different parts is done, it is necessary to repair the improper holes of the aircraft components. The damages that occur may be of geometry, an oversized hole, or an improper alignment. Basically, a chopped fiber-reinforced epoxy resin is used to cure the holes to solid, and then the new drilling operation is used.

13.5.3   Plug Repair

Plug repair is primarily used for components such as wood, pitch, glue, and wadding in the marine industry. In this technique, plugs are placed with adequate coating in the improper hole.

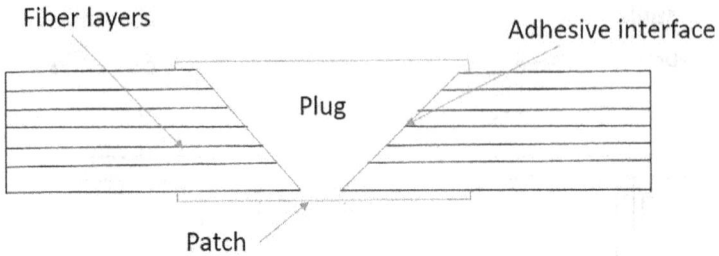

FIGURE 13.5 Plug repair configuration.

This is a cost-effective practice. For external patch repair, this approach is less employed. The configuration diagram is shown in Figure 13.5.

13.5.4 SCARF REPAIR PATCH

Bonded repair shown in Figure 13.6 is mostly preferred to repair the primary and secondary aircraft structures to influence the better strength, smooth aerodynamic surface, and load transfer. It requires more time and costly equipment for repairing function. A taper ratio of 50:1 is required for better result and operation performance. It is basically designed to attain the shear force to double of tensile strength. Material is removed from one side and dressed into the damage by which the depression region is created. Highly trained individuals is a necessary requirement. This technique is not suitable for in situ repairs [39–40].

13.5.5 EXTERNAL PATCH REPAIR

It's the most commonly accepted repair method and is very simple and cost-effective; the least amount of planning is needed. It can also be used for in situ repairs. This repair can be permanent and temporary. In this technique, the damage site in parent laminate is dressed by cutting a hole after this dressed area is filled with the neat resin, and then FRP patches are wrapped over it. This process is performed on the material in order to enhance the properties and increase the

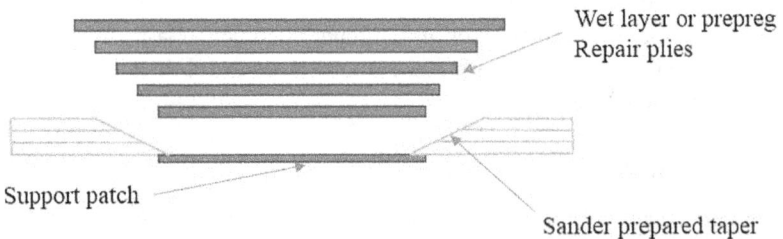

FIGURE 13.6 Scarf repair configuration.

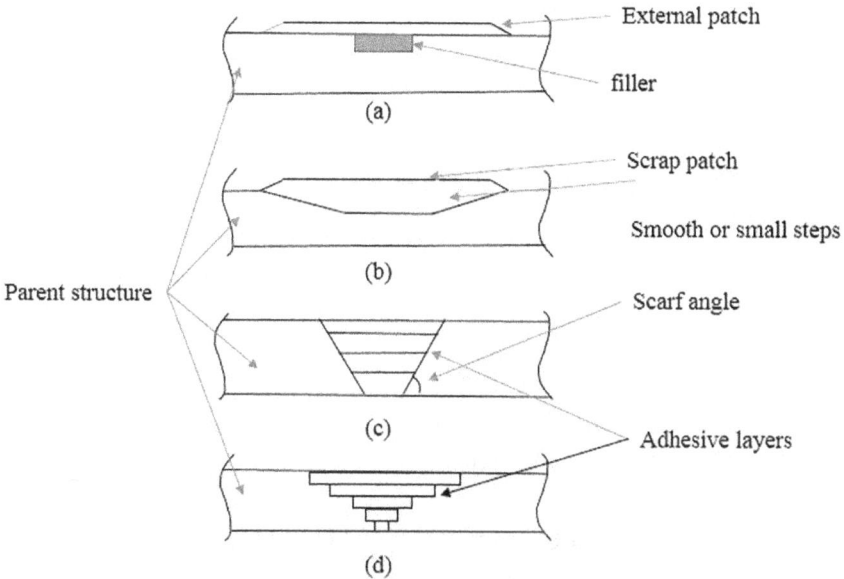

FIGURE 13.7   External patch repair configuration.

strength of the material [41–42]. The systematic external patch configuration is shown in Figure 13.7.

13.5.6   STRUCTURAL MECHANICALLY FASTENED REPAIR

It is used to attach the rigid structure of composite materials that are heavily filled. No aerodynamically smooth property surface exhibits this form of repair. Basically, they move the load to the corners and the free edges from the actual damage location.

13.5.7   TYPICAL REPAIR PROCEDURE

For performing the scarf repair, the primer should be removed first. The damaged area is machined using a portable grinder into a circular shape. Two methods are used, either scarfed or stepped.

The vacuum bagging technique shown in Figure 13.8 is used to repair the fabrication in situ. Basically, this process is done under the vacuum condition to enhance the bonding and for proper distribution of the resin. This process can be changed from one configuration to another according to the need of requirement.

In some special cases like the flat laminates, the vacuum bag system is used with the heat produced by the electric heating blanket. The control panel must be used in a vacuum bag to validate the repair process and curing simultaneously. After

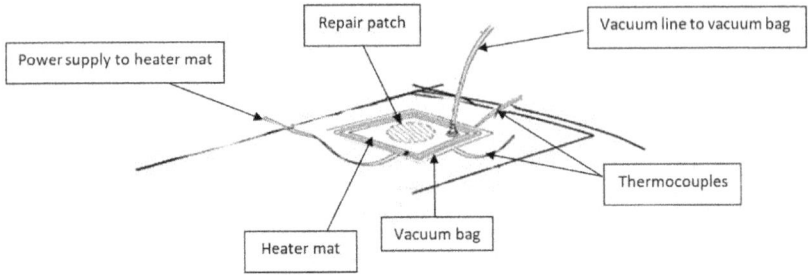

FIGURE 13.8   Vacuum bagging repair configuration.

the completion of the procedure by the control process panel, the repair process can be checked. For improving the performance of the repairing, some additional layers are added; this basically decreases the efficiency of aerodynamics [43].

13.6   SELF-HEALING OF POLYMER MATERIALS

Self-healing means that a material can regain its standard form when external forces acting on them are removed. Polymers are like those materials which gain their conventional properties back after damage. This is an excellent property of these materials, which is essential for engineering applications [44]. We can interpret these self-healing properties by examining the data of their changes in actual dimensions of a system parameter. These properties, like self-mending, are mainly inspired by living organisms as they retain their properties by evolution. The research has shown that the example of healing by self has inspired industrialists to copy natural materials and make polymer and their composites as having self-mending properties. Research has shown that polymer has the most self-healing property by their functional is action and change of polymeric coordination. There are some disadvantages of this property as very minute braking, scratches, and surface degradation are seen in products. However, these are most secure and have a longer life as compared to other materials [45–46].

13.7   CONCLUSIONS

The sandwich process with progressive literature on sandwich boards was discussed in this paper. Sandwich panels have excellent mechanical properties than traditional materials. Sandwich structures are light materials in weight and possess high strength-to-weight ratio. For complicated structures, it is required to perform machining operations on them, such as drilling. Drilling with different drill materials and drill tool geometries is well understood. Damages in polymer composites during service occur due to fatigue, which is generated by repeated loading. The repair of composites is intricate. Maintenance organizations have data and knowledge deficient in repairing composites. This chapter has initiated the sandwich panel review with various types of damage and various types of repair methods.

Finally, the discussion on the repair of damaged portion and self-healing has been done. Polymers are similar to those materials that gain their conventional properties back after damage. These properties, like self-mending, are mainly inspired by living organisms.

REFERENCES

1. Campbell, F.C. (2006) *Manufacturing Technology for Aerospace Structural Materials*, Elsevier, London.
2. Krzy'zak, A., and Vali˘s, D. (2015) Selected safety aspects of polymer composites with natural fibres. in *Safety and Reliability: Methodology and Applications*, T. Nowakowski, M. Mły´nczak, A. Jodejko-Pietruczuk, and S.Werbi´nska-Wojciechowska, Eds. pp. 903–909, Taylor & Francis Group, London.
3. Landowski, M., Budzik, M.K., and Imieli´nska, K. (2001) Wpływ metody wytwarzania na wła´sciwo´sci laminat´ow poliestrowo/ szklanych do budowy małych jednostek pływających. *zynieria Materiałowa* 5: 868–872.
4. Campbell, F.C. (2010) *Structural Composite Materials*, ASM International, Novelty, OH.
5. Muc, A., and Nogowczyk, R. (2005) Formy zniszczenia konstrukcji sandwiczowych z okładzinami wykonanymi z ompozyt´ow. *Composites* 5(4): 31–36.
6. Ochelski, S. and Niezgoda, T. (2007) Kompozytowe konstrukcje pochłaniające energię uderzenia. *Przegląd Mechaniczny* 1: 21–28.
7. Campbell, F.C. (2004) *Manufacturing Processes for Advanced Composites*, Elsevier, London.
8. Dweib, M.A., Hu, B., O'Donnell, A., Shenton, H.W., and Wool, R.P. (2004). All natural composite sandwich beams for structural applications. *Compos Struct* 63(2): 147–157.
9. J˙ungert, A. "Damage detection in wind turbine blades using two different acoustic techniques," in Proceedings of the 7th fib PhD Symposium, Journal of Nondestructive Testing, Stuttgart, Germany, September 2008.
10. Mouritz, A.P., and Gibson, A.G. (2006). *Fire Properties of Polymer Composite Materials*, Springer: Netherlands.
11. Abrate, S. (1997) Machining of composite materials. In *Composites Engineering Handbook*; Mallick, P.K., Ed. Marcel Dekker: New York, pp. 777–809.
12. Koplev, A., Lystrup, A., and Vorm, T. (1983) The cutting process, chips, and cutting forces in machining CFRP. *Composites* 14: 371–376.
13. Davim, J.P., and Reis, P. (2003) Drilling carbon fibre reinforced plastics manufactured by autoclave Experimental and statistical study. *Mater Des* 24: 315–324.
14. Hocheng, H., and Tsao, C.C. (2006) Effects of special drill bits on drilling-induced delamination of composite materials. *Int J Mach Tools Manuf* 46: 1403–1416.
15. Hocheng, H., and Tsao, C.C. (2003) Comprehensive analysis of delamination in drilling of composite materials with various drill bits. *J Mater Process Technol* 140: 335–339.
16. Bosco, M.A.J., Palanikumar, K, Durga, P.B., and Velayudham, A. (2013) Influence of machining parameters on delamination in drilling of GFRP-armour steel sandwich composites. *Procedia Eng* 51: 758–763.

17. Zitoune, R., El-Mansori, M., and Krishnaraj, V. (2013). Tribo-functional design of double cone drill implications in tool wear during drilling of copper mesh/CFRP/ woven ply. *Wear* 302: 1560–1567.
18. Mohan, N.S., Ramachandra, A., and Kulkarni, S.M. (2005) Influence of process parameters on cutting force and torque during drilling of glass-fiber polyester reinforced composites. *Compos Struct* 71: 407–13.
19. Caprino, G., and Tagliaferri, V. (1995) Damage development in drilling glass fi ber reinforced plastics. *Int J Mach Tools Manuf* 35: 817–29.
20. Singh, I., Rakesh, P.K., and Malik, J. (2012) Predicting forces and damage in drilling of polymer composites: Soft computing techniques. *Woodhead Publishing Rev: Mech Eng Series*, 227–258 Doi:10.1533/9780857095893.227.
21. Singh, I., and Bhatnagar, N. (2006) Drilling of uni-directional glass fiber reinforced plastic (UD-GFRP) composite laminates. *Int J Adv manuf Technol* 27: 870–6.
22. Palanikumar, K., Rubio, J.C., Abrao, A.M., Correia, A.E., and Davim, J.P. (2008) Influence of drill point angle in high speed drilling of glass fiber reinforced plastics. *J Compos Mater* 42: 2585–97.
23. Ramulu, M., Branson, T., and Kim, D. (2001) A study on the drilling of composite and titanium stacks. *Compos Struct* 54: 67–77.
24. Shyha, I.S., Aspinwall, D.K., Soo, S.L., and Bradley, S. (2009) Drill geometry and operating effects when cutting small diameter holes in CFRP. *Int J Mach Tools Manuf* 49: 1008–14.
25. Murphy, C., Byrne, G., and Gilchrist, M.D. (2002) The performance of coated tungsten carbide drills when machining carbon fiber-reinforced epoxy composite materials. *PI MECH ENG B–J ENG* 216: 143–52.
26. Quan, Y., and Zhong, W. (2009) Investigation drilling grinding of CFRP. *Front Mech Eng China* 4: 60–3.
27. Abrao, A.M., Rubio, J.C., Faria, P.E., and Davim, J.P. (2008) The effect of cutting tool geometry on thrust force and delamination when drilling glass fiber reinforced plastic composite. *Mater Des* 29: 508–13.
28. Ei-Sonbaty, I., Khashaba, U.A., and Machaly, T. (2004) Factors affecting the machinability of GFR/epoxy composites. *Compos Struct* 63: 329–38.
29. Mathew, J., Ramakrishnan, N., and Naik, N.K. (1999) Investigations into the effect of geometry of trepanning tool on thrust and torque during drilling of GFRP composites. *J Mater Process Technol* 91: 1–11.
30. Di-Paolo, G., Kapoor, S.G., and Devor, R.E. (1996) An experimental investigation of the crack growth phenomenon for drilling of fiber reinforced composite materials. *J Eng Ind Trans ASME* 118: 104–10.
31. Andrew, J., and Ramesh, C. (2015) Residual strength and damage characterization of unidirectional Glass–Basalt hybrid/epoxy CAI laminates, *Arab J Sci Eng* 40 (6): 1695–1705.
32. MIL-STD-1530B. (2004) *Standard Practice: Aircraft Structural Integrity Program (ASIP)*, United State Air Force Aeronautical Systems Division, WPAFB, Dayton, OH, MIL STD 1530B.
33. Baker, A., Dutton, S., and Kelly, D. (2004) *Composite Materials for Aircraft Structures*, AIAA Education Series, AIAA, Reston, VA.
34. Hess, M.S. (1969) The end problem for a laminated elastic strip—II. Differential expansion stresses, *J Compos Mater* 3: 630–641.
35. Jefferson Andrew, J., Arumugam, V., and Dhakal, H.N. (2018) *Repair of Polymer Composites Methodology, Techniques, and Challenges* Woodhead Publishing Series in Composites Science and Engineering

36. Jefferson Andrew, J., and Arumugam, V. (2017) Effect of patch hybridization on the tensile behaviour of patch repaired glass/epoxy composite laminates using acoustic emission monitoring, *Int J Adhes Adhes* 0143–7496, 74: 155–166. Doi: 10.1016/j.ijadhadh.2017.01.014.

37. Azouaoui, K., and Azari, Z. (2010) Evaluation of impact fatigue damage in glass/epoxy composite laminate. *Int J Fatigue* 32 (2): 443–452.

38. Baker, A.A., Chester, R.J., Hugo, G.R., and Radtke, T.C. (1999) Scarf repairs to highly strained graphite/ epoxy structure. *Int J Adhes Adhes* 19: 161–171.

39. Sabelkina, V., Malla, S., Hansena, M.A., Vandawakerb, R.M., and Derrisob, M. (2007) Investigation into cracked aluminum plate repaired with bonded composite patch. *Compos Struct* 9 (1): 55–66.

40. Mansberger, L.L. (2009) *General Aviation Composite Repair*, The University of Texas, Arlington, TX. http://mansbergeraircraft.com/.

41. Baker, A.A. (2006) *Development of a Hard-Patch Approach for Scarf Repair of Composite Structure*, Report TR-1892, 19, Defence Science and Technology Organisation, Canberra.

42. Jefferson Andrew, J., and Arumugam, V. (2016) Effect of patch hybridization on the compression behavior of patch repaired glass/epoxy composite laminates using acoustic emission monitoring. *Polym Compos* Doi: 10.1002/pc.24149.

43. Jefferson Andrew, J., Arumugam, V., Ramesh, C., Poorani, S., and Santulli, C. (2017) Quasi-static indentation properties of damaged glass/epoxy composite laminates repaired by the application of intra-ply hybrid patches. *Polym Test* 0142–9418, 61: 132–145. Doi: 10.1016/j.polymertesting.2017.05.014.

44. Mansberger, L.L. (2009) *General Aviation Composite Repair*, The University of Texas, Arlington, TX. http://mansbergeraircraft.com.

45. Trask, R, Williams, H, and Bond, I. (2007) Self-healing polymer composites: Mimicking nature to enhance performance. *Bioinspir Biomim* 2: 1–9.

46. Dong, Y., Wu, S., and Meure, D.S. (2008) Self-healing polymeric materials: A review of recent developments. *Prog Poly Sci* 5: 479–522.

14 Composite Sandwich Structures in the Marine Applications

Thuy Thi Thu Nguyen, Tuan Anh Le, and Quang Huy Tran
Phenikaa University

CONTENTS

14.1 Introduction 277
14.2 Materials... 280
    14.2.1 Materials for Skins 280
    14.2.2 Core Materials 281
        14.2.2.1 Foam Core Sandwich.. 282
        14.2.2.2 Honeycomb Core 283
    14.2.3 Skin–Core Adhesive Materials.. 284
14.3 Response of CSS in Marine Applications 285
    14.3.1 Effect of Temperature.. 285
    14.3.2 Effects of Seawater 286
    14.3.3 Effects of Underwater Blast Loading 287
14.4 Challenges in Marine Applications 288
References... 288

14.1 INTRODUCTION

Systems operating in the marine environment suffer from inclement conditions, i.e. stresses arising from wind, waves, and tides; corrosions due to sea salts; erosion caused by sand and pebble; high atmosphere humidity; variations of temperature from arctic conditions to over 45°C; marine borers; and complex loads. Marine structures are subjected to exposure to water conditions and hydrodynamic loading caused by waves occurring on the sea. Besides, the submerged structures encounter the additional effects of hydrostatic compression. Therefore, the materials used in shipbuilding must preserve some most important characteristics as follows:

1. Lightweight structure enables buoyancy with high cargo loading, less fuel consumption, high acceleration, and increased ship stability.

DOI: 10.1201/9781003143031-14

2. High stiffness avoids buckling, high strength resists tensile loads, and high fracture toughness resists brittle failure. As a result, high mechanical properties provide benefits of augmented payload, lengthened range, and enhanced speed.
3. High resistance to corrosion, moisture, and fatigue in a seawater environment reduces lifecycle costs.
4. Low acquisition cost is of prime importance as shipbuilders require what balances by performances and maximises the returns for the initial investment and running cost.
5. Non-magnetic property, high stealth, and blast loading are specific characteristics needed for naval ships.

Historically, shipbuilders have used steel as the primary material for ship structure for over 150 years due to its low cost and high mechanical properties. The main disadvantage of using steel in shipbuilding is the high weight of steel, which typically accounts for 35%–45% of the total weight of the ship [1,2]. This fact causes the reduction in payload, signatures, and combat system effectiveness. Moreover, ships primarily constructed using steel require continual maintenance to prevent corrosion. An estimated 52% of manpower working on ship is spent on maintenance due to corrosion [2].

New materials based on composites are employed for various components and structures used in all areas of the marine sector, such as hull, deckhouse, propellers, railings, hatch covers, and other subsea structures. The widely used type of composite materials is the sandwich structure, which is typically fabricated by adhering thin stiff face-sheets (or skins) on the two opposite sides of a lightweight and thicker core [3,4]. The stiff skins provide high bending stiffness to hold up loads in tensile or compressive stresses. Meanwhile, the core bears the shear and compression loads and keeps stabilising the sandwich structure under loading. The sheets of glass or carbon fibre-reinforced polymers [unsaturated polyesters and epoxies (EPs)] or metals (steel and aluminium) are used as skin materials. Based on the configuration, the core is classified into foam cores, corrugated cores, honeycombs, fold cores, and lattice cores. Amongst them, composite sandwich panels using foam cores are designed for the development of lightweight structures applied in some engineering fields, such as aerospace, automobile, and marine [4].

The use of composite sandwich structure (CSS) has led to decreased weight and enhanced mechanical performance of marine vessels in comparison with steel material. The lightweight CSS provides benefits of buoyancy, speed, cargo payload, manoeuvrability, and reduction in fuel consumption [5]. The Visby class of stealth corvettes is built for the Swedish Navy by using CSS entirely with carbon- and glass-reinforced vinyl resin. An estimated 30% of hull weight was reduced, without spending additional manufacturing cost when using carbon fibre-reinforced polymer for the hull construction [1].

The great advantage of using composite panels in marine vessels is that they have higher stiffness and strength per unit weight than their metal counterparts

[2]. The mechanical properties of CSS strongly depend on three factors [6,3]. (1) The skins have high stiffness and strength that can carry almost all the bending loads. When the sandwich is stressed by means of a force in the middle of the beam, the bending moment and shear forces are introduced in the material. (2) The lightweight core is placed between the two skins and provides the CSS with high bending stiffness with overall low density. The rigidity of CSS could be improved by increasing the thickness of the core. (3) For the strength of core-to-skins bonding, the adhesive layer between the core and the skins undertakes some degree of shear force transferred from the core and the skins. The CSS delaminates if the adhesive bond is weak.

Nowadays, due to the development of materials science, the number of advanced materials has remarkably enlarged with the addition of more functional characteristics. Therefore, the design of CSS is based on material selection and geometrical sizing, which are best suited for marine fields. In marine applications, composite materials reinforced by aramid, carbon, or glass fibre are mainly used as skins. Lightweight core materials are commonly designed by polymeric foams and honeycombs. CSSs containing glass fibre-reinforced plastics (GFRPs) are mostly chosen for stiffness, whereas carbon fibre-reinforced composites (CFRPs) are used for their strength. The CSS made of aramid fibre-reinforced skins and polyvinyl chloride (PVC) foam provided a fivefold enhancement in tensile strength of the hull in comparison with steel and reduced the noise [1,7].

Another unique feature of fibre-reinforced plastic (FRP) is its non-magnetic property, which is necessary for naval ships, especially the minesweepers. Since 2011, the Integrated Technology Mast (ITM) consisting of shape-designed CSS has been installed on ships. The ITM improved electromagnetic shielding, stealth capability, and 10%–20% weight reduction with respect to conventional steel masts. In addition, FRP provided resistance to corrosion in the marine environment, and hence lessened maintenance [1].

Given the above-mentioned advantages of FRP, a large number of small marine vehicles and some naval vessels are built using this material. For instance, Brødrene Aa Company, a Norwegian shipbuilder, switched to composite materials when building the first FRP sandwich vessel approved by DNV. In the early 2000s, this company became the manufacturer of the world's first commercial passenger vessel made of carbon fibre. Besides, GFRPs are used worldwide to manufacture small boats, barges, and lifeboats. The superstructures of several naval vessels have been constructed by composites, namely two Zumwalt-class destroyers for the US Navy, the Steregushchiy-class corvettes, and a stealth frigate for the Russian Navy.

However, the development of composite materials is limited because of higher fabrication costs than conventional metals and the associated problems of fire, smoke, and toxicity. The mechanical properties of sandwich composites could also be significantly reduced due to thermal softened and degraded skins and core. Therefore, shipbuilders have to deliberate the use of FRP holistically regarding its safety. The International Maritime Organizations' Safety of Life at Sea Convention (SOLAS) gave regulations of fire prevention, detection, and extinguishing in separate sections to reduce the use of potentially flammable materials in shipbuilding. Anjang et al. [8] developed a model for calculating the post-fire mechanical properties of a CSS made up of E-glass/vinyl ester skins and a balsa wood core. Their testing showed that the thermal decomposition of skins mainly resulted in a rapid decrease in the tension and compression properties of CSS after fire exposure. The model revealed that the tension properties are affected by decomposition to the whole CSS, whereas the compression properties depend on carbonisation of the front skin; hence, the decrease in stiffness and strength is more rapid than that in the tension properties. Kim et al. [9] coated an aramid/phenolic composite layer on the skin surface and used carbonised phenolic resin instead of phenolic resin to improve the post-fire characteristics and fire resistance of composites. The tensile strength of the modified composite after fire exposure increased by 95%. The temperature of the fire-exposed side could be maximised to 1,000°C, and the other side was maintained at 60°C for 1 h.

For the applications of FRP and other new materials into building ship's structure and components, further research and detailed analyses are needed in terms of fire safety engineering.

14.2 MATERIALS

14.2.1 MATERIALS FOR SKINS

The main requirements for the skins of CSS used in marine applications are high tensile and compressive strength, stiffness, impact resistance, and corrosion resistance. Historically, metal alloys (namely stainless steel, aluminium, and titanium) have served as skin materials used in ships and submarines [10]. Steel has the benefits of high stiffness, high ductility, isotropy, and ease of manufacturing and outfitting. Steel structures typically contribute to 35%–45% of the overall weight of a ship, leading to limits in payload, speed, and combat effectiveness for battleships. Moreover, steel skins require an estimated 52% of a ship's manpower for regular maintenance due to the rapid rate of corrosion in the sea environment. Aluminium alloy has been used instead of steel to reduce the topside weight of large ships. However, the structures made of aluminium alloy show several disadvantages, such as high thermal conductivity and low temperatures of softening and melting.

Nowadays, FRP has emerged as a lightweight skin material for CSS applied in aerospace and marine fields. FRP skins consist of fibre-reinforced laminates with either a thermosetting polymer matrix (polyester resin, EP resin, or vinyl ester) or a thermoplastic polymer matrix [polypropylene (PP), polyamide (PA), polyetherimide (PEI), or polyether ether ketone (PEEK)] [11]. Polymer matrices have been reinforced by glass, carbon, and aramid fibres; natural fibres; and nanoparticles. E-glass fibre is commonly used as a reinforcing material for composites in the marine applications because of its high tensile strength and tensile strain (maximum of 2,200 MPa and 2.5%, respectively), and moisture and chemical resistance. Carbon fibres have higher tensile strength (maximum of 4,000 MPa), but they are much more expensive than glass fibres, thereby limiting their usage in applications [1].

Most CSS skins are currently based on fibre-reinforced thermoset polymer matrices. The vinyl ester matrix is characterised by high chemical resistance due to lessened reactive sites and its ability to prevent hydrolytic attack. Vinyl esters are harder and more flexible than polyesters. Therefore, fibre-reinforced vinyl ester composites could well afford the effect of fatigue resulting in failure of hull ships [1]. However, the manufacturing process of vinyl ester-based sandwiches is laborious and time-consuming. In addition, it could not fully meet the demands for people's health and environmental sustainability due to the high levels of styrene emission. Indeed, EPs provide significant improvements in performance and manufacturing. They are 20%–30% stronger than vinyl ester in terms of elongation, tensile strength, and modulus features. The fibre content in the EP matrix could be increased, whilst the number of layers of the laminates could be decreased, leading to weight reduction in EP-based composites. EPs could be cured at low temperature without dangerous emissions and produce a smooth surface due to the low cure shrinkage (< 2%). A shipbuilder, Scout Boats, Inc. (Summerville, SC), has used closed-mould EP infusion for hull production of 42-foot sport fishing boats. This EP-based sandwich composite reduces weight and maintenance cost and improves the hydrodynamic performance, structural stability, speed capability, and fuel economy [12]. The outstanding characteristic of phenolic resins is the ability to preserve fire resistance at around 200°C with the low release of smoke and toxic gas. However, phenolic resins have lower mechanical properties than polyester resin [13].

In general, thermoset-based composites have a little damage tolerance because of the relatively low toughness of the resins. The usage of thermoplastic-based composites could overcome the above-mentioned disadvantages. The composites of glass fibres and PP are commercialised by many suppliers, such as Fiber Glass Industries, Inc. (USA), Quadrant Composites (Switzerland), and Saint-Gobain Vetrotex (France). This composite has the advantages of low price and ease of processability of PP.

14.2.2 Core Materials

Some researchers reported that the characteristics of the core affect more the fracture mechanism of CSS than those of skins [14]. The lightweight core mainly bears compression and shear loads whilst reserving the whole sandwich structure of the panel under stresses. Brittle cores cause shear breaking across the thickness, whilst tough cores lead to buckling failure on the skins. A sandwich core can be classified into several types on the basis of their architecture, including corrugated, foam, lattice, and honeycomb cores. Amongst them, foam and honeycomb cores are commonly used in CSS applied in the marine field because of their light weight and performance.

14.2.2.1 Foam Core Sandwich

Foam cores have been utilised in naval ships since the 1970s. Hulls of small marine vehicles have also been manufactured by foam core sandwiches. Foam cores provide higher stiffness in the thickness direction than other core structures. However, some foam cores show the capacity to absorb large amounts of energy caused by shock and blast loadings. Therefore, they may offer possibilities for shock mitigation, which is necessary for mine countermeasure vessels. The underwater blast mitigation effect of core crushing is recorded when using PVC foam core in composite sandwich panels [15]. Foam cores also exhibit other useful properties, such as acoustic damping and insulation.

Based on cellular structure, polymer foams are divided into two categories of open and closed cells. Foams with closed-cell structure feature higher compressive mechanical strength than open-cell foams. The disadvantage of closed-cell foams is the accumulation of moisture inside the pore holes, which could increase the foam density and degrade the foam material after a long-term exposure [7]. PVC and polyurethane (PU) foams in densities ranging from 21 to 400 kg/m^3 are the two most common core materials for CSS. The compressive properties of a PVC foam depend on its density, cell size, and wall thickness [16]. In the measurement of the compressive stress–strain behaviour, PVC has a high sensitivity to strain rate because of its viscoelastic material. As a result, when a failure appears at the weakest cell walls of foams, time is not enough for the stress concentration to spread out before reaching the target strain, causing the initiation of failure at multiple sites. PU foam core having low density is recommended for buoyancy and thermal insulation for CSS. Foam cores made of thermoplastics, such as PP and PET, are also increasingly being applied for CSS because of their availability, cost–benefit, thermoforming, and recyclability [11]. Cabrera et al. [17] reported 100% PP sandwich panels manufactured by the same polymer for the skins and the cores without reducing the mechanical properties of all-PP sheets. The bonding of foam and skins was homogeneous because of a high contacting surface area. The possibility to produce all-PP face panels in incremental layers of thin plies provides a designer with further freedom to optimise lightweight sandwich structure. Thermoformed PEI sandwich structures were manufactured successfully using a two-step heating cycle by Rozant et al. [18]. First, the sandwich structure was heated between two hot plates at the lower forming temperature of the foam. Then, only the skins were heated to the temperature that skins are formed whilst ensuring that the temperature was not over the forming temperature of the foam through its entire thickness.

Syntactic foam is one type of closed-cell foams, in which thin-walled hollow particles (such as cenosphere, glass microballoon, and EP hollow sphere) are incorporated in a polymer material (such as PU, PP, vinyl ester, and EP resin). Syntactic foam exhibits specific features in comparison with other types of foams and promises its high demand in marine industry [19]. Most studies related to syntactic foam core-based CSS dealt with the influences of hydrothermal conditions, environmental temperature, hydrolytic ageing, and ingress of moisture on the degradation process and mechanical strength. Kumar et al. [20] developed sandwich composites by using glass/EP laminates as skins and a syntactic foam core. The core was fabricated by combining the fly ash cenosphere and phenolic resin and reinforced by integrating it with honeycomb structure. The resistance to moisture absorption was important to its sandwich construction. The post-aged flatwise compressive strength of sandwich composites which use syntactic foam core is higher than that of the unaged ones. Meanwhile, the sandwich composites showed marginally lowered edgewise compressive properties after ageing. The interfacial bonding is poor due to the moisture adsorption in syntactic foam. The influence of moisture ingression depend on wall thickness, and volume fraction of the glass particles in the vinyl ester syntactic foam was studied by Tagliavia et al. [21]. The moisture absorption reduced Young's modulus, but increased the flexural property of the syntactic foams. The results indicated that the volume fraction of particles significantly affects the diffusion properties of syntactic foams, whilst the wall thickness of particles is a secondary parameter.

Sandwich composites constructed by syntactic foam core and laminated glass fibre skin [22], carbon fibre skin [23], and interleaved syntactic foam cores [24] have been investigated for their impact properties. These structures kept at minimum damage caused by collision and enhance the remaining strength of the composites. The change in particle structure and volume proportion could be used to create novel syntactic foams with novel functions as well as high ability of energy absorption [25]. The disadvantage of this solution is that a certain composition of syntactic foams gives the highest benefit of energy absorption in a limited range of impact.

14.2.2.2 Honeycomb Core

Recently, honeycomb cores have been paid much attention, generally because of their greater performance in terms of strength and stiffness compared to foam cores. The honeycomb cores are made of thermoplastic or aluminium strips welded to each other to form hexagonal cells. Baral et al. [26] reported that honeycomb core CSS, which are widely used for manufacturing the hull of racing yachts, started to be damaged because the core was crushed at impact energies around 550 J. Corex Honeycomb Company manufactures aluminium and PP honeycomb cores used in boat and ship construction. They ensure the sandwich made from PP honeycomb has valuable characteristics, such as lower weight with increasing pressure, impact and bending stability, low water absorption, and corrosion resistance. The impact response of aluminium sandwiches with foam and honeycomb cores for lightweight ship structure was investigated by Crupi et al. [27]. They demonstrated that the collapse of the CSS with honeycomb core structure occurs for continuous crumpling of the cell wall and this phenomenon was significantly affected by the cell size. Meanwhile, the foam core CSS collapsed due to the foam crushing and the nature of the foam core material strongly influenced the CSS's capacity of energy absorption.

FRP is an effective material for developing an ultralight honeycomb sandwich structure with outstanding mechanical properties. Florence et al. [28] used hybrid FRP to fabricate skins and honeycomb core filled with energy-absorbing materials, such as Rohacell, wheat husk, and PU foam. The CSS with PU foam-filled honeycomb core enhanced the three-point bending, compression, and impact behaviours satisfactorily. This result opened a new approach to the fabrication of hybrid honeycomb CSSs, which are typically meaningful for marine sector.

14.2.3 SKIN–CORE ADHESIVE MATERIALS

Core–skin delamination is a serious failure of CSS. This failure mode is related to the properties of the adhesive to form an effective bonding between the skins and the core. The adhesive used in CSS must have some important characteristics, which are good bonding pressure, filet forming, adaptability, bond line control, and toughness [6]. However, the choice of proper adhesive is based on many factors, including the type of core and skin, cure method, stresses, and environments. In particular, for the adhesive bonding of CSS in marine applications, the certification for adhesive materials needs to be approved by classification societies. This certification ensures conformity with specified requirements by standard tests, including lap shear testing, cleavage peel testing, fatigue testing, immersion tests, and glass transition temperature test [29]. Then, the marine durability of the sample is tested by weathering under different conditions involving loading. The prolonged exposure of the adhesive under seawater at temperature in the range from −30°C to 70°C often leads to changes in the chemical and physical properties, such as plasticisation, swelling, and degradation.

PUs, methacrylates (MAs), and EPs are three kinds of adhesives commonly applied in the marine applications. These adhesives have significant differences in terms of stiffness, strength, and ductility. PUs are known for their excellent adhesion, flexibility, low-temperature performance, resistance to fatigue, impact resistance, and high cure speeds at room temperature (RT) [30]. They effectively wet the surface of most substrates and easily permeate porous substrates due to their small molecular size. EP adhesive is frequently used in marine industry because of its high adhesive performance. The properties of EP adhesive depend on cure temperature at which the reaction of the EP and hydroxyl groups of resin molecule with reactive groups of the hardeners takes place. The shear strength of adhesives cured at RT is lower than that of adhesives cured at elevated temperature (above 100°C). The adhesives cured at elevated temperature also exhibit resistance to the impact of elevated temperature. EP solvent-based adhesive is characterised by low viscosity, and it requires pre-drying prior to the actual curing process. This adhesive may also lead to a certain amount of solvent being remained, which could have an adverse effect on its strength. These advantages could be laminated by using EP solvent-free adhesives [18].

The thermoplastic hot melt film is another choice for adhesive bonding in CSS. This film is sandwiched between the skin and the core, molten by heating and ultimately solidified after cooling to form an adhesive layer. The bonding technique is controlled by the conditions of temperature, pressure, and consolidation time. Kumar et al. [31] manufactured an all-olefin thermoplastic sandwich system, in which a glass fibre PP composite and a high-density PE (HDPE) foam with closed cells were used as skins and the core, respectively. Ethylene–propylene copolymer (EPC) and HDPE/elastomer films were used as hot melt adhesives to bond the substrates. A higher bond strength of the all-olefin sandwich system was observed in the case of EPC tie layer (27.4 kg/cm^2) than that of the HDPE/ elastomer blend (19.7 kg/cm^2). The bonding between the substrates occurs by interdiffusion of polymer segments and the mechanical interlocking caused by local crystallisation at the interface. Ning et al. [32] reported a design of CSS with E-glass fibre/PP skins and PP honeycomb core as main components that gave the advantages of low weight, high strength, and energy absorption. They suggested the use of hot melt adhesives developed by Dow Automotive for bonding PP laminates with low-energy surface and methacrylate adhesives for providing high strength and high elongation. The destruction of this CSS appeared via adhesive failure when reaching the load of 11.7 kN. The body panel of the sandwich composite successfully met the static loading criterions of the American Public Transportation Association and saved a weight of more than 55% compared with using aluminium skin.

Besides the extensive availability of standard commercial products, adhesive suppliers could design special formulations to fully satisfy the requirements of marine applications. Other types of adhesive can be listed, such as phenolics blended with vinyl rubbers or EPs, EPs modified with nylon or other polyamides, nitrile rubber modified epoxies, and highly specialised adhesives.

14.3 RESPONSE OF CSS IN MARINE APPLICATIONS

14.3.1 EFFECT OF TEMPERATURE

In the case of marine applications, CSS often suffered extreme and varied conditions, including variable temperature ranging between −30°C and 70°C. Therefore, the fatigue behaviour of CSS must be investigated at different temperatures to assess the ability of using these materials in the hull structures of ships in different regions.

The effects of temperature on the fatigue behaviour of foam core CSS were investigated by Kanny et al. [33] at RT, 40°C, and 80°C. They used linear high-density PVC (HD130) foam and slightly cross-linked PVC (H130) foam to make cores. The skins were made of S2-glass-reinforced vinyl ester. At RT, the use of HD130 structures significantly prolonged fatigue life than the use of H130 structures, whilst at elevated temperature, the opposite trend was observed. Fatigue life and mechanical strength of both structures decreased when the temperature raised. The failure of CSS that use H130 core at RT and 40°C was caused by a fatigue crack in the core. Meanwhile, the gradual collapse of the foam core was the main reason leading to failure at 80°C.

Singh et al. [34] determined the influences of sub-zero temperature, seawater immersion, and low-velocity impact on the flexural strength, stiffness, and life of CSSs. A decrease in temperature from 20°C to −20°C caused an average increase in static strength of 24% and their fatigue life. At the same level of load, the low-temperature specimens prolonged their lives an average of 6.3 times those of the RT specimens. The results suggested that accurate design of materials and structural response are needed in order to be used in a wide range of environmental conditions.

The influences of temperatures in the range of 22°C and −60°C on the failure behaviours of CSS constructed by carbon fibre/Rohacell foam core (CF/RC) and glass fibre/Rohacell foam core (GF/RC) were studied by Soni et al. [35] under different loading conditions. They concluded that the stiffness and brittleness increased for GF/RC and CF/RC CSSs when temperature decreased, in which the skin materials influenced the strength of any sandwich beam at each temperature. No failure was found after 10^5 cycles in the case of GF/RC CSS at −60°C and 80% load condition, thus confirming the great improvement of fatigue life at this low temperature. Core shear was observed as the only cause leading to failure when the flexural static loads were applied at different temperatures. However, monitoring the failure occurrences in CSS and controlling the safety process similar to those for their metal counterparts are difficult at a low temperature.

14.3.2 Effects of Seawater

Most polymers could absorb moisture under long-time exposure in water environment. The moisture absorption is accompanied by a decrease in the glass transition temperature, expansional strains, and degradation of polymers. In case of CSS used in marine applications, an enhanced understanding of their dimensional stability, mechanical behaviour, fatigue response, and face–core debonding fracture under the effects of seawater is important.

Siriruk et al. [36] reported the influences of seawater on the interfacial bonding between foam and skin of the sandwich composite, which was composed of a closed-cell H100 Divinycell PVC foam core placed between CFR and GFR composite skins (Devold AMT AS, Sweden). Long-term exposure to seawater led to a 30% decrease in delamination fracture toughness. The crack spreads near the interface in the wet case, whereas it remains within the foam in the dry case.

Li et al. [37] studied seawater-induced damage, weight gains and expansional strains, and the possible degradation of foam materials under sustained exposure. They used two closed-cell PVC foams (H100 and H200) and their polymeric composite skin. A large amount of seawater penetrated and filled in the cells in the exterior regions of those materials, leading to the microstructural and irreversible damage to the foam and inducing swelling strains. The absorption of seawater in the fracture regions enhanced the toughness level of the H100 and H200 PVC foams by 31% and 8%, respectively.

Foam core CSSs were tested to creep failure and cyclic creep in seawater to examine the response of CSS in an actual service scenario [38]. The sandwich specimens consisted of PU foam core and GFR skins. The skins were bonded to the foam core with EP resin. Creep failure test was performed by placing a constant load on the specimens in air or seawater until reaching failure. Cyclic creep test kept constant loading at 24 h, whilst the relaxation periods were varied at 6, 12, and 24 h. Approximately 15% larger deflection and more than 50% decrease in the whole lifetime were obtained in the specimens tested in seawater than in those in air. This finding could be attributed to the various possible paths to crump the cell wall of foam. Cyclic creep led to significantly reduced lifetime and increased damage in sandwich specimens. Core compression and skin indentation were modes of failure in all cases.

Davies [39] reported the effect of seawater ageing on sandwich composites for building marine structures, such as boats and tidal turbines. With regard to environmental problems, the author suggested using recyclable materials, such as Elium range of acrylic resins from Arkema. Unreinforced and CFR acrylic specimens were immersed in seawater for 1-year ageing. In case of unreinforced specimens, the water saturation occurred after approximately two months and a mitigation of mechanical properties appeared during 1-year exposure in seawater at 60°C. The tensile properties of the resin and composite could be recovered after drying at 60°C, suggesting that the main ageing mechanism was matrix plasticisation. Using natural fibres (flax fibres) instead of glass fibres to reinforce the polyester matrix provided the same weight gains after exposure in seawater under similar conditions. On the basis of these results, a 6 m long bio-composite multihull constructed by a flax-reinforced polyester sandwich composite was used by Kairos in 2013. It showed very promising materials for marine applications.

14.3.3 Effects of Underwater Blast Loading

Underwater blast loading is a complex phenomenon, which is of particular interest to the naval field. When an underwater explosion occurs, it produces blast waves and high-pressure gas bubbles. Therefore, the effects of these factors on CSS during underwater blast need to be well investigated to prevent potential dangers in service. The response of CSS to strong blast loads resulting in core crushing and failure of the facings has been reported [40–42]. Arora et al. [40] investigated the mechanism of failure observed within GFRP sandwich composite specimens due to underwater blast shocking. The core suffered significant damage (up to 50%), whilst the skins got very large strains, leading to fibre rupture on both faces when the back of targets was air. With water as the back medium, the surface strains reduced as the core crushing increased. The authors underlined the significance of boundary conditions on the structural response when designing the blast load resistance. Using a novel experimental set-up for HP100 PVC foam core–GFRP skins sandwich structure, Avachat et al. [41] observed multiple failures caused by core compression, delamination between core and skins, translaminar cracking, matrix, and fibre breakage. The cylindrical sandwich structures showed more blast resistance than cylindrical monolithic structures with the same mass.

The blast-resistant capacity of CSS depends on the type of core and skin materials. The CSS with CFRP skins experienced more severe damage due to its lower strain to failure than that with GFRP skins [43]. In the comparison of blast resistance of different core materials (PVC, polymethacrylimide (PMI), and styrene acrylonitrile (SAN)), it was found that SAN showed the least damage under the same loads, minimising the plane displacement and pull-out [15]. Rolfe et al. [44] demonstrated the advantages of hybrid composite sandwich panels for enhancing blast resistance. The combination of GFRP and CFRP layers in the skins reduced the normalised deflection compared to the use of GFRP and CFRP skins by up to 41% and 23%, respectively.

14.4   CHALLENGES IN MARINE APPLICATIONS

In the last several decades, CSS has been optimised in design and materials to encounter harsh conditions in the time of its service. The first GFR vessels have been utilised in the naval field in 1972. However, the development of composite materials into a navy ship has been slow because of the limitation of experience of shipbuilders in using composite materials instead of metals. Accompanying with the advancement of composite materials, it is required that the technologies for CSS manufacture must also be developed to ensure its performances with complex shape, large size, and being hazardless to labour and environment. Many design parameters, such as material and structure of core and skins, manufacturing process, joining technique for connecting structures, for large marine vessels need to be carefully considered. The design for a ship structure becomes a more difficult task because of the requirements of low weight and cost. Higher acquisition costs and issues relating to fire are still big challenges in the production of large marine vessels. For new CSS, the confirmation testing of its structure and performance would be a significance step in monitoring its suitability with specifications for applying in marine applications. However, reliable analysis tools for determining the in-service performance of CSS, particularly the response to impacts, collisions, fire, and explosions, still do not meet the requirements.

In the future, the flexibility in the design of CSS will allow innovative structural developments from composite materials to apply in building a range of ship lengths and loads with efficient weights and costs. According to the results of research, development, and implementation, CSS has the potential to be the first-choice material for marine applications because of its unique characteristics and performances.

REFERENCES

1. Rubino, F., Nisticò, A., Tucci, F., and Carlone, P. 2020. Marine application of fiber reinforced composites: A review. *J. Mar. Sci. Eng.* 8: 1–26.
2. Liang, R., and Hota, G. 2013. Fiber-reinforced polymer (FRP) composites in environmental engineering applications. In *Developments in Fiber-Reinforced Polymer (FRP) Composites for Civil Engineering*, ed. Uddin, N., 410–468, Cambridge: Woodhead Publishing.
3. Xiong, J., Du, Y., Mousanezhad, D., Eydani Asl, M., Norato, J., and Vaziri, A. 2019. Sandwich structures with prismatic and foam cores: A review. *Adv. Eng. Mater.*, 21: 1–19.

4. Calabrese, L., Di Bella, G., and Fiore, V. 2016. Manufacture of marine composite sandwich structures. In *Marine Applications of Advanced Fibre-Reinforced-Composites*, ed. Graham-Jones, J., and Summerscales, J., 57–78, Cambridge: Woodhead Publishing.

5. Graham-Jones, J., and Summerscales, J. 2016. Introduction. In *Marine Applications of Advances Fiber-Reinforced Composites*, ed. Graham-Jones, J., and Summerscales, J., 1–15, Cambridge: Woodhead Publishing.

6. Ramakrishnan, K.V., and Architect, N. 2016. Applications of sandwich plate system for ship structures. *Int. Conf. Emerg. Trends Eng. Manag.*, 83–90.

7. Summerscales, J. 2019. Materials selection for marine composites. In *Marine Composite: Design and Performance*. ed. Pemberton, R., Summerscales, J., and Graham-Jone, J. 4–30 Cambridge: Woodhead Publishing.

8. Anjang, A., Chevali, V.S., Lattimer, B.Y, Case, S.W., Feih, S., and Mouritz, A.P. 2015. Post-fire mechanical properties of sandwich composite structures. *Compos. Struct.*, 132: 1019–1028.

9. Kim, M., Choe, J., and Lee, D.G. 2016. Development of the fire-retardant sandwich structure using an aramid/glass hybrid composite and a phenolic foam-filled honeycomb. *Compos. Struct.*, 158:227–234.

10. Vinson, J.R. 2005. Sandwich structures: Past, present, and future. In *Sandwich Structures 7: Advancing with Sandwich Structures and Materials*. ed. Thomsen, O.T., Bozhevolnaya, E., Lyckegaard, A. 3–12, Dordrecht: Springer.

11. Grünewald, J., Parlevliet, P., and Altstädt, V. 2017. Manufacturing of thermoplastic composite sandwich structures: A review of literature. *J. Thermoplast. Compos. Mater.* 30:437–464.

12. Hoge, J. and Leach, C. 2016. Epoxy resin infused boat hulls. *Reinf. Plast.* 60:221–223.

13. Mougel, C., Garnier, T., Cassagnau, P. and Sintes-Zydowicz, N. 2019. Phenolic foams: A review of mechanical properties, fire resistance and new trends in phenol substitution. *Polymer* 164:86–117.

14. Manalo, A., Aravinthan, T., Fam, A., and Benmokrane, B. 2017.State-of-the-art review on FRP sandwich systems for lightweight civil infrastructure. *J. Compos. Constr.* 21:04016068.

15. Kelly, M., Arora, H., Worley, A., Kaye, M., Del Linz, P., Hopper, P.A., and Dear J.P. 2016. Sandwich panel cores for blast applications: Materials and graded density. *Exp. Mech.* 56:523–544.

16. Luong, D.D., Pinisetty, D., and Gupta, N. 2013. Compressive properties of closed-cell polyvinyl chloride foams at low and high strain rates: Experimental investigation and critical review of state of the art. *Compos. Part B Eng.* 44:403–416.

17. Cabrera, N.O., Alcock, B., and Peijs, T. 2008. Design and manufacture of all-PP sandwich panels based on co-extruded polypropylene tapes. *Compos. Part B Eng.* 39:1183–1195.

18. Rozant, O., Bourban, P.E., Manson, J.A.E., and Drezet, J.M. 2000. Pre-heating of thermoplastic sandwich materials for rapid thermoforming. *J. Thermoplast. Compos. Mater.*, 13:510–523.

19. Salleh, Z., Islam, M., and Ku, H. Study on compressive properties of syntactic foams for marine applications. *J. Multifunct. Compos.* 2:21–27.

20. Kumar, S.J.A., and Ahmed, K.S. 2016. Effects of ageing on mechanical properties of stiffened syntactic foam core sandwich composites for marine applications. *J. Cell. Plast.*, 52:503–532.

21. Tagliavia, G., Porfiri, M., and Gupta, N. 2012. Influence of moisture absorption on flexural properties of syntactic foams. *Compos. Part B Eng.* 43:115–123.

22. Kim, B., Ahn, K., and Lee, C. 2014. Low Velocity Impact Behaviors of a Laminated Glass. *Smart Sci.* 2:209–213.

23. Scarponi, C., Briotti, G., Barboni, R., Marcone, A., and Iannone, M. 1996. Impact Testing on Composites Laminates and Sandwich Panels. *J. Compos. Mater.* 30:1873–1911.

24. Ishai, O. and Hiel, C. 1992. Damage tolerance of a composite sandwich with interleaved foam core. *J. Compos. Technol. Res.* 14:155–168.

25. Gupta, N. 2007. A functionally graded syntactic foam material for high energy absorption under compression. *Cit. Inf. Mater. Lett.* 61:979–982.

26. Baral, N., Cartié, D.D.R, Partridge, I.K., Baley, C., and Davies, P. 2010. Improved impact performance of marine sandwich panels using through-thickness reinforcement: Experimental results. *Compos. Part B Eng.* 41:117–123.

27. Crupi, V., Epasto, G., and Guglielmino, E. 2011. Impact Response of Aluminum Foam Sandwiches for Light-Weight Ship Structures. *Metals* 1: 98–112.

28. Florence, A., Jaswin, M.A., Prakash, M.D.A.A., and Jayaram, R.S. 2020. Effect of energy-absorbing materials on the mechanical behaviour of hybrid FRP honeycomb core sandwich composites. *Mater. Res. Innov.* 24: 244–255.

29. Lucas, R.D.A., da Silva, F.M., Ochsner, A. 2011. Introduction to adhesive bonding technology. In *Handbook of Adhesion Technology*, 2–7, Verlag Berlin Heidelberg: Springer.

30. Banea, M.D., and da Silva, L.F.M., 2009. Adhesively bonded joints in composite materials: An overview. *Proc. Inst. Mech. Eng. Part L J. Mater. Des. Appl.*, 223:1–18.

31. Kumar, G., Ramani, K., and Xu, C. 2002. Development and characterization of an all-olefin thermoplastic sandwich composite system. *Polym. Compos.* 23:647–657.

32. Ning, N., Janowski, G.M., Vaidya, U.K., and Husman, G. 2007. Thermoplastic sandwich structure design and manufacturing for the body panel of mass transit vehicle. *Compos. Struct.* 80:82–91.

33. Kanny, K., Mahfuz, H., Thomas, T., and Jeelani, S. Temperature effects on the fatigue behavior of foam core sandwich structures. *Polym. Polym. Compos.* 12:551–559.

34. Singh, A., and Davidson, B. 2011. Effects of temperature, seawater and impact on the strength, stiffness, and life of sandwich composites. *J. Reinf. Plast. Compos.* 30:269–277.

35. Soni, S.M., Gibson, R.F., and Ayorinde, E.O. 2009. The influence of subzero temperatures on fatigue behavior of composite sandwich structures. *Compos. Sci. Technol.* 69:829–838.

36. Siriruk, A., Penumadu, D., and Weitsman, Y.J. 2009. Effect of sea environment on interfacial delamination behavior of polymeric sandwich structures. *Compos. Sci. Technol.* 69:821–828.

37. Li, X., and Weitsman, Y.J. 2004. Sea-water effects on foam-cored composite sandwich lay-ups. *Compos. Part B Eng.* 35:451–459.

38. Figueroa, E., Shafiq, B., and de la Paz, I. 2013. Creep to failure and cyclic creep of foam core sandwich composites in seawater. *J. Sandw. Struct. Mater.* 15:657–670.

39. Davies, P. 2016. Environmental degradation of composites for marine structures: new materials and new applications. *Philos. Trans. R. Soc. A Math. Phys. Eng. Sci.* 374: 20150272.

40. Arora, H., Hooper, P.A., and Dear, J.P. 2012. The effects of air and underwater blast on composite sandwich panels and tubular laminate structures. *Exp. Mech.* 52:59–81.

41. Avachat. S., and Zhou, M. 2014. Response of cylindrical composite structures to underwater impulsive loading. *Procedia Eng.* 88:69–76.
42. Bahei-El-Din, Y.A., and Dvorak, G.J. 2007. Wave Propagation and Dispersion in Sandwich Plates Subjected to Blast Loads. *Mech. Adv. Mater. Struct.* 14:465–475.
43. Rolfe, E., Kelly, M., Arora, H., Hooper, P.A., and Dear, J.P. 2017. Failure analysis using X-ray computed tomography of composite sandwich panels subjected to full-scale blast loading. *Compos. Part B Eng.* 129:26–40.
44. Rolfe, E., Quinn, R., Sancho, A., Kaboglu, C., Johnson, A., Liu, H., Hooper, P.A., Dear, J.P., and Arora H. 2018. Blast resilience of composite sandwich panels with hybrid glass-fibre and carbon-fibre skins. *MultiscaleMultidiscip. Model. Exp. Des.* 1:197–210.

15 Composite Sandwich Structures in Aerospace Applications

Erik Vargas-Rojas
Polytechnic Institute, Mexico
Franche-Comté University

CONTENTS

15.1 Introduction 294
15.2 Sandwich Structures Used in Aerospace 295
    15.2.1 Analogy with Composites.. 295
    15.2.2 Cellular Solids 297
15.3 Applications in Primary Structures.. 297
    15.3.1 Cantilever Wing.. 298
    15.3.2 Ultralight Monoplane 299
    15.3.3 Jet Engines 300
15.4 Mechanical Behavior 300
    15.4.1 Specific Mechanical Properties and Cost 300
    15.4.2 Equivalent Mechanical Properties... 301
    15.4.3 Equivalent Flexural Stiffness 302
    15.4.4 Failure Modes of Sandwich Panels Subjected to Flexural
        and Compressive Loading 302
    15.4.5 Equivalent Flexural Strength.. 304
        15.4.5.1 Facings Yielding 304
        15.4.5.2 Facings Buckling 305
        15.4.5.3 Core Shear Failure 305
    15.4.6 Index Lines 306
15.5 Case Study: Sandwich Materials Employed in Spacecraft Structures 306
    15.5.1 Microsatellite 306
    15.5.2 Alternative Materials 307
    15.5.3 Mechanical Properties Evaluation... 308
        15.5.3.1 Equivalent Flexural Stiffness..................................... 309
        15.5.3.2 Equivalent Flexural Strength 309
        15.5.3.3 Specific Mechanical Properties and Cost 309
        15.5.3.4 Discussion 314

DOI: 10.1201/9781003143031-15

        15.5.4 Buckling...314
                15.5.4.1 Facings Critical Compression Stress...........................315
                15.5.4.2 Facings Critical Compression Load............................315
                15.5.4.3 Facings Critical Stress before Wrinkling....................315
                15.5.4.4 Cells Instability Critical Stress 315
15.6 Maintenance 316
15.7 Conclusions..316
15.8 Note...318
Acknowledgments...318
References...318

15.1 INTRODUCTION

It is common practice in aerospace to class structural elements into primary, secondary and tertiary (Wang and Duong, 2016, p. 5). The primary structure carries flight, ground or pressurization loads. The secondary structure carries air or inertial loads generated on or within itself. The tertiary structure is defined according to the consequences of a failure scenario. Failure of the primary structure would reduce the structural integrity of the aircraft. If the secondary structure was to fail, it would affect the operation of the aircraft, but not lead to its loss. Failure of the tertiary structure would not significantly affect the operation of the aircraft. This classification has been adopted for spacecraft with a different nuance. The primary structure is the backbone, too; secondary structures include support beams, booms, trusses and solar panels; tertiary structures are the smallest structures such as boxes that house electronics or brackets that support electrical cables (Moreland 2009, p. 552).

Composite materials and structures are playing an increasingly important role as the aerospace engineer's choice in each of the classifications given above. For instance, the long-haul Airbus A350 and Boeing 787 jetliners employ advanced fiber composites as more than 50% of their total mass (Wang and Duong, 2016, p. ix). Back in 2005, Herrmann et al. (2005) stated, "The application of sandwich structures in commercial aviation is currently restricted to secondary structures." In the early years of the century, the German Aerospace Center (DLR) and Airbus explored the possibility to incorporate foam sandwich as part of the fuselage structure (Kolesnikov and Herbeck, 2004; Wilmes et al., 2002). They aimed at the implementation of more functions into fewer parts in addition to thermal insulation, noise damping, impact protection and mass reduction. However, despite the advantages of composite sandwich structures related to their high specific mechanical properties, they continue to shape radomes, fairings, nacelles, landing gear doors, control surfaces, leading and trailing edge panels, wing tips and floor panels (see Figure 15.1). The main use of composites in the above-mentioned aircraft remains as monolithic CFRP (carbon fiber-reinforced polymer, particularly carbon–epoxy or C-Ep) structures. In contrast, small aircraft designers have made use of sandwich as part of major primary structures.

Materials
■ Carbon laminate ■ Carbon sandwich ▨ Other composites ▨ Aluminum ▨ Titanium ▨ Titanium/Steel/Aluminum

FIGURE 15.1 Composites employed in the structure of Boeing 787. (Lewis, 2014. Courtesy of the Boeing Company.)

Why are composite sandwich structures so widely used in the construction of aerospace structures? In order to answer this fundamental question, the topics to be reviewed comprise the arguments that justify the use of composite materials and structures; their failure modes; cost and maintenance aspects. The concept of the sandwich structure is reviewed based on these arguments. Two applications to be inspected include the construction details of an ultralight monoplane fabricated mostly with CFRP and the comparison of the specific mechanical properties of several materials thought to be used as structural sandwich panels of an artificial satellite.

15.2 SANDWICH STRUCTURES USED IN AEROSPACE

15.2.1 Analogy with Composites

Sandwich structures behave analogously to composites. Composites are hybrid materials because fibers and resins are combined together resulting in the superposition of their properties, creating a material whose resultant attributes cannot be predicted by simply summing those of the constituents. Broadly speaking, high-strength or high-stiffness fibers exert an influence on the overall mechanical properties, whereas the matrix contributes to the physical properties, enhances fracture toughness, transfers the applied forces onto the embedded fibers and protects them from chemical environments. Porosity is an undesirable gaseous phase that takes the form of voids or bubbles.

Volume fractions are the ratios of the volumes v_x of the constituent materials to the total volume v. The volume fractions V_x of reinforcement (r), matrix (m) and porosity (p) are given with Expression (15.1). The concept of volume fractions is used in the analysis of the mechanical behavior of sandwich structures.

$$V_r = \frac{v_r}{v}; V_m = \frac{v_m}{v}; V_p = 1 - V_r - V_m \qquad (15.1)$$

Similarly, as with composite materials, each of the constituents of a sandwich structure has a well-defined role. A sandwich structure consists of thin load-bearing skins or facings bonded to a foam or honeycomb interior (Ashby, 2011, p. 244). The sandwich structure is compared to an I-beam with the facings and core corresponding to the flanges and the web, respectively. The facings bear axial tensile and compressive stresses, while the core carries shear and compressive stresses normal to the panel and stabilizes the outer skins, preventing wrinkling and buckling of the facings under axial compressive forces (Megson, 2018, p. 382; Hackman, 1973). The core-to-facing bond must be adequate enough to transfer the stresses from the facings to the core. Facings, core and the interface are the substructures of the sandwich panel.

Two classifications of sandwich structures are adopted from the literature and depicted together in Figure 15.2a. The first one classifies sandwich structures as structural composites, i.e., the combination of composites and homogeneous materials (Callister, 2001, p. 164). Consequently, sandwich structures are by definition hybrid materials, too. In the second classification, sandwich structures fall into two categories: symmetrical and asymmetrical. The definitions given in the previous paragraph corresponds indeed to the symmetric panel (see Figure 15.2b, left). The hybrid asymmetric panel (see Figure 15.2b, right) consists of two sections and one nucleus configured such that the primary section bears a majority of the applied operating load. When used in commercial jetliners, the primary section comprises relatively thick load-carrying outer plies made out of high-strength, high-modulus toughened epoxy unidirectional tape applied by an automated machine (e.g., automated tape laying). The core is positioned next and then covered with a limited number of inner fabric plies that can be laid down by hand. Thus, the hybrid asymmetric panel combines automated with manual processes (Ackermann and Gleason, 2008).

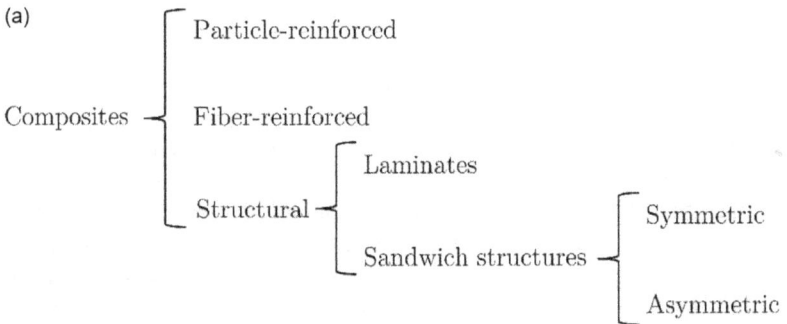

(a)

Composites —
- Particle-reinforced
- Fiber-reinforced
- Structural —
  - Laminates
  - Sandwich structures —
    - Symmetric
    - Asymmetric

FIGURE 15.2 (a) Classification of sandwich structures; (b) symmetric panel (left) and asymmetric panel (right) [leaders: ① sandwich facings; ② sandwich nucleus; ③ adhesive]; (c) classification of cellular solids.

(Continued)

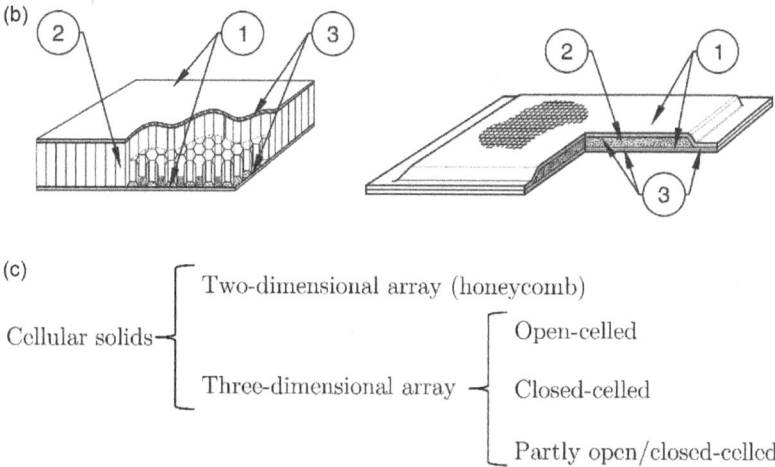

FIGURE 15.2 (*CONTINUED*) (a) Classification of sandwich structures; (b) symmetric panel (left) and asymmetric panel (right) [leaders: ① sandwich facings; ② sandwich nucleus; ③ adhesive]; (c) classification of cellular solids.

15.2.2 CELLULAR SOLIDS

Two are the most common materials used to shape the nucleus of composite sandwich structures used in aerospace: the honeycomb core made of aluminum alloys or aramid paper, and foam cores, either metallic or polymeric. All these materials are cellular solids. A cellular solid is made up of an interconnected network of solid struts or plates which form the edges and faces of cells (Gibson and Ashby, 1997). They are classified as shown in Figure 15.2c. The simplest structure is the honeycomb. In contrast, foams are polyhedral cells which pack in three dimensions to fill space. In open-celled foams the cells connect through open faces because the material is contained in the cell edges only. As for closed-celled foams, the cells are solid and each cell is sealed off from its neighbors. The relative density $\rho*/\rho_s$, i.e., the ratio of the density of the cellular material $\rho*$ to the density of the solid from which the cell walls are made, ρ_s is a parameter of comparison. As the parameter increases, the cell walls thicken and the pore space shrinks. At $\rho*/\rho_s = 0.3$, a transition from a cellular structure to a solid containing isolated pores is observed. By way of example, the average relative density of the aluminum honeycombs listed by Hexcel (2014) is 0.031 (standard deviation, $s = 0.014$).

15.3 APPLICATIONS IN PRIMARY STRUCTURES

An introduction to the structural elements that constitute a wing are given first, together with a brief explanation of their functionality. Several construction details of an ultralight monoplane are given afterward. A mention to the use of sandwich structures employed in primary structural components of jet engines is also treated.

15.3.1 CANTILEVER WING

The cantilever wing is a framework made up of spars and ribs covered with aluminum sheets, composite laminates or sandwich panels. The spars run along the wingspan, being the main structural members. The shape of the wing section or airfoil defines the aerodynamic characteristics of the wing. When an airfoil moves through the air, each surface element is subjected to a normal pressure stress p and a tangential viscous shear stress τ; their integration around the perimeter of the airfoil gives a resultant aerodynamic force $\mathbf{A}$ and a pitching moment $\mathbf{M_o}$. The ribs, which give the airfoil its shape, transmit the air loads from the wing covering to the spars. The spanwise flow of each airfoil produces a three-dimensional flow which occurs over the finite wing. This lift distribution provokes the wing to bend upward, taking up the contour of a smile. Thus, size and configuration of the wing structure are dictated by flexure and torsion acting together. In addition, wing bending due to lift induces compression of the extrados (wing top surface) and tension of the intrados (wing bottom surface), so additional reinforcing elements are added in order to stabilize the outer skins. Resultants $\mathbf{A}$ and $\mathbf{M_o}$ are transmitted to the fuselage through the spars and rib attachment points at the wing root.

The standard construction practice of wing extrados and intrados consists of sheet metals stiffened with riveted, bonded or co-cured slender stiffeners also known as stringers. They are aimed at resisting in-plane compressive and pressure loads. Plain sheet metals require to be folded in order to support compressive loads (this action increases the second moment of area I, together with the radius of gyration and the bending stiffness). Stringers are used to achieve higher stress levels with relative thin skin thicknesses. Therefore, the design criteria dictate that stiffeners should accomplish sufficient rigidity to remain straight; that the whole stiffened surface must not buckle as it would for a plain sheet; and that the stringers do not fail due to proper local buckling modes (Davies, 2003). When thicker skins are necessary (for instance, due to high surface pressures which imply high wing loads), one alternative is the use of sandwich panels.

Figure 15.3 illustrates a typical structural layout of the outer wing panel for a small airplane. It consists of a one-cell beam with a front and rear spar and

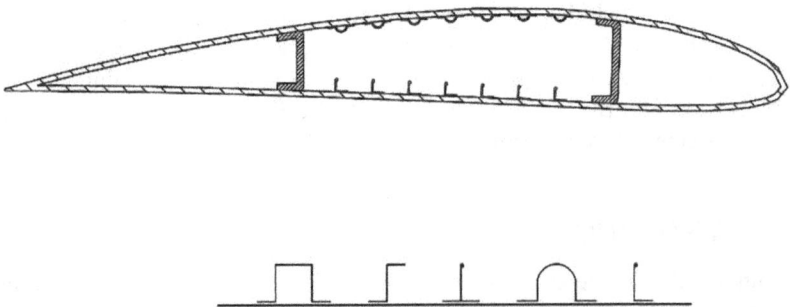

FIGURE 15.3 Typical cantilever wing section breakdown (a); sample of stringers used to avoid buckling of wing skins: hat stiffener, Z-section, blade, channel and angle (b).

multiple spanwise stingers placed between them. The leading and trailing edge cells are considered ineffective to load bearing.

15.3.2   ULTRALIGHT MONOPLANE

The aircraft depicted in Figure 15.4a is a two-seat ultralight braced monoplane fabricated mostly with composite materials. It is powered by a four-stroke, flat-six, air-cooled engine (90 kW @ 3,300 rpm) and a ground-adjustable, two-blade

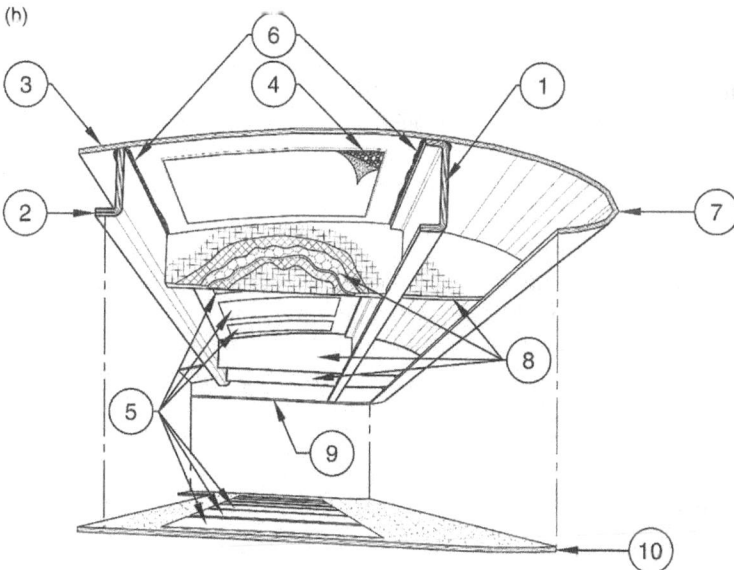

FIGURE 15.4   (a) Monoplane; (b) wing structure details and materials: ① front spar (C-Ep), ② rear spar (C-Ep), ③ wing extrados (C-Ep), ④ and ⑤ skin stiffeners (asymmetric sandwich, C-Ep skins/foam nucleus), ⑥ epoxy adhesive, ⑦ leading edge, ⑧ ribs (symmetric sandwich, C-Ep skins/honeycomb core), ⑨ wing tip and ⑩ wing intrados.

tractor propeller. Its MTOW (maximum takeoff weight, weight at which the aircraft is certified for takeoff due to structural limits) is limited to 680 kgf.

A brief description of the main structural elements is given next. The wing is constructed with symmetric and asymmetric sandwich and monolithic CFRP laminates (see Figure 15.4b). Asymmetric sandwich panels are employed instead of the traditional monolithic skin panels reinforced with stringers depicted in Figure 15.3. They provide stabilization of the outer skins of the wing extrados and intrados between the spars and ribs. The main structural element of the wing is the front spar which consists of a C-channel. The rear spar is an L-beam that provides fixation to the ailerons and flaps. The wing ribs are fabricated with symmetric sandwich panels. In contrast, the spars, the leading and trailing edges and the wing tip are constructed with monolithic CFRP layups. The control surfaces are fabricated from symmetric sandwiches with the skins made of CFRP and a foam or honeycomb core.

The fuselage consists of a semi-monocoque structure reinforced with transverse frames. It is produced in two halves via wet layup. Once cured, they are bonded together with strips of CFRP. Each half of fuselage includes the vertical tail fins and the firewall. Monolithic CFRP laminates predominate in the construction of these elements. The transverse frames are fabricated with symmetric sandwich panels and bonded to the fuselage in subsequent operations, together with the floor framing. Epoxy is the main adhesive employed in bonding. The undercarriage, engine mount and propeller are metallic. The wing and main landing gear are attached to the fuselage with metallic hardware.

15.3.3 Jet Engines

The ducted fan of a high bypass ratio turbofan engine maximizes the propulsive force generated from the power supplied by the turbine. Its blades are major structural components.

Over three decades ago, Rolls-Royce developed hollow, wide-chord titanium alloy fan blades subjected to fatigue, high centrifugal loads and impacts. They consist of skins that are separated and supported by a thin-walled small cell honeycomb. High-strength and good-quality finish joints (skin-to-skin and honeycomb-to-skin) are fused together by diffusion bonding (Fitzpatrick and Broughton, 1988). Recent developments comprise stainless steel foam sandwich or aluminum honeycomb sandwich with 60%C-Ep facings (Min et al., 2015; Coroneos and Reddy, 2012).

15.4 MECHANICAL BEHAVIOR

15.4.1 Specific Mechanical Properties and Cost

Metals in aerospace applications are being substituted by materials with higher specific mechanical properties (e.g., specific strength or specific stiffness).

The strength (σ_{max}) of metals corresponds to the maximum tensile stress. As for composites, it agrees with the tensile stress at rupture. Concerning honeycombs and foams, strength stands for the maximum compressive stress of stabilized test probes (i.e., specimen with facings).

The term specific strength stands for the strength per unit mass of the material. It means that although a composite may have lower tensile strength than denser materials such as aluminum, titanium or steel, on a weight basis it will sustain larger loads (Callister, 2001, p. 418). Specific mechanical properties are quantified by the respective mechanical property–density ratio: specific strength, (σ_{max}/ρ); specific stiffness, (E/ρ). Apart from their advantageous structural performance, composite materials and structures must be cost-effective. Material cost per unit of mass can also be introduced into the equation with the specific ratios $(\sigma_{max}/\rho)/\cos t$ and $(E/\rho)/\cos t$. In closing, the greater the result, the better the performance the material will show with reduced weight and cost.

15.4.2 EQUIVALENT MECHANICAL PROPERTIES

Sandwich structures are composed of high-stiffness or high-strength facings and low-stiffness or low-strength but lighter core materials giving as result extremely high flexural stiffness-to-weight and high flexural strength-to-weight ratios (Lee and Suh, 2006, p. 15). From the standpoint of selection of materials, these are the main arguments—together with cost—that justify the use of composite sandwich structures in aerospace applications. Ashby proves the aforementioned properties with numbers. He proposes an approximate analytic method that computes the flexural stiffness and strength of an equivalent beam to that of a symmetric sandwich structure (see Figure 15.5). The equivalent beam is made of a homogeneous material and shares the same thickness, width and length as the sandwich. Expression (15.2) defines its equivalent flexural strength.

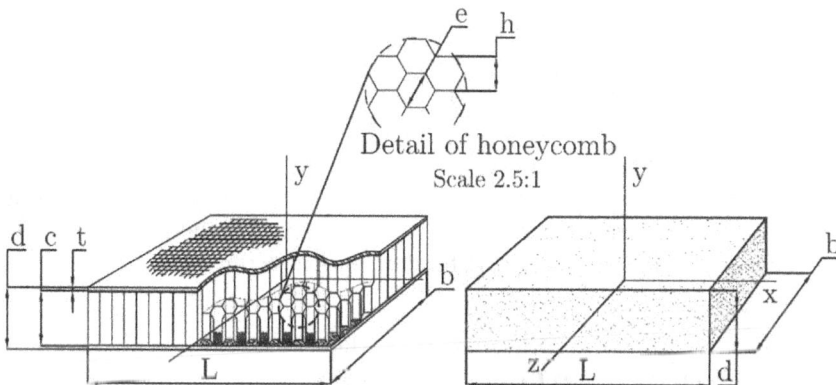

FIGURE 15.5 Parameters of a symmetric sandwich structure (a) and equivalent homogeneous material (b): d, sandwich thickness; c, core thickness; t, facing thickness; L, panel length between supports; and b, panel width.

$$\sigma_{\text{eq flex}} = \frac{M_f}{Z} = \frac{6M_f}{bd^2} \tag{15.2}$$

where M_f, bending moment; $Z = I/y$, section modulus; $I = bd^3/12$, second moment of area of a rectangular section about the neutral axis; and $y = d/2$, normal distance from the neutral axis of bending to the outer surface or any other intermediate point.

This approach allows consideration of the sandwich as a material rather than as a structure. The most advantageous outcome of this procedure is the possibility to compare the equivalent properties of the sandwich structure with those of conventional constituent materials of composite structures.

Ashby's (2011, §11.5) approach is implemented in Section 15.5 in order to compare the specific mechanical properties of a range of materials used in aerospace. Among them, a combination of materials intended to be used in the structure of a satellite is included. The following subsections present the required information to compute the equivalent properties of sandwich structures.

15.4.3 Equivalent Flexural Stiffness

The path to compute the equivalent flexural stiffness of the symmetric sandwich shown in Figure 15.5 is straight and narrow with only two expressions needed: the equivalent density ρ_{eq}, Expression (15.2), and the equivalent flexural modulus E_{eq}, Expression (15.3). The former follows the rule of mixtures, that is, an arithmetic average of the properties of the substructures of the sandwich weighted by their volume fractions: f for both facings and $(1-f)$ for core. E_{eq} represents the flexural modulus of the homogeneous material.

$$\rho_{\text{eq}} = \frac{2t}{d}\rho_f + \frac{c}{d}\rho_c = f\rho_f + (1-f)\rho_c \tag{15.2}$$

$$E_{\text{eq}}^{-1} = \frac{1}{E_f}\left\{\left[1-(1-f)^3 + \frac{E_c}{E_f}\right](1-f)^3\right\}^{-1} + \frac{B_1}{B_2}\left(\frac{d}{L}\right)^2\frac{(1-f)}{G_c} \tag{15.3}$$

where c, core thickness; d, sandwich thickness; t, facing thickness; L, sandwich length between supports; ρ_c, density of core; ρ_f, density of facings; E_c, elastic modulus of core; E_f, elastic modulus of facings; G_c, shear modulus of core; B_1/B_2, ratio given in Table 15.1; and $(d/L)^2$, bending-to-shear term.

15.4.4 Failure Modes of Sandwich Panels Subjected to Flexural and Compressive Loading

Size and configuration of most of primary aircraft structures is dictated by the combination of flexural, compressive and torsion loading. Such is the case of wing extrados which is called upon to resist high compression loads as the wing bends

TABLE 15.1

Ratios Describing Loading Modes and Boundary Conditions of Sandwich Beams Subjected to Flatwise Bending Loads (Ashby, 2011, p. 321)

Type of Beam	Loading Mode (flatwise)	B_1/B_2	B_3/B_4
Cantilever	Point load at free edge	3	1
	Uniformly distributed load	4	2
Simply supported	Point load at half-length	12	2
	Uniformly distributed load	9.6	4
Built-in	Point load at half-length	48	4
	Uniformly distributed load	48	6

upward due to lift distribution. When an aircraft is on the ground, the roles invert and the wing intrados is under compression due to the weight of the wing itself, engines and fuel comprised.

Sandwich structures are prone to fail in numerous different ways depending on the type of loading and fixation of the panel ends; the mechanical properties of the materials; and the sandwich geometric dimensions. The analysis of the failure modes of sandwich panels is by itself a complex topic owing to the elaborate interactions of single failure modes, and the nonlinearities and inelastic behavior of the materials of the substructures. A particular failure mode may initiate and interact with other modes, and the final failure may follow another failure mode. For instance, *shear crimping* is considered as a local mode of failure of facings and core that leads to *global buckling* (see Figure 15.6). *Shear crimping* is due to *core shear instability* and *interface shear failure* at the crimp (Hackman, 1973). First, the overall buckle appears, and then the crimp occurs attenuating the overall buckle. The examination of a failed sandwich may indicate *crimping* or *core shear instability* (see Figure 15.7), but failure could have started as *global*

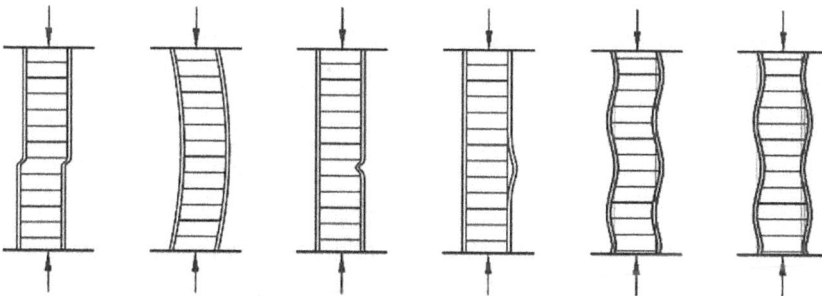

FIGURE 15.6 Possible modes of failure of sandwich structures under edgewise loading (from left to right): *shear crimping, global buckling, intracell buckling, facing wrinkling* away from the core (with *separation*), antisymmetrical *wrinkling* and symmetrical *wrinkling* (with core *crushing*).

FIGURE 15.7   Cardboard core shear instability due to bending fatigue (Hernández, 2002).

buckling that provoked *crimping* in the end. Names of failure modes adopted for sandwich structures are written in italics.

The nature of the core influences significantly the failure mode. If the core is a honeycomb, facings may dimple or buckle into the spaces between the cells. If the amplitude of the dimples is large enough, *facings wrinkling* is prone to occur. Thus, *dimpling* may imply plastic deformation of thin metallic facings.

The experimental results listed in Table 15.2 show that different loading types give rise to similar failure modes. Through the previous examination of several failure modes of sandwich, the great depths of the subject become evident.

15.4.5   EQUIVALENT FLEXURAL STRENGTH

Three specific modes of failure occurring under bending loading are reviewed as follows: *facings compressive failure* (yielding and buckling) and *core shear failure*.

15.4.5.1   Facings Yielding

$$\sigma_{\text{eq yielding}} = \sigma_f \left(1 - \left(1 - f\right)^2\right) + \sigma_c \left(1 - f\right)^2 \qquad (15.4)$$

where $\sigma_{\text{eq yielding}}$, equivalent flexural facing yielding strength; σ_f, yield stress of facings; and σ_c, yield stress of core. Expression (15.4) takes the form of the rule of mixtures. It is obtained by postulating that facings and core reach yielding simultaneously.

TABLE 15.2

Failure Modes Observed of Sandwich Structures according to Loading

Loading	Substructures of Sandwich				
	Facings	Core	Interface	Global	Source
Axial compression	Compressive failure, wrinkling	Compressive failure, shear instability	N/A	Buckling	a*
Pure bending	Compressive failure, wrinkling	N/A	N/A	N/A	a*
Combined bending and shear	Compressive failure, wrinkling (subsequent to core failure)	Shear due to combined loading (critical), shear crimping, yielding, plastic deformation	N/A	N/A	a*
Possible failure modes	Tensile or compressive failure, indentation	N/A	Debonding	Buckling	a
Fatigue	N/A	Shear instability	Separation	N/A	b

Sources: (a) Daniel et al. (2003); (b) Hernández (2002).
NB N/A is employed when failure modes are not reported by the respective source.
*Indentation is avoided intentionally.

15.4.5.2 Facings Buckling

$$\sigma_{eq\ buckling} = 1.71 f \left(1 - \frac{f}{2}\right)^2 \left(E_f E_c^2\right)^{\frac{1}{3}} \tag{15.5}$$

where $\sigma_{eq\ buckling}$, equivalent flexural facing buckling strength. Expression (15.5) is obtained by substituting expressions (15.6) and (15.7) into (15.1).

$$\sigma_{cr} = K \left(E_f E_c^2\right)^{\frac{1}{3}} \tag{15.6}$$

$$I_f = y^2 A = bt \frac{(t+c)^2}{2} \tag{15.7}$$

where σ_{cr} is the compressive critical stress in the facings; K is computed in terms of Poisson's ratio of the core ν_c (if $\nu_c = 0.3$, $K = 0.57$; refer to Allen (1969, p. 159)); and I_f agrees with the moment of inertia of the cross-sectional area A of the facings obtained with the expression of the radius of gyration only. The influence of the moment of inertia of the core is omitted, and the facings are considered too thin.

15.4.5.3 Core Shear Failure

$$\sigma_{eq\ core\ shear} = \left[\sigma_f f^2 + 4\tau_c \left(\frac{d}{L}\right)^{-1} (1-f)\right] \left(\frac{B_3}{B_4}\right)^{-1} \tag{15.8}$$

where $\sigma_{eq\ core\ shear}$, equivalent flexural core shear strength; B_3 / B_4, ratio given in Table 15.1; and τ_c, shear strength of core material.

The first value of flexural strength reached ($\sigma_{eq\ min}$) of either facings or core dominates over the rest. It can be identified for each specific case under analysis with criterion (15.9).

$$\sigma_{eq\ min} = \min\left(\sigma_{eq\ yielding}, \sigma_{eq\ buckling}, \sigma_{eq\ core\ shear}\right). \tag{15.9}$$

15.4.6 INDEX LINES

A series of parallel straight lines, called index lines, are drawn for each plot. They serve as a reference of the properties of the materials that are common to those crossed by the same line. Besides, they aid in the optimization of the sandwich proportions given by f, and as a useful reference in the search of new materials that equate or exceed the properties of the existing materials. They take the index form $(Q^{1/m} / \rho) = k$, where Q is the material property; m is the slope of the family of parallel lines belonging to that index; and $m \log_{10} k$ is the intercept of each line given with Expression (15.10). For instance, concerning the E–ρ plot, a material with $k = 1$ stands for a beam with 1/10 of the mass of one with $k = 0.1$.

$$\log_{10} Q = m \log_{10} \rho + m \log_{10} k \tag{15.10}$$

15.5 CASE STUDY: SANDWICH MATERIALS EMPLOYED IN SPACECRAFT STRUCTURES

This section is dedicated to the comparison of a range of materials based on specific mechanical properties and cost. These criteria are employed in the selection process of a sample of materials intended to be used in the primary structure of a microsatellite. The first topic deals with structural requirements.

15.5.1 MICROSATELLITE

The case study consists of the structure of an experimental microsatellite. The prefix "micro" indicates a mass range between 10 and 100 kg. Its payload consists of telecommunication experiments. Its primary structure houses the payload and is designed to transmit mechanical and thermal loads that will be produced during the integration, certification, launch, injection and operation in orbit. Besides, the structure must withstand the effects of severe spatial environment (extreme temperature variations, ultra-high vacuum outgassing, atomic oxygen flux etching and erosion, ultraviolet degradation, and impacts from meteoroids and man-made space debris). A redesign process of the original all-metal structure, motivated by a compulsory reduction in the total mass, seeks to reduce the mass of the structure by using alternative construction materials and side panels concepts (Vargas and Nocetti, 2020).

The primary structure consists of a central cylindrical column, three horizontal plates and four side panels (see Figure 15.8). The side panels ($d = 12.7$ mm)

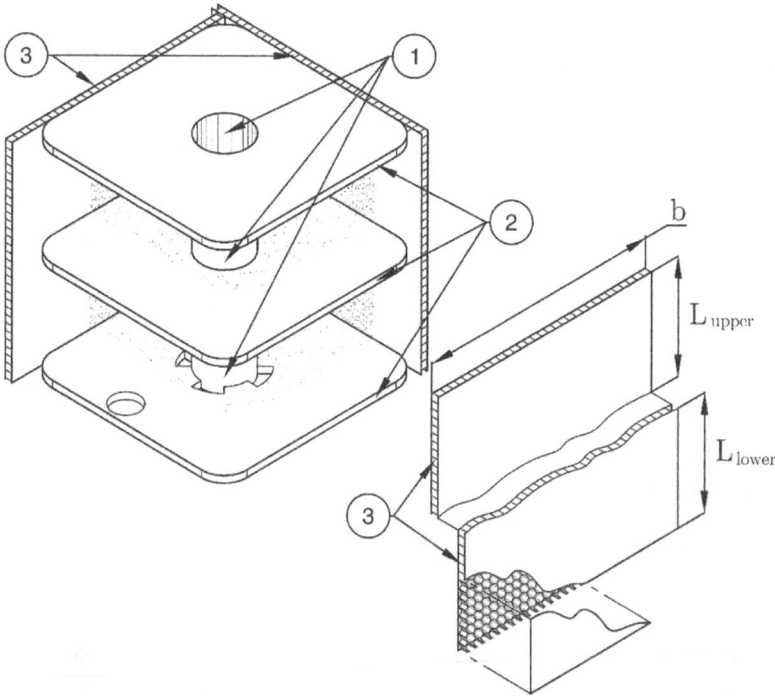

FIGURE 15.8 Primary structure: ① central column; ② horizontal plates; and ③ side panels.

are fabricated as sandwich structures made of aluminum alloy (7075-T73) skins ($t = 0.4$ mm; $f = 0.063$) bonded to an Aluminum 5052 honeycomb core. The side panels are clamped around their perimeter and at half-length ($b = 437$ mm; $L_{upper} = 211$ mm; $L_{lower} = 213$ mm).

15.5.2 ALTERNATIVE MATERIALS

The following analysis is about the comparison of the specific mechanical properties of the combination of C-Ep, aluminum alloys and foams. Their possible combinations are listed in Table 15.3. They are used in the constitution of alternative

TABLE 15.3
Material Combinations

Material Combination	Specification (Facings–Core)
#1	Al 7075-T73–Al 5052 honeycomb
#2	Al 7075-T73–Alporas foam
#3	60%C-Ep laminate–Al 5052 honeycomb
#4	60%C-Ep laminate–Alporas foam

TABLE 15.4

Mechanical Properties of Materials under Comparison

Material	Relative Cost	Material Properties				
		Density [kg/m³]	Relative Density ×100	Young's Modulus [MPa]	Strength [MPa]	Source
Al 7075-T73	1.00×	2,710*	100	71,110	386	a
HexWeb™ CR-PAA™ 3/8-5052-0.005 honeycomb	10.8	104	3.84	1,827**	2**	b
Alporas closed-cell foam	N/A	256	9.44	500	2	c, d
60%C-Ep quasi-isotropic laminate	4.2××	1,530	N/A	50,380	320	a
63%C-Ep AS4/3501-6 unidirectional	4.3××	1,580	N/A	142,000	2,280	e
65%C-Ep IM6/SC1081 unidirectional	4.4××	1,600	N/A	177,000	2,860	
Carbon fiber AS4	6.4×	1,810	N/A	235,000	3,730	
Al 2024-T3	1.0×	2,800	100	73,000	414	
Steel AISI 1025	0.4×	7,800	N/A	207,000	394	
Titanium Ti-6Al-4V	3.4×	4,450	N/A	108,000	896	
Al foam AlSi12	N/A	600	22.14	4,900	16	f
Al foam AlSi12	N/A	800	29.52	8,400	29	
Al foam AlMg1Si	N/A	600	22.14	3,900	20	
Epoxy resins	0.8×	N/A	N/A	N/A	N/A	

Sources: (a) Vargas and Nocetti (2020); (b) Hexcel (2014); (c) Mohan et al. (2005); (d) Mohan
       et al. (2011); (e) Daniel and Ishai (1994, Ch. 1); (f) Gutiérrez and Oñoro (2008).
(×) After Infosys (2018).
(××) Computed with the rule of mixtures.
(*) Considered as ρ_s for all the aluminum foams.
(**) Plate shear in W-direction (lowest).

sandwich structures (Materials #2 to #4) to the original one (Material #1). Other materials are included as reference in Table 15.4. Metallic foams are considered as space debris shock absorbers.

15.5.3 MECHANICAL PROPERTIES EVALUATION

The material combinations listed in Table 15.3 are evaluated via their specific mechanical properties with the method shown in Figure 15.9.

For a finer interpretation of these charts, the larger the index $(Q^{1/m}/\rho) = k$ is, the better the material performance. Concerning the E_{eq}/ρ_{eq}–σ_{eq}/ρ_{eq} chart, the best choice is found above and to the right.

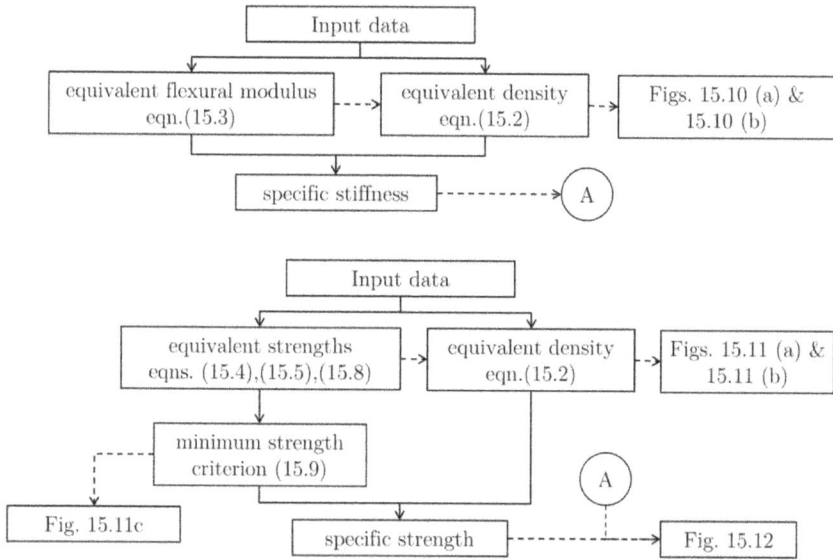

FIGURE 15.9 Method implemented to evaluate the specific mechanical properties of the material combinations under study.

15.5.3.1 Equivalent Flexural Stiffness

Figure 15.10a plots the equivalent flexural modulus of the equivalent homogeneous material according to its density for different relative thicknesses of faces (f), one boundary condition ($B_1 / B_2 = 48$) and two ratios d/L. The analysis is conducted over the whole panel ($d / L = 0.03$) and either the upper or lower half-panel ($d / L = 0.06$) assuming flatwise loading. The constants of the respective index lines are $m = 3$ and $0.001 < k < 0.013$ GPa/(kg/m^3). In Figure 15.10b, the four combinations listed in Table 15.3 are plotted. Half-panels ($d / L = 0.03$) are the only ones inspected hereinafter.

15.5.3.2 Equivalent Flexural Strength

Figure 15.11a illustrates simultaneously the three modes of failure for Material Combination #1, while Figure 15.11b is the resultant envelope after accounting for the competition among them with criterion (15.9). Facings yielding dominates for $f < 0.1$, while facings buckling does for $f > 0.2$. A transition zone is found in between. Figure 15.11c shows the resultant envelope for all the combinations of materials. Mind $m = 2$ and $0.003 < k < 0.050$ MPa/(kg/m^3).

15.5.3.3 Specific Mechanical Properties and Cost

The procedure permits to plot the specific modulus–specific strength chart of the combination of materials of Table 15.3 in Figure 15.12a, together with some reference materials found in Table 15.4. Sandwich material combinations are limited to $f < 0.13$. The straight lines with unitary slope agree with strain relations

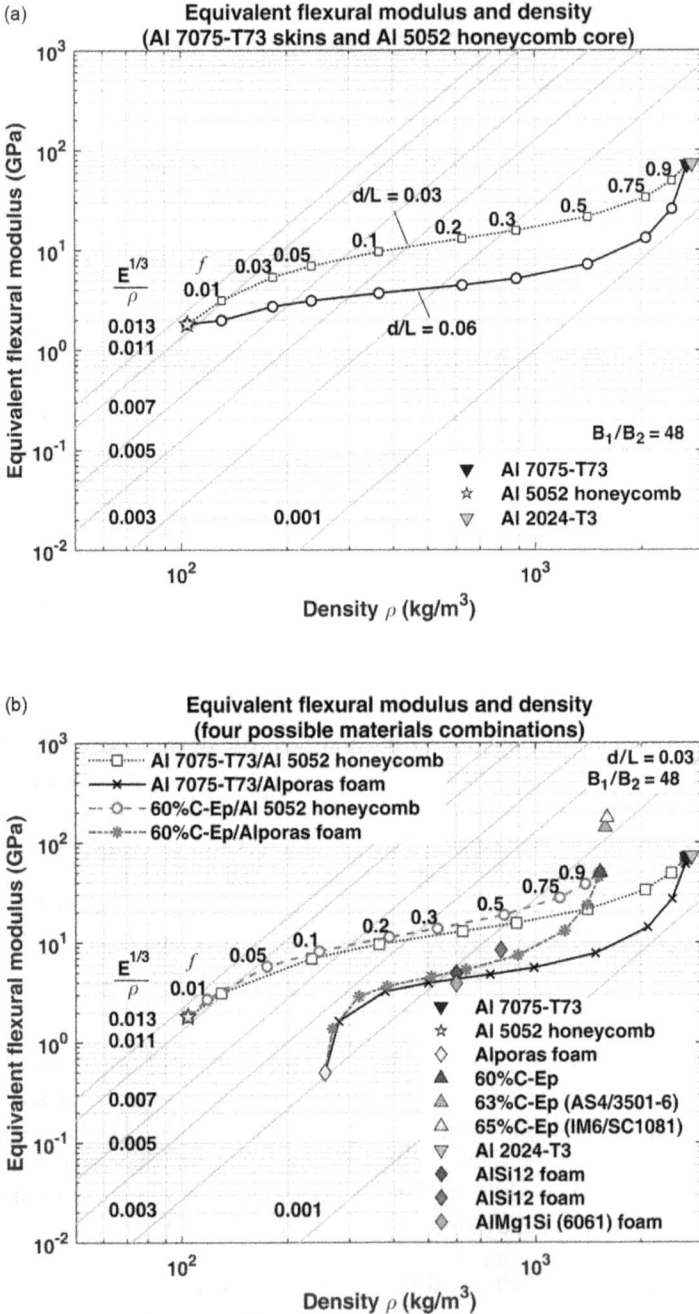

FIGURE 15.10 Log–log plots of equivalent flexural stiffness and density: (a) Material Combination #1; (b) all material combinations.

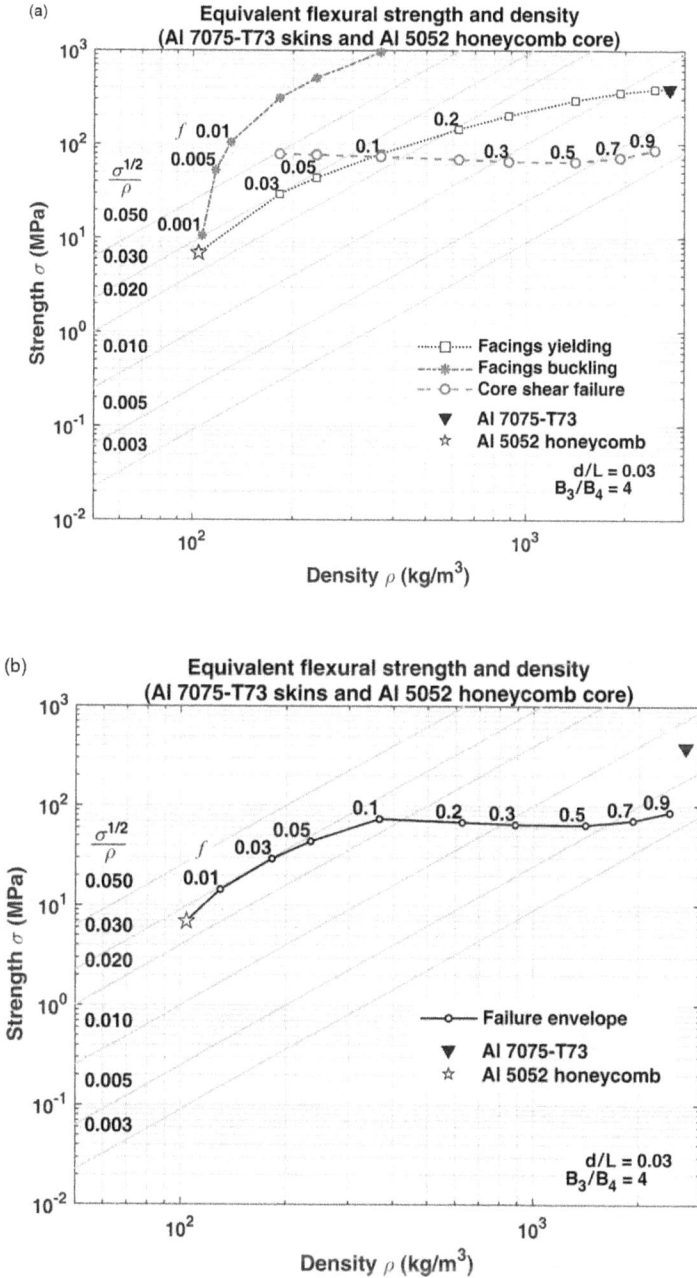

(a) Equivalent flexural strength and density
(Al 7075-T73 skins and Al 5052 honeycomb core)

(b) Equivalent flexural strength and density
(Al 7075-T73 skins and Al 5052 honeycomb core)

FIGURE 15.11 Log–log plots of equivalent flexural strength and density: (a) Material Combination #1; (b) failure envelope of Material Combination #1; (c) failure envelopes of all combinations.

(Continued)

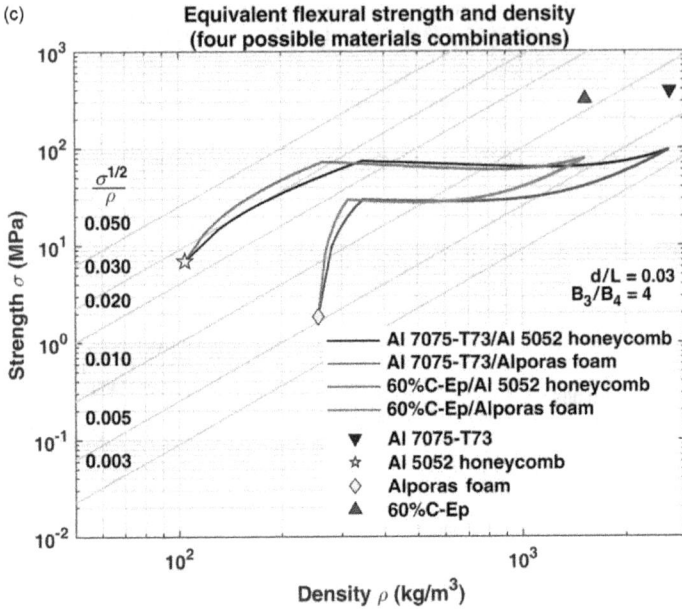

FIGURE 15.11 (*CONTINUED*) Log–log plots of equivalent flexural strength and density: (a) Material Combination #1; (b) failure envelope of Material Combination #1; (c) failure envelopes of all combinations.

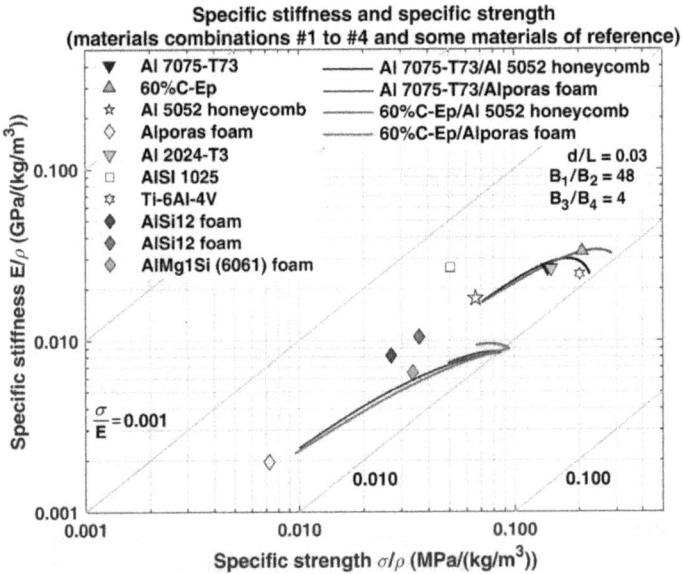

FIGURE 15.12 Log–log plot of the specific mechanical properties for all combinations and materials of reference.

defined with Expression (15.11), where E_{eq} and σ_{eq} are given in MPa and $k \in \mathbb{N}$. Process-based costing is not considered.

$$\frac{\sigma_{eq}}{E_{eq}} = 10^{-k} \tag{15.11}$$

Costs are included in Figure 15.13 by considering the $(E/\rho)/\text{cost}$ and $(\sigma_{max}/\rho)/\text{cost}$ criteria.

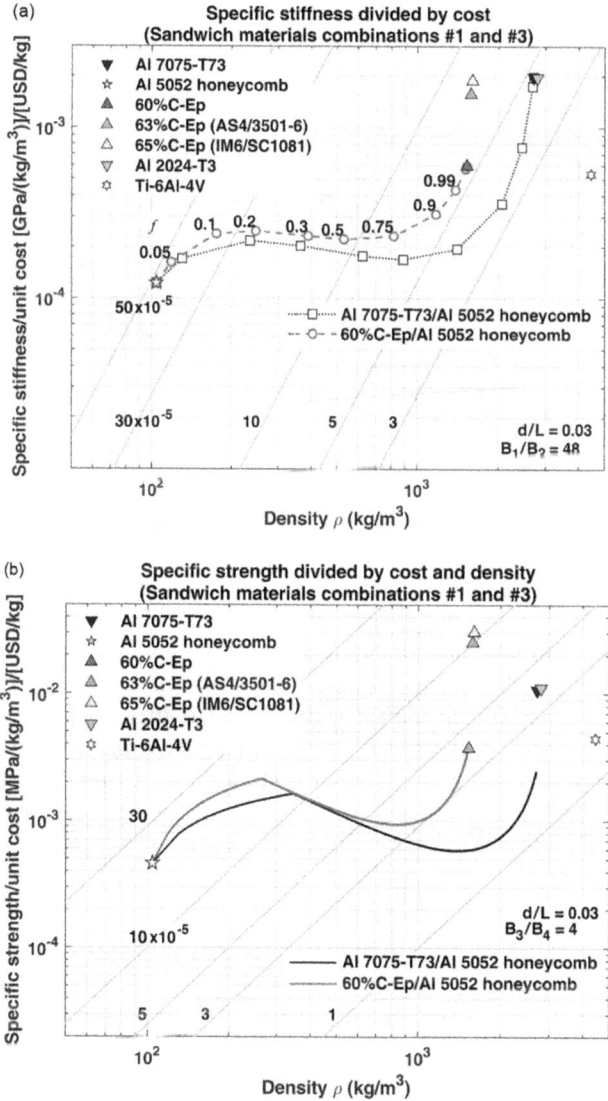

FIGURE 15.13 Log–log plot of specific mechanical properties divided by cost: (a) specific stiffness/cost; (b) specific strength/cost.

15.5.3.4 Discussion

Figures 15.10–15.13 illustrate how sandwich panels with different proportions of facings–core create new sets of materials offering complementary performances, sometimes better with respect to the existent monolithic ones. In particular, Figure 15.10a shows that each combination of input variables (B_1/B_2, B_3/B_4 and d/L) creates a line representing a new set of materials. Rather than a line drawn on each graph, each of the four material combinations in Table 15.3 could be seen as an area. Concerning strength of any material combination, the chart in Figure 15.11a evinces that the dominating failure modes are facings yielding and core shear failure, resulting in the envelopes plotted in Figure 15.11b and c. For its part, Figure 15.12 shows that Material Combinations #1 and #3 outperform the Al and Ti alloys and the AISI 1025 steel for the highest values of f within the inspected domain. In contrast, Al foam sandwich leaves much to be desired in all aspects. Further attempts with metallic foams with higher relative densities are recommended.

Cost-related charts show that the 60%C-Ep laminate–Al 5052 honeycomb is less expensive than its all-metal counterpart, but without being cheaper than traditional monolithic materials. Notice that although the 60%C-Ep laminate has similar unit cost to its $6x$%C-Ep unidirectional counterparts, it reflects a lower profit without considering process-based costing yet.

The effects of mechanical behavior of the sandwich as structures and of space environment still need to be evaluated.

15.5.4 Buckling

A crucial aspect in the analysis of sandwich structures is buckling. The previous approach focused on considering the panel as a homogeneous material. However, "the main difference in design procedures for sandwich structural elements as compared to design procedures for homogeneous material is the inclusion of the effects of core shear properties on deflection, buckling, and stress" (Hackman, 1973). A core with low stiffness makes the effects of transverse shear critical, being the predicted global buckling load higher if they are not accounted for (Kassapoglou, 2013). This effect is visible through the comparison of Euler's formula for pin-ended columns, Expression (15.2), with Expression (15.13) which accounts for the influence of core.

$$P_{cr} = \pi^2 \frac{E_f I_f}{L_e^2} \tag{15.12}$$

$$P_{cr} \approx \kappa \pi^2 \frac{E_f I_f}{L^2 + \kappa \pi^2 \dfrac{E_f I_f}{G_c b (c + 2t)}} \tag{15.13}$$

where L_e, column effective length and κ, coefficient according to column ends condition (see Table 15.5). Notice the absence of the term $E_c I_c$.

TABLE 15.5

Effective Length L_e and Coefficient κ of Columns

Column Ends Condition	$L_e{}^a$	κ^b
Fixed–free	$2L$	0.25
Pinned–pinned	L	1
Fixed–pinned	$0.7L$	2.04
Fixed–fixed	$0.5L$	4

Sources: (a) Beer and Russell (2012, p. 642); (b) Gay (1997).

The following formulas aid in pre-sizing of honeycomb sandwich panels subjected to edgewise loading.

15.5.4.1 Facings Critical Compression Stress

$$\sigma_{cr} = 3 \left[\frac{E_f E_c^2}{\left[12(3 - v_c)^2 (1 + v_c)^2 \right]} \right]^{\frac{1}{3}} \tag{15.14}$$

15.5.4.2 Facings Critical Compression Load

$$P_{cr} \approx 1.64 E_f t \sqrt{\frac{E_c}{E_f} \frac{t}{c}} \tag{15.15}$$

15.5.4.3 Facings Critical Stress before Wrinkling

$$\sigma_{cr} = 1.03 \left[\frac{t_{cell}}{h} E_f E_c G_c \right]^{\frac{1}{3}} \tag{15.16}$$

15.5.4.4 Cells Instability Critical Stress

$$\sigma_{cr} = 2 E_f \left(\frac{t_{cell}}{e} \right)^2 \tag{15.17}$$

where t_{cell}, cell (foil) thickness; h, cell width (across the flats); and e, diagonal length (see details in Figure 15.5).

15.6   MAINTENANCE

The goal of structural maintenance is to maintain airworthiness throughout the life of an aircraft in an economical manner; it is based on the assessment of design information, fatigue and damage tolerance evaluations and test results, and service experience with similar equipment (Akdeniz, 2003). The design methods of structures of commercial jet airliners in the 1950s and 1960s drew upon the *fail-safe* concept (attribute of the structure that permits to retain its required residual strength after failure of a principal structural element). However, failure modes were not always predicted accurately to ensure that structural failures would be obvious and within safety limits. In the 1970s, the implementation of adequate inspection planning led to the timely detection of structural discrepancies (e.g., barely detectable cracks) before they become critical. Consequently, *damage tolerance* was embraced as a design philosophy.

Concerning composites and sandwich structures, fail-safe design is complicated to produce and validate owing to the need to demonstrate a capability for load redistribution after failure (Backman, 2008, p. 81, 84). This is particularly true for compression due to complex failure mechanics. In such case, fail-safe design requires to lean on design development testing, design data testing and compliance testing. The preservation of limit load capability is a design objective for composite and sandwich structures. Ongoing research copes with damage tolerance, fail-safe design criteria, requirements for maximum damage sizes, and categorization of damage and its origins. Once in service, walk-around inspections (detailed, surveillance and special detailed) that seek for external damage are effective tools in preserving safety.

Damages in composites and sandwich structures cannot be detected by visual inspection as it is possible for aluminum components. Detection of imperceptible damages requires nondestructive inspection technologies such as ultrasonic or X-ray procedures (see Figure 15.14). Once detected, the surrounding area has to be inspected to make sure that no fluid ingress (water, deicing fluids and hydraulic fluid), delamination or debonding occurred. Injection repairs provide a moisture entry path, too. Besides, sandwich and laminate composites are sensitive to impacts. These occurrences provoke high maintenance costs because of the need of frequent inspections. In such case, maintenance costs can diminish the positive aspects of the high specific mechanical properties of sandwich to the extent that heavier constructions could be more economical over the life cycle of the component. Hence, it becomes imperative to take into consideration the life cycle when assessing the economic viability of sandwich structures (Kleineberg et al., 2002).

15.7   CONCLUSIONS

Several technical details of the sandwich structure construction are briefly reviewed throughout this text: applications in aerospace, mechanical behavior under flexure, failure modes under compression, and maintenance. A comparison of the specific mechanical properties of monolithic and cellular solids and four

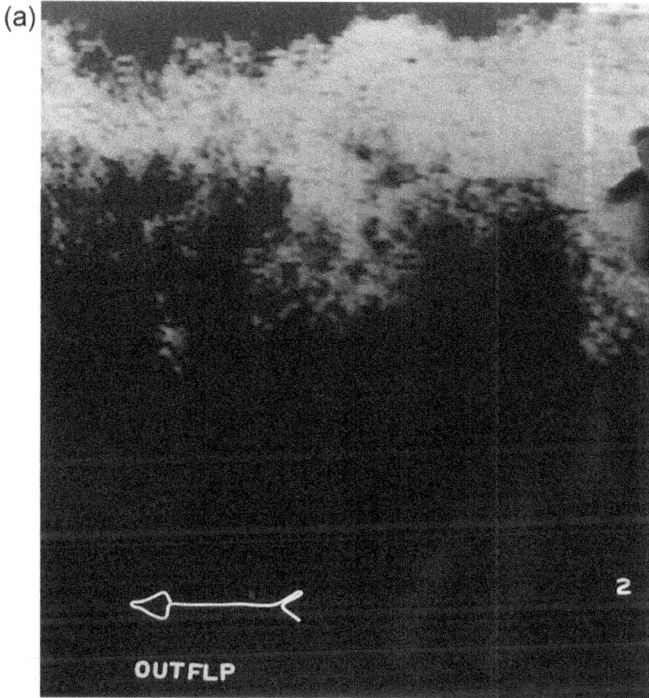

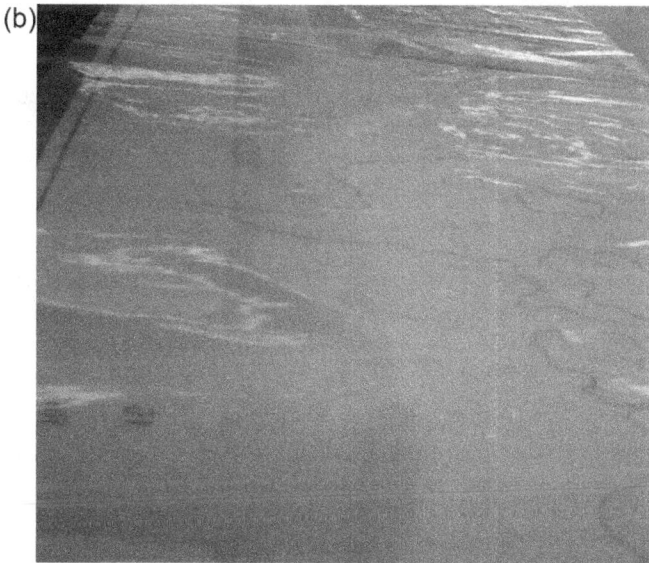

FIGURE 15.14 Special detailed inspection: (a) X-ray of an aileron showing water ingress; (b) location of the water in the aileron according to radiography in the workshop.

proposed own combinations is conducted. Results show that the convenience of the materials in a sandwich structure can exceed aluminum and titanium alloys of aerospace grade. However, when the issue of raw materials cost is put on the table, it is observed that sandwich structures begin to lose ground, and even more so when maintenance cost is factored in.

Composite sandwich structures are broadly discussed in the literature because they show good performance where high specific stiffness is required; however, there is a lack of data regarding the failure mechanics of sandwich subjected to torsion, a circumstance that contrasts notably with bending, buckling, fatigue or impact. Recent research has focused mainly on analytical or numerical models. Being wings and fuselages the structural components that are subjected to simultaneous flexural, column and torsional loads, it would be of interest to delve further into this matter.

The asymmetric sandwich makes progress in terms of its use in typical lifting surfaces with respect to its counterpart. Its use in higher-demand applications relies on solving manufacturing, joining, inspection, maintenance and cost-related challenges.

15.8   NOTE

Mention of commercial or trade names does not imply any endorsement of the products by the author. They are used for identification purposes only.

ACKNOWLEDGMENTS

I wish to thank Dr. K. Senthilkumar for inviting me to participate in this endeavor. I express my gratitude to Professor Carlos Nocetti-Cotelo for his encouragement over time.

REFERENCES

Ackermann, J.F. and Gleason, G.R. 2008. Hybrid composite panel systems and methods. US Patent 2008/0277531A1:1–9 Publication date: November 13th, 2008.

Akdeniz, G. 2003. Section 18: Aircraft maintenance/Part 6: Airframe maintenance. In *The Standard Handbook for Aeronautical and Astronautical Engineers*, ed. Davies, M., 18.52–18.61. New York: McGraw-Hill.

Allen, H.G. 1969. *Analysis and Design of Structural Sandwich Panels*. Oxford: Pergamon Press.

Ashby, M.F. 2011. *Materials Selection in Mechanical Design*, 4th ed. New York: Butterworth-Heinemann.

Backman, H.F., 2008. *Composite Structures Safety Management*, 2nd ed. Oxford: Elsevier.

Beer, F.P. Russell-Johnston, E.R. et al. 2012. *Mechanics of Materials*, 6th ed. New York: McGraw-Hill.

Callister, W.D. 2001. *Fundamentals of Materials Science and Engineering*, 5th ed. New York: John Wiley & Sons.

Coroneos, R.M and Reddy-Gorla, R.S. 2012. Structural analysis and optimization of a composite fan blade for future aircraft engine. *Int. J. Turbo Jet Engines* 29(3):131–164.

Daniel, I.M. and Ishai, O. 1994. *Engineering Mechanics of Composite Materials.* Oxford: Oxford University Press.

Daniel, I.M., Gdoutos, E.E., et al. 2003. Deformation and failure of composite sandwich structures. *J. Thermoplast. Compos. Mater.* 16:345–364. Doi: 10.1177/089270503030670.

Davies, G. 2003. Section 9: Aerospace structures/Part 3: Structural considerations. In *The Standard Handbook for Aeronautical and Astronautical Engineers*, ed. Davies, M., 9.15–9.43. New York: McGraw-Hill.

Fitzpatrick, G.A. and Broughton, T. 1988. The diffusion bonding of aeroengine components. In *Sixth World Conference on Titanium*, France, 1451–1456.

Gay, D., 1997. *Matériaux Composites*, 4ème éd. Paris: Éditions Hermès.

Gibson, L.J. and Ashby, M.F. 1997. *Cellular Solids, Structure and Properties*, 2nd ed. Cambridge: Cambridge University Press. Doi: 10.1017/CBO9781139878326.

Gutiérrez-Vázquez, J.A. and Oñoro, J. 2008. Aluminum foams. Manufacture, properties and applications. *Rev. Metal.* 44 (5):457–476. Doi: 10.3989/revmetalm.0751.

Hackman, L. 1973. Sandwich construction and design. In *Analysis and Design of Flight Vehicle Structures*, authors Bruhn, E. and Bollard, R., 12.1–12.52. Carmel: Jacobs Publishing.

Hernández-Moreno, H. 2002. *Development of a Composite Material for Structural Application* (in Spanish). MSc report. Mexico City: National Polytechnic Institute.

Herrmann, A.S., Zahlen, P.C. and Zuardy, I. 2005. Sandwich structures technology in commercial aviation, present applications and future trends. In *Sandwich Structures 7: Advancing with Sandwich Structures and Materials*, ed. Thomsen, O., Bozhevolnaya E., Lyckegaard A. : 13–26. Dordrecht: Springer. Doi: 10,1007/1-4020-3848-8_2

Hexcel, 2014. HexWeb® CR-PAA™ Phosporic oxide anodized aluminum honeycomb. *Product data sheet:* 1–5.

Infosys. 2018. Carbon composites are becoming competitive and cost effective [White paper]. Retrieved from: www.infosys.com/engineering-services/white-papers/documents/carbon-composites-cost-effective.pdf.

Kassapoglou, C. 2013. *Design and Analysis of Composite Structures with Applications to Aerospace Structures.* New York: John Wiley & Sons.

Kleineberg, M., Wenner, U. and Hanke, M. 2002. Cost-effective CFRP fuselage manufacturing with liquid resin infusion technologies. In: *Proceedings of Institut für Faserverbundleichtbau und Adaptronik Workshop*, Braunschweig: 1–10.

Kolesnikov, B.Y. and Herbeck, L. 2004. Carbon fiber composite airplane fuselage: concept and analysis. In *Proceedings of ILA Conference (Russia in European Research Programs on Aeronautics)*, Berlin: 1–11.

Lee, D.G. and Suh, N.P. 2006. *Axiomatic Design and Fabrication of Composite Structures: Application in Robots, Machine Tools and Automobiles.* New York: Oxford University Press.

Lewis, A. 2014. Making composite repairs to the 787. *AeroMagazine* 56 (04):5–13. Seattle.

Megson, T.H.G. 2018. *Introduction to Aircraft Structural Analysis*, 3rd ed. New York: Butterworth-Heinemann.

Min, J.B., Ghosn, L. and Lerch, B.A. 2015. A study for stainless steel fan blade design with metal foam core. *J. Sandwich Struct. Mater.* 17(1):56–73. Doi: 10.1177/1099636214554181

Mohan, K., Hon, Y.T., et al. 2005. Failure of sandwich beams consisting of alumina face sheet and aluminum foam core in bending. *Mater. Sci. Eng.: A* 409 (1): 292–301. Doi: 10.1016/j.msea.2005.06.070

Mohan, K., Hon, Y.T., et al. 2011. Impact response of aluminum foam core sandwich structures. *Mater. Sci. Eng., A* 529:94–101. Doi: 10.1016/j.msea.2011.08.066.

Moreland, D.W. 2009. Mechanical systems safety. In *Safety Design for Space Systems*, ed. Musgrave, G.E.: 549–606. New York: Butterworth-Heinemann.

Vargas-Rojas, E. and Nocetti-Cotelo, C.-A. 2020. Alternative proposal, based on systems-engineering methods, aimed at substituting with carbon-epoxy laminates the load-bearing aluminum sandwiches employed in the structure of a small satellite. *Adv. Space Res.* 66 (2020): 193–218. Doi: 10.1016/j.asr.2020.04.004.

Wang, C.H. and Duong, C.N. 2016. *Bonded Joints and Repairs to Composite Airframe Structures*. New York: Academic Press.

Wilmes, H., Kolesnikov, B. and Fink, A. 2002. New design concepts for a CFRP fuselage. In: *Proceedings of Institut für Faserverbundleichtbau und Adaptronik Workshop*, Braunschweig, 1–7. Braunschweig: DLR (Deutsches Zentrumfür Luft- und Raumfahrt)

16 Crashworthiness Applications of the Composite Sandwich Structures

S. Hassouna, M. Janane Allah, and A. Timesli
Hassan II University of Casablanca

CONTENTS

16.1 Introduction 321
16.2 Metrics Used in Crashworthiness Studies 323
    16.2.1 Specific Energy Absorption.. 323
    16.2.2 Mean Crushing Force 324
16.3 Crushing Modes.. 324
16.4 The Axial Compression of Sandwich Composite Tubes 327
16.5 The Three-Point Bending of Sandwich Composite Tubes 332
16.6 Automotive Sandwich Composite Structures for Crashworthiness:
    Applications Review 339
16.7 Marine Sandwich Composite Structures for Crashworthiness:
    Applications Review 342
16.8 Concluding Remarks 343
References... 345

16.1 INTRODUCTION

Crashworthiness is the absorption capacity of high impact energy of a material during collision in an irreversible, progressive and controlled manner. A high crashworthiness value is desirable for a material when it is used to reduce the impact force on transportation cargo and passengers. Given that, in the last two decades although much works [Ramakrishna et al. 1998, Mahdi et al. 2001, 2003, 2005, Khalid et al. 2002, Warrior et al. 2003, Abosbaia et al. 2003, Elgalai et al. 2004, Mamalis et al. 2004] have been done on composite materials' crashworthiness capabilities, the goal of most of these works is the understanding of the crashworthiness of tube-like geometries. They have shown that the good designation of composite materials to crush in a stable manner can provide good energy absorption capabilities. On the other hand, little works have been realized to explain the effectiveness of composite sandwich structures (CSSs) as energy absorption devices. Tube-like structures should not be neglected. Because cylindrical sandwich tubes have high strength-to-weight ratio and extremely high flexural stiffness-to-weight ratio, their applications are of larger importance unlike other structures.

Among the researches in the field of crashworthiness of CSS are Goldsmith and Sackman (1992), Found et al. (1997), Mamalis et al. (2002, 2005), Tarlochan et al. (2005), Ji et al. (2011), Ji et al. (2012), Kalhori et al. (2015) and Nguyen et al. (2005). These works show that early buckling can be avoided by adjusting the designation of sandwich panels, which can demonstrate progressive crushing, giving excellent energy absorption capacity, as shown by Tarlochan et al. (2007). Among the recently published papers in this research area are Mamalis et al. (2013), Sun et al. (2014, 2019), Jing et al. (2019) and Ha and Lu (2020). Mamalis et al. (2013) investigated the effects of mechanical and geometrical characteristics on the energy absorption of CSS panels under in-plane compression. In this paper, they used the LS-DYNA3D code based on an explicit finite element formulation. Sun et al. (2014) analyzed crushing of functionally graded thin-walled structures using a multi-objective optimization. Sun et al. (2019) performed numerical and experimental studies on the crashworthiness of metal foam in hybrid tubular CSS. Jing et al. (2019) investigated the three-point bending of layered-gradient foam core sandwich beams under low-velocity impact. This study was conducted using experimental and numerical approaches. Zhu et al. (2019) did the same study, but of composite sandwich laminated box beam and by three methods: analytical, numerical (finite element method) and experimental. This work discusses the influence of stiffness performance, ultimate load and failure modes, and their influencing factors. Ha and Lu (2020) presented a review on crashworthiness of thin-walled corrugated structures; this work is devoted to crashworthiness designs and energy absorption characteristics. Haug et al. presented a new material model made entirely of composite sandwich material to predict the crash behavior of a prototype passenger car compartment structure. The components of this model are made of continuous fiber-reinforced composite as carbon fiber, which allows to increase the rigidity of materials [Timesli 2020a, 2020b] because of its exceptionally high elastic modulus [Timesli 2020c].

The study of crashworthiness is very important in different fields such as aerospace, automotive, transportation and defense industries. This paper focuses on automotive safety as a field of study, which requires the occupant safety in vehicle design objectives [Mulla Salim et al. 2013]. Accident statistics show that frontal collisions are the reasons of serious injuries for drivers [Sen et al. 2013]. The passengers and vehicle must be saved by ensuring that vehicle structures are stiff enough and not be so soft to protect pedestrians, drivers and vehicles [Srinivas et al. 2017]. The two requirements mentioned above are opposed. The safety of a vehicle requires an optimum design, which imposes that the vehicle has the right amount of stiffness to absorb right amount energy at the moment of crash and ensure non-penetration into the passenger compartment.

This chapter is outlined as follows: Section 16.2 presents metrics used in crashworthiness studies. Section 16.3 explains crushing modes. Sections 16.4 and

16.5 show the axial compression of sandwich composite tube and the three-point bending of sandwich composite tube, respectively. Some crashworthiness applications of sandwich composite structures are presented in Sections 16.6 and 16.7, and this chapter ends with concluding remarks in Section 16.8.

16.2 METRICS USED IN CRASHWORTHINESS STUDIES

16.2.1 Specific Energy Absorption

The specific energy absorption (SEA) is the energy absorbed per unit mass of material, and it is defined as:

$$SEA = \frac{W}{m} \tag{16.1}$$

Where m is the mass and W is the total energy absorbed in crushing; W is given by:

$$W = \int_{\Delta displacement} F_{Av} dx = F_{Av}\, d \tag{16.2}$$

With d being the displacement and F_{Av} being the average load over the crush zone, which is defined as:

$$F_{Av} = \int_0^L F(x)\, dx \tag{16.3}$$

Based on these precedents definitions, we can rewrite SEA in the following form:

$$SEA = \frac{F_{Av}\, d}{\rho A L} \tag{16.4}$$

where ρ is the density of the material, A is the cross-sectional area, and L is the length of the crushed part of the tube. In general, it's difficult to deduct the average crush load due to its volatility. In practice, it is common to calculate the total energy absorbed W corresponding to the area under the force–displacement curve. Thus, the SEA per mass unit is obtained by Eq. (16.1). The average force on the crush zone can be deducted using Eq. (16.4).

As the specimens are all crushed in a brittle manner, the term collapsibility is mostly used compared to elastic behavior problems. The ratio d/L is a metric of the collapsibility of the structure. Replacing $c = d/L$ in Eq. (16.4), we obtain the SEA as a function of collapsibility:

$$SEA = \frac{F_{Av}\, c}{\rho A} = \frac{\sigma\, c}{\rho} \tag{16.5}$$

where σ means crush stress.

16.2.2 Mean Crushing Force

Mean crushing force (MCF) is the total energy absorbed divided by the crushed structure distance obtained from the force–displacement curve. It is expressed by:

$$\text{MCF} = \frac{1}{d} \int_0^d F_{Av}\, dx \qquad (16.6)$$

Major researches have been conducted on the energy absorption capability of structures in tube shapes, because this geometry is easy to manufacture. Other studies were performed on a three-point flexural beam with square or rectangular cross section. In all crashworthiness studies, these metrics are mainly used to determine the absorption capability, in the case of metal/steel materials as well as composite sandwich specimens.

16.3 CRUSHING MODES

To analyze the role of composite faceplates in the energy absorption of sandwich panels, Gary and Robert (1989) grouped the crushing mechanisms into four crushing modes: "brittle fracturing," "transverse shearing," "local buckling" and "lamina bending," resulting in different energy absorption efficiencies. Many experiments, as per Tarlochan et al. (2007), have proven that the first three modes resulted from "brittle fiber-reinforced composites," and not from "ductile fiber-reinforced composite," as Kevlar. Local buckling can be faced by both composites: ductile and brittle, as per the case of edge compression. Most experiments showed a combination of modes, such as the "edge compression crushing experiments" (see Section 16.4) and "three-point bending" (see Section 16.5), which are related to the composite design and mechanical properties. Let us discuss the description of each mode in the following.

- Transverse Shearing Crushing Mode

    The "transverse shearing" crushing mode is described by the formation of lamina bundles made by laminate cross section with one or more short inter-laminar and longitudinal cracks on the faceplates. The load is not transferred uniformly across the surface of the tube. This is a consequence of the scalloped cross surface tube. The main energy absorption mechanism is the transverse shearing of the edges of the lamina bundles (Figure 16.1).

    The lamina bundles act as columns to resist the applied compression load. As the load applied increases, the inter-laminar cracks grow until the edges of the column are sheared. The length of lamina bundles of specimen depends on the structure and constituent material properties and is less than the laminate thickness. The short length of the failed material results in an efficient crushing mode.

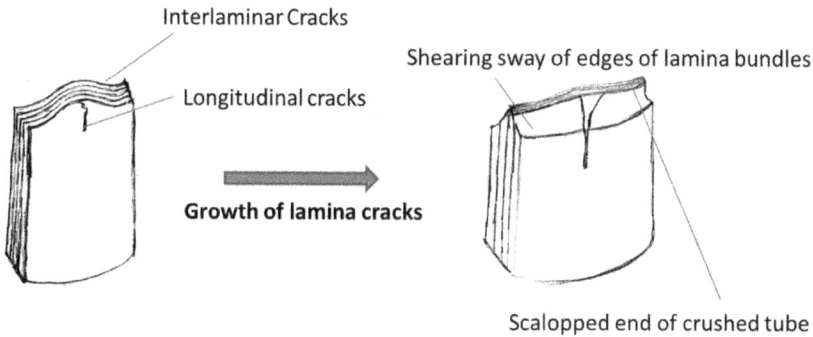

FIGURE 16.1 Crushing mode characteristics of transverse shearing crushing mode.

- Brittle fracturing crushing mode:

    The brittle fracturing crushing mode displays similar crushing characteristics to the transverse shearing mode: inter-laminar and longitudinal cracks, scalloped crushing surface and principal energy absorption mechanism (failure of the lamina bundles). The length of the interlaminar cracks is equivalent to 1–10 laminate thicknesses, which is one to ten times the lamina bundles length for the transverse crushing mode. The longer the fractured lamina bundle, the less efficient the crushing mode (Figure 16.2).

- Lamina bending crushing mode:

    The lamina bending crushing mode is similar to the brittle crushing mode; lamina bundles are one to ten times the length of the laminate thickness and are subjected to significant bending deformation, but lamina bundles are not fractured. The parallel-to-fiber cracks propagate within a ply or within several adjacent laminas that have common fiber orientation.

    The long length and the nonexistence of fractures result in inefficient energy absorption (Figure 16.3).

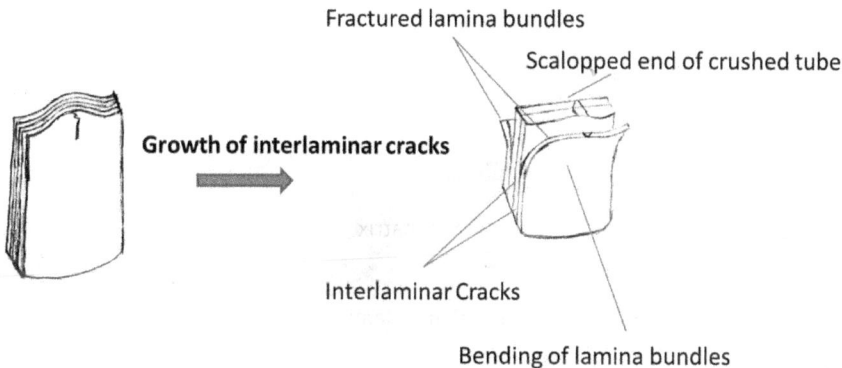

FIGURE 16.2 Crushing characteristics of brittle crushing mode.

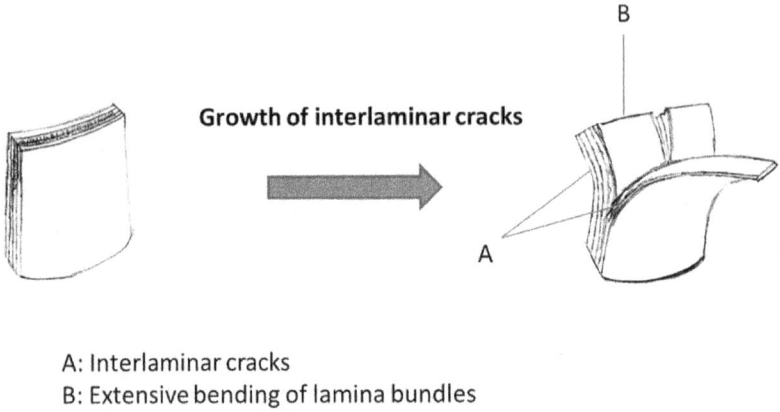

A: Interlaminar cracks
B: Extensive bending of lamina bundles

FIGURE 16.3   Crushing characteristics of lamina bending crushing mode.

- Local buckling crushing mode:
  The local buckling mode can occur in both ductile fiber-reinforced composite materials such as Kevlar and brittle fiber-reinforced materials made of carbon and glass (Tarlochan 2007). The crushing behavior is similar to that observed in ductile metals. This mode is described by the formation of local buckles through plastic deformation of the material. Ductile composite materials deform plastically at the buckle site along the compression side of the buckled structure. Delamination can also occur between plies because of fibers split along the buckled side, but the composite demonstrates post-crushing integrity. In contrast to the first two crushing modes described above, the buckling crushing mode does not show any bundles fracture and results in an inefficient energy absorption capability; a small part of energy is absorbed owing to plastic deformation and bundles formation (Figure 16.4).

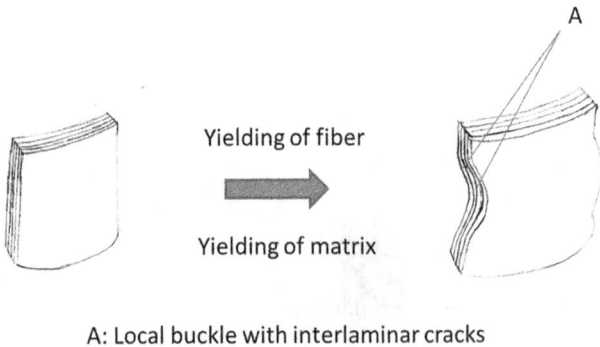

A: Local buckle with interlaminar cracks

FIGURE 16.4   Crushing characteristics of local buckling crushing mode.

16.4   THE AXIAL COMPRESSION OF SANDWICH COMPOSITE TUBES

The energy dissipation mechanisms are directly dependent on the collapse mode observed and can indicate how the energy is absorbed while the structure is crushing. Understanding the basic mechanics of energy dissipation mode could lead to the determination of the corresponding mode for a particular CSS configuration.

In case of compression, fibers play an important role as a column resistant to the compression load. In CSSs, the core material is the determinant of the structure stability and integrity (Mamalis 2013). A good choice of core can contribute to the sandwich integrity and avoid early buckling, leading to good energy absorption. In this section, experimental results are described to better understand the absorption mechanism of sandwich composite tubes subjected to axial compression and the role of the materials in good energy efficiency. The fibers used are glass in epoxy resin matrix, with different foam cores: polymethacrylimide (PMI), polyurethane (PUR) and polyvinylchloride (PVC).

Three modes of failure have been observed, each resulting in a form of energy absorption. Stable and controlled deformation (progressive collapse) is preferable, due to high part absorbing of the impact energy. All specimens start crushing after an initial elastic phase, but after a peak load, each tube follows a crushing mode. Mamalis et al. (2013) showed that we can obtain the same crushing mode as experimental results using the explicit finite element code LS-DYNA3D. Experimental results are described below separately.

- First crushing mode:

    The crushing of the sandwich specimen tube starts by a linear elastic compression that results in an increase in the stress load, which is followed by a decrease in loading and a stabilization of the force along the displacement of the moving crosshead. The initial phase is called peak load and is a metric to assess the efficiency of energy absorption of a material.

    This mode is characterized by an early global column buckling of the CSS associated with debonding of face sheets and foam core, in the compressed side of the tube, after an initial elastic compression. This is due to the "shear failure" of the core material, resulting in a frictional resistance at the interface between face sheets and core material. At an early stage, the flexural rigidity of the panel is reduced by the partial failure of the sandwich structure debonding, which leads to the bending and then to the fracture of the faceplate laminates. This fracture has contributed to a localized debonding of the faceplate from the sandwich core and a rapid drop in the CSS strength leading to bending, which decreased also the energy absorption capability as shown in Figure 16.5.

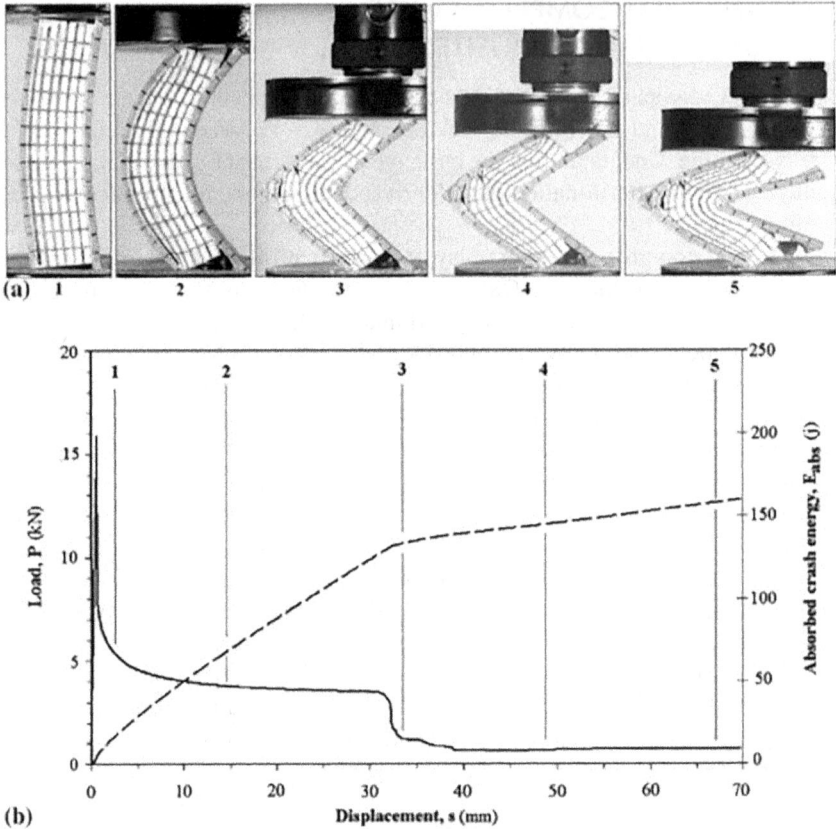

FIGURE 16.5 (a) Failure mode I and (b) load–displacement curve [Mamalis et al. 2005].

From that point on, the resistance of the CSS panel to deformation became irrelevant in compression, which leads to plastic deformation and densification of the foam core.

The different steps of failure are listed below:

- Initial peak load owing to the elastic compression of the CSS (foam core and FRP face sheets);
- Global column buckling, followed by bending of the sandwich panel and the debonding of FRP faceplates after shear failure of the foam core;
- Face sheets delamination in the compressed side;
- Sliding of the foam core on the debonded FRP faceplate under high frictional resistance;
- FRP faceplates fracture; and
- Foam core permanent plastic deformation.

- Second crushing mode:

  As per the first failure mode, the second crushing mode is defined by an initial elastic compression followed by an unstable collapse mode, which will be detailed below:

  The failure starts by the delamination of both the FRP faceplates, and a layer of the foam core is disbanded and remains on the faceplate laminates. This is due to its weak strength to tension that is not sufficient to keep it bonded to FRP. The CSS starts to deflect in buckling, reducing the resistance to compression and energy absorption. After facing debonding, the foam core has been subjected to compressive buckling leading to shear failure and local buckling occurrence around a weakened section of the core. As the crosshead continues to move downward, buckling of the faceplates causes more propagation of delamination of the FRP faceplates, followed by relative sliding between the bundles of laminate plies resulted from delamination, sliding, bending and flexural damage of the delaminated bundles of plies.

  The factors that contribute to energy dissipation in this mode may be listed as follows:
  - Elastic compression of the FRP faceplates and the foam core;
  - Debonding of both the face sheets laminates from the foam core of CSS;
  - Buckling of the delaminated FRP face sheets;
  - Delamination and cracks transmission between the plies of the FRP laminate;
  - Slipping between delaminated FRP layers yielding in high frictional resistance;
  - Sliding of delaminated FRP while the moving press crosshead is moving down;
  - Bending and flexural damage of the delaminated FRP layers;
  - Permanent deformation of the bent fiber layers;
  - Shear failure of the foam core; and
  - Permanent and plastic deformation of the foam core.

  Figure 16.6 shows the different steps of mode II failure of the sandwich tube, as well as the load–displacement graph.
- Third crushing mode:

  It is called progressive end-crushing mode. Like the first two collapse modes I and II, this mode starts with an elastic compression followed by a symmetric debonding of both the sandwich face sheets close to pressing crosshead area. This failure mode is characterized by a bonding to the foam core of the faceplate and resistance to tension in the region between them. The starting of the inter-laminar fracture of the faceplates reduces the stiffness of the sandwich composite. This causes a sudden drop in the load stress, as marked in Figure 16.7. But, in general, the resistance to compression is higher than the two other modes and the composite maintains a part of the original rigidity. This "progressive

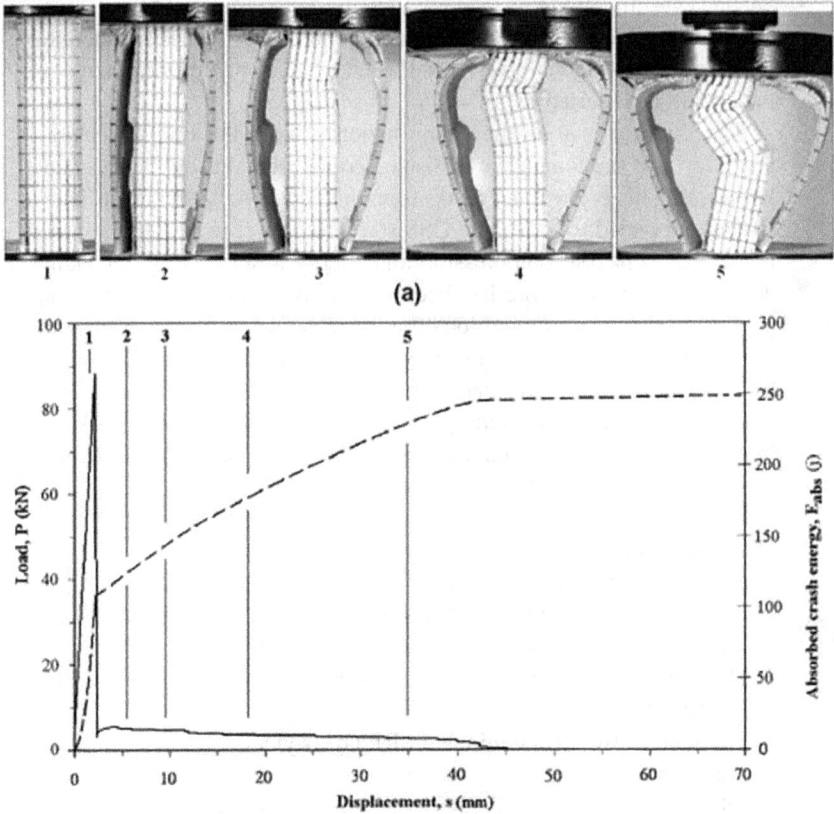

FIGURE 16.6   (a) Crushing mode II and (b) load–displacement curve [Mamalis et al 2005].

end-crushing" mode results in high average crushing load, P, and in low compressive load uniformity index, in contrast to the other first two failure modes. This is due to the large number of factors contributing to energy absorption. The factors that play an important role in crash energy absorption in the stable progressive end-crushing mode of sandwich panels are the following:

- Elastic compression of the FRP faceplates and the foam core;
- Delamination and cracks transmission between the plies of the FRP laminate;
- Elastic bending, permanent deformation and flexural damage of the delaminated faceplates;
- Sliding of delaminated FRP while the moving press crosshead is moving down;
- Sliding of delaminated bundles of plies on the moving press crosshead and on the debris wedge under high friction;

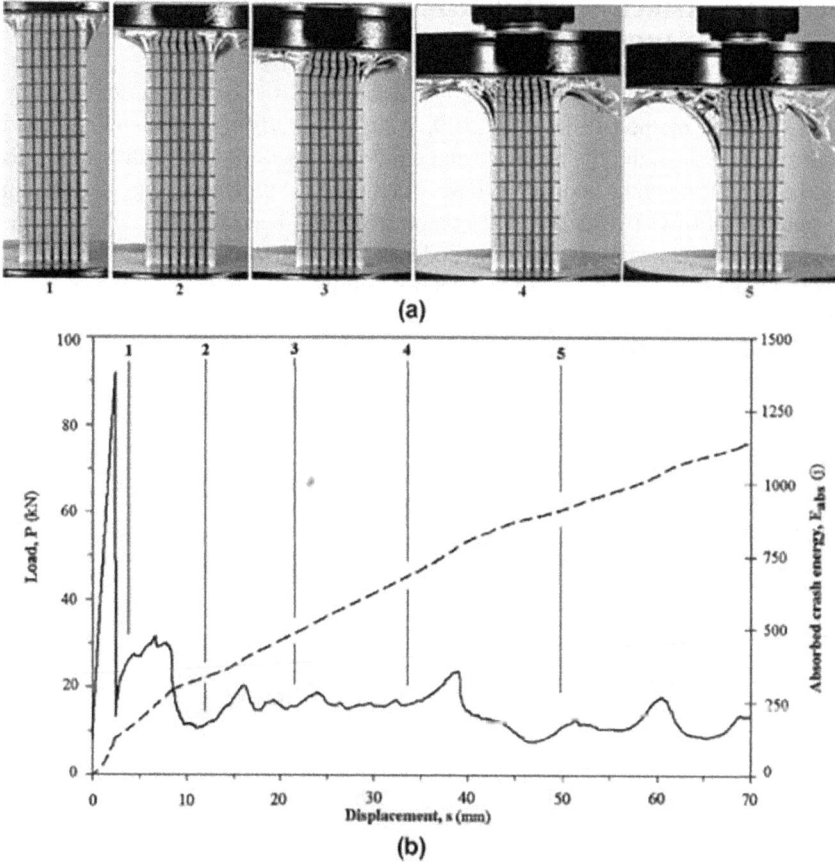

FIGURE 16.7   (a) Crushing mode III and load–displacement curve [Mamalis et al. 2005].

- Debonding of both the FRP laminates from the CSS foam core; and
- Permanent plastic deformation and densification of the foam core.

Note that for a CSS panel in edgewise compression, the strength of the foam core and the mechanical properties are the most influencing factors that control the collapse mode and the response to crushing. Experiences obtained by Mamalis et al. (2005) showed the importance of choice of the core in the achievement of crashworthy CSS panels to maintain a progressive integrity of the specimen while crushing and to avoid unstable failure mode. Thus, high-performance properties and strength of foam core could lead to mode III failure, which corresponds to high crashworthiness capabilities as shown in Figure 16.7. This progressive end-crushing mode is owing to the several micro-collapse mechanisms that occur after the initial elastic loading.

16.5 THE THREE-POINT BENDING OF SANDWICH COMPOSITE TUBES

Many experiments [Mamalis et al. 2008, Zhu et al. 2019] and numerical tests by finite element method [Zhu et al. 2019, Wang et al. 2019] have been conducted in order to understand the behavior and crashworthiness of sandwich composites in the case of bending impact. To have a comparative view of energy absorption capabilities in the two cases, i.e., compression and bending, the same materials (foam core and fibers) in the precedent case were studied by Mamalis et al. (2008). As shown in Figure 16.8, a lateral loading was applied on the plane to investigate the flexural capabilities of sandwich specimens. The crosshead speed is 5 mm/min, and the materials used are detailed in Table 16.1.

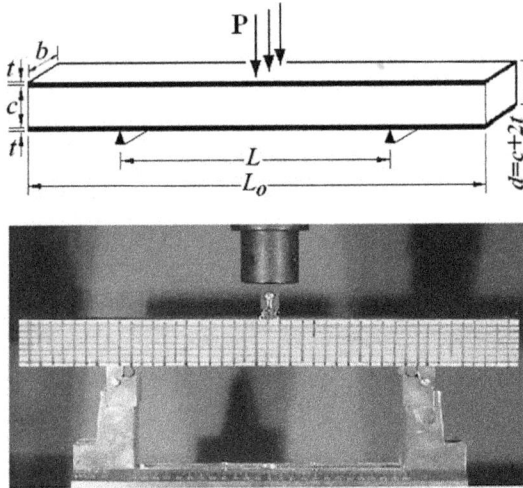

FIGURE 16.8 Sandwich test experimental setup [Mamalis et al. 2008].

TABLE 16.1
Mechanical Properties of Foam Core Materials

	Unit	PMI "Rohacell 71s"	PVC Cross-Linked "Herex C70.90"	Linear PVC "Airex R63.80"
Density	kg/m³	75	100	90
Compressive strength	N/mm²	1.5	2	0.9
Tensile strength	N/mm²	1.9	2.7	1.4
Tensile modulus	N/mm²	90	84	50
Elongation at break	%	3.5	23	75
Shear strength	N/mm²	1.2	1.7	1
Shear modulus	N/mm²	34	40	21

Figure 16.9 shows the results of the load–deflection curve grouped according to the fibers of faceplate laminates. C and D faceplate laminates are fiber glass layers combining modified acrylic resin with different fiber contents, 45% and

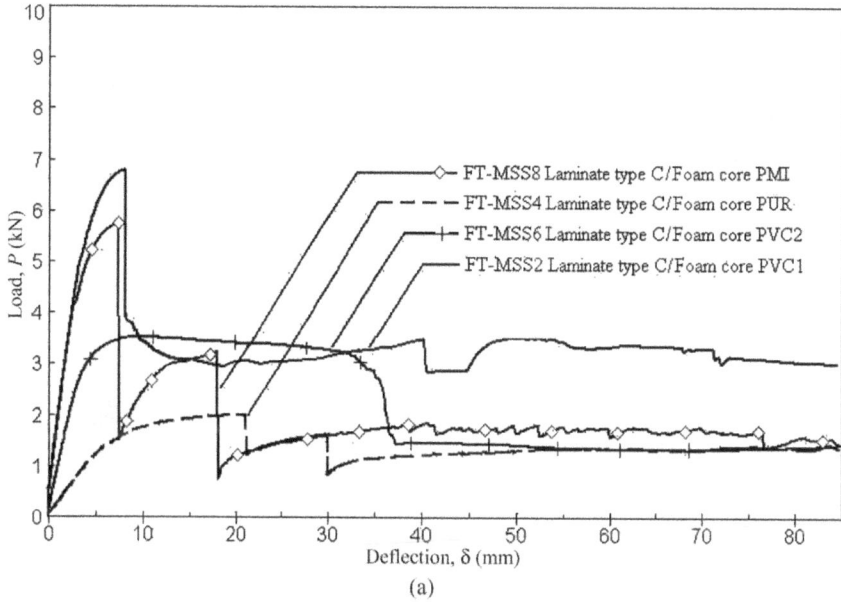

(a)

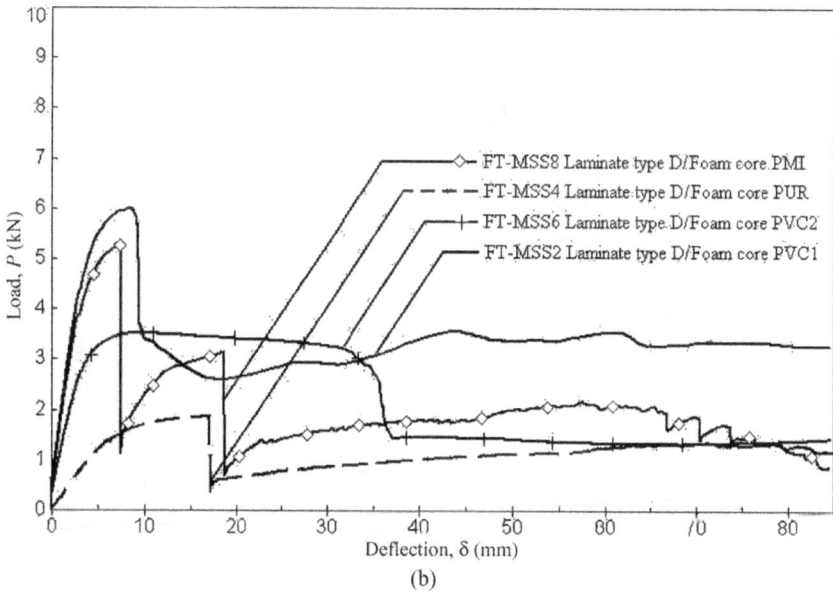

(b)

FIGURE 16.9   Load–deflection curve for faceplate type D (left) and C (right) [Mamalis et al. 2008].

41% for C and D types, respectively. In both the cases, i.e., C and D face laminate fibers, it is clear that energy absorption is dominated by the foam core material. Thus, the PVC1 (HEREX C 70.90) curve shows high crashworthiness capabilities and the PU foam core recorded the lowest SEA (Figure 16.10).

The experiences also showed that specimens crushed in two different manners. The authors have classified the results into two modes. The first one is unstable and less efficient in terms of crashworthiness capabilities in contrast to the second one that is more stable and more energy absorbing as depicted in Figure 16.11.

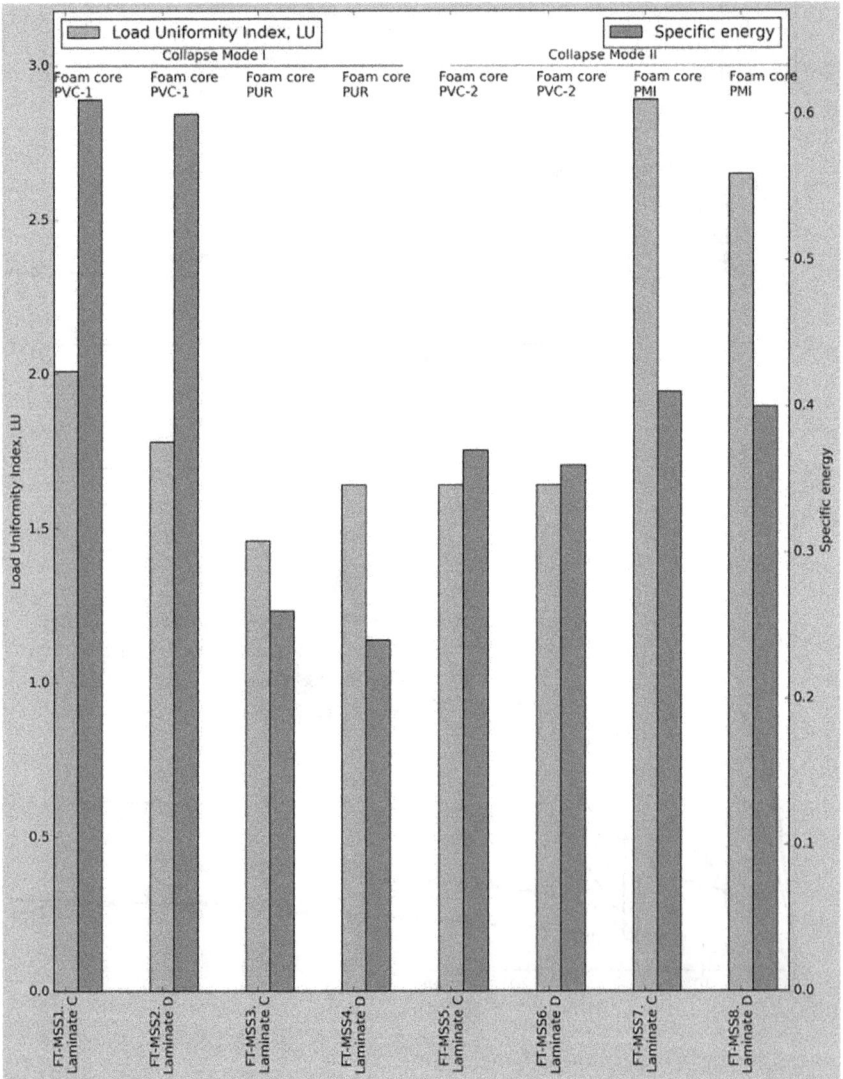

FIGURE 16.10 Load uniformity and SEA by collapse mode [Mamalis et al. 2008].

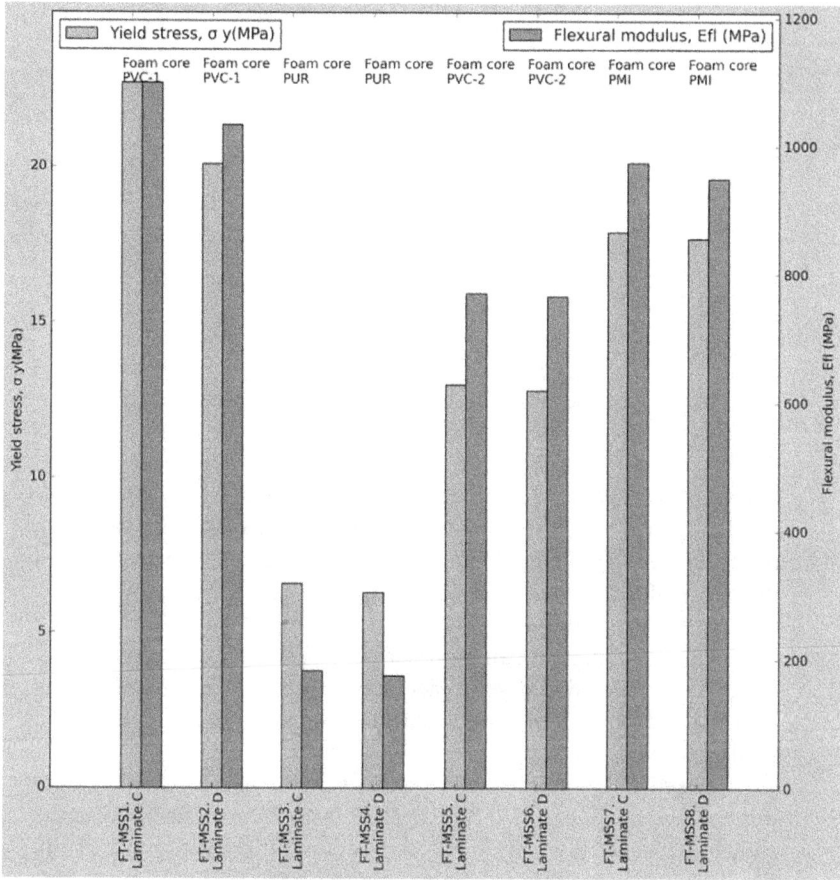

FIGURE 16.11   Mechanical properties of tested sandwich panels [Mamalis et al. 2008].

- Effects of faceplate laminate material:

    The face sheets carry the bending loads. Compared with the low-density core material, the face sheets are strong and stiff in compression and tension.

    In the majority of the cases, the energy absorption of laminates of type C is significantly higher than that of faceplates made by type D fibers (see Figure 16.9). This is due to the uniaxial-reinforcing fiber layers that have higher mechanical properties than FRP with the chopped-strand mat layers contained within type D laminate, and also, the fibers of type D have a volume content lower than the type C laminate. Therefore, the faceplate laminates of type C exhibit a higher Young's modulus, resulting in an efficient strength of CSS panel bending. Figure 16.12 summarizes the different flexural properties of different sandwich specimens.

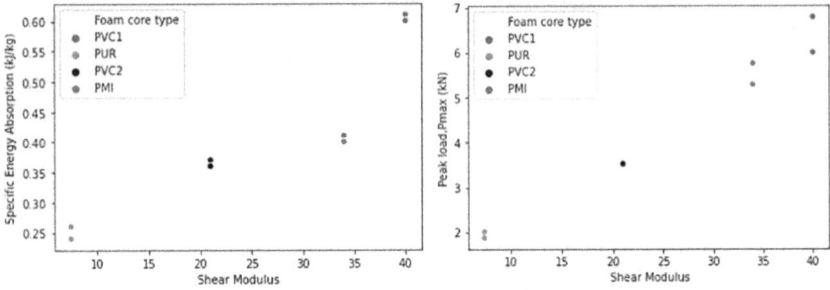

FIGURE 16.12 Specific energy absorption versus shear modulus of foam core.

According to the finite element method simulation carried out by Donga (2011), introducing angled plies in the laminate is a method for optimizing sandwich panels. He used the plies configuration [(0/30/-30/-60/60/90/0/S] for optimizing a sandwich beam. A comparison between a validated sandwich beam and the optimized sandwich beam showed that the mesh quality is the same and the contact laws in the two models have not changed. As shown by Donga (2011), load–displacement curve profiles were obtained by numerical simulations. From these numerical data, it can be observed that the peak load carried by the optimized sandwich beam increases approximately by 900%. The area covered by the load–displacement curve proves that the optimized sandwich beam gives higher energy absorption.

To improve the interfacial properties of the sandwich during crushing, the face sheets can be tufted together using "aramid fibers" over the entire CSS panel assembly [Dell'Anno et al. 2016, Henao et al. 2010]. For example, aramid threads were used by Blok et al. (2017) as reinforcement strategy throughout the thickness of the sandwich panels.

- Effects of foam core material:

    In sandwich materials, the core carries the shear loads. As depicted in Figures 16.10 and 16.11, foam core materials play an important role in the crushing mechanism. Mode II crushing mode, corresponding to PVC foam, has higher absorption capabilities. This is due to the stable and symmetric crushing that mainly involves the mechanism of crashworthiness and the mechanical strength of the foam material.

    The stable mode is also characterized by high elastic and shear properties of the foam core as per Gibson and Ashby's (1997) classification. These sandwich specimens made by PVC exhibit the highest shear strength as depicted in Table 16.1. Based on the data of Mamalis et al. (2008), we plot Figure 16.12, which shows SEA in terms of shear modulus of the foam core. It's clear from the curve that SEA is almost linearly dependent on shear properties. For the peak load, it is increasing linearly with the shear modulus of the foam. Thus, the crashworthiness of a CSS panel can be enhanced mainly by the selection of the adequate mechanical properties of the foam.

To understand the mechanism of energy dissipation, the two main dissipation modes are detailed hereafter.

- Crushing mode I:
  Mode I is characterized by non-symmetrical bending of the sandwich. As the deflection continues to grow, delamination of the face sheets and the core continues and fractures take place on the core material. Local damages, fracture and permanent deformation of the FRP face sheets lead to very low absorption of crash energy as depicted in Figure 16.13.

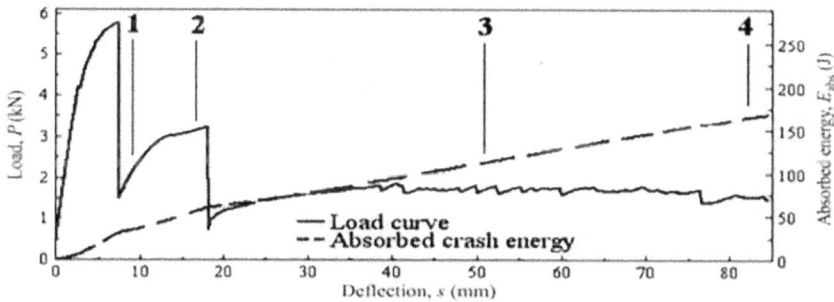

FIGURE 16.13   Crushing mode I results [Mamalis et al. 2008].

- Crushing mode II:

  In contrast to unstable mode I, mode II is characterized by uniform and symmetrical bending of the CSS structure. The deflection occurs without separation between the FRP and the core, generating stable local indentation and flexural damage of the face sheets. The high elongation capacity of the foam core leads to elastic deflection followed by local plastic deformation of the foam core. Due to the structural integrity during crushing, the energy absorption recorded is higher. Figure 16.14 depicts the different steps of crushing in mode II [Mamalis et al. 2008].

(a)

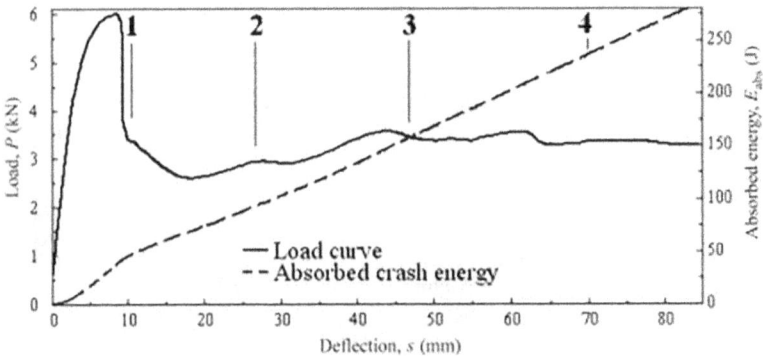

FIGURE 16.14   Crushing mode II results [Mamalis et al. 2008].

- Energy absorption:

  The SEA recorded for the four specimens increases linearly with the shear properties of the core material. As depicted in Table 16.1, the flexural peak load of cross-linked PVC has the highest value followed by panels made of PMI, linear PVC and finally PU. This descending order corresponds to the same order of shear strength and shear modulus. But this is not the case for the energy absorption which is dependent on the stability of the crushing mode.

16.6 AUTOMOTIVE SANDWICH COMPOSITE STRUCTURES FOR CRASHWORTHINESS: APPLICATIONS REVIEW

According to several research papers [Wekezer et al. 1993, Simms et al. 2009, Han et al. 2012, Jurewicz et al. 2016, Mujeeb et al. 2020], researchers have carried out a wide range of studies and analyses on vehicle impacts. They have largely succeeded in reducing parameters such as injury to the occupant of the vehicle. Figure 16.15 explains the different car parts made of sandwich materials.

Among the researches in this field which studied the use of CSS are Aly et al. (202), Lukaszewicz et al. (2016), Khan et al. (2017), Lukaszewicz et al. (2017) and Donga (2011). Considering motor vehicle crash, there are different types of crash events, such as frontal, lateral and oblique, side, rear and rollover. Donga (2011) used LS-DYNA FE software to show the impact effects of both steel and

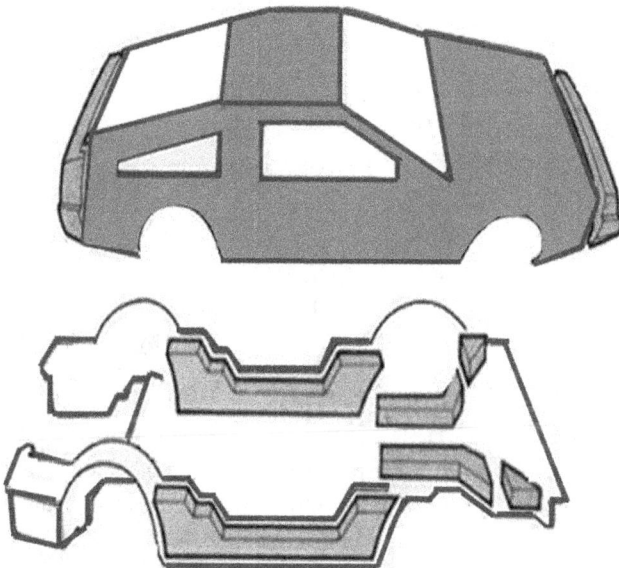

FIGURE 16.15 Different car parts made of sandwich materials.

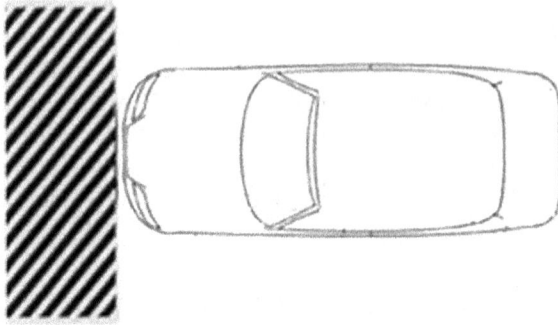

FIGURE 16.16 Frontal impact configuration.

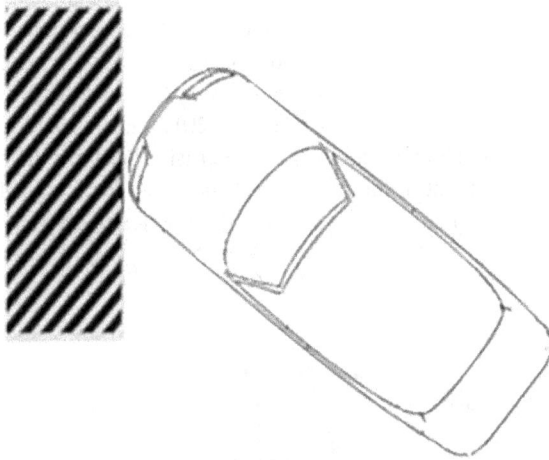

FIGURE 16.17 Oblique impact configuration.

composite sandwich models; Figures 16.16 and 16.17 represent the frontal and oblique impacts, respectively.

Table 16.2 summarizes the results obtained in this work. By comparing the two beams, we observe that the vehicle bumper made by a sandwich material is less deformed with an increase in the contact force and then an increase in the acceleration.

Based on the data from Donga (2011), the force versus displacement relation for oblique impact shows high absorption energy for the sandwich beam compared to steel beam in the case of oblique impact of 30°. The results show that the sandwich materials present a high structural stability and a smaller overall deformation. On the other hand, this kind of material is more crashworthy for frontal impact.

TABLE 16.2
Mechanical Properties of Foam Core Materials

		Front Impact 56 km/h	Front Impact 16 km/h	Oblique Impact 56 km/h
Displacement	Steel	500	310	930
	Sandwich	450	200	900
Force (KN)	Steel	700	140	500
	Sandwich	810	225	680
Acceleration (g)	Steel	38	20	6
	Sandwich	44	25	20

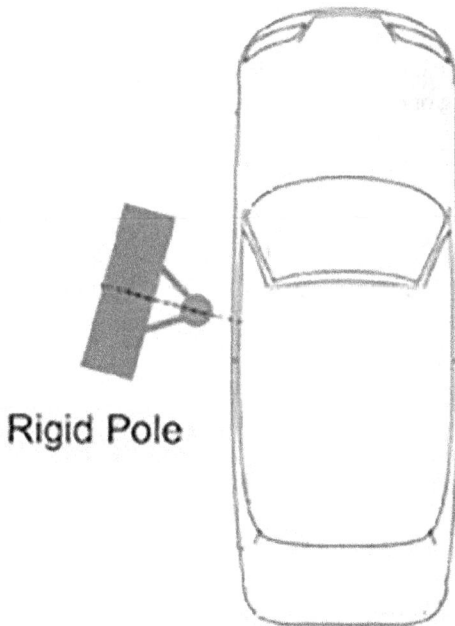

Rigid Pole

FIGURE 16.18   Lateral impact configuration.

Lukaszewicz et al. (2017) studied automotive crashworthiness during lateral impact (see Figure 16.18). In this research paper, the application of plastic tufted sandwich structures reinforced by aramid fiber is being studied through crash testing and numerical modeling.

According to this study, a CSS panel was produced with a "Rohacell" core (a 75 kg/m³ PMI foam core) and carbon FRP C with aramid tufting through the face sheets and the core. They tested a "floor panel demonstrator" with bonded door sill, and they performed a numerical modeling for crush simulation using

ABAQUS, commercial FEA software with a composite crushing add-in CZone. The crush simulation of the sandwich panel is in accordance with the experiments.

16.7 MARINE SANDWICH COMPOSITE STRUCTURES FOR CRASHWORTHINESS: APPLICATIONS REVIEW

During the 1980s and 1990s, consistent efforts have been made to develop sandwich composite materials for marine applications. PVC has been used in military and passengers vessels in marine manufacturing. First, glass facing was used and then the carbon-reinforced skins were used. Face sheets improve bending stiffness, and the core resists the shear and compressive stresses and keeps the skins bonded, preventing the structure from early damages and instabilities (Mamalis et al. 2004 and Mamalis et al. 2008). Today, sandwich composite is the best choice of material in the marine construction industry. Sandwich composites offer many advantages over traditional materials (metal, aluminum, etc.). Mechanical properties, such as resistance to impact by absorbing energy and by deformations such as shearing, bending or crack growing, still need more researches to design better crashworthy marine structures. The maritime classification society closely controls the sandwich panel construction in order to enable the marine craft to obtain certification.

Bulla and Edgrenb (2004) investigated the influence of low-velocity impact on sandwich panels made of CFRP skins and PVC foam core with different thicknesses and densities to find the effect of each parameter on energy-absorbing capabilities of marine hulls. The main metric used by the authors to study the response of sandwiches to drop weights is the residual strength. The purpose of the experiments was to explore failure in sandwich panels with carbon FRP faces and a PVC foam core to enhance the mechanical properties of hull vessels. The authors used digital speckle photography; the equipment consisted of two industrial-grade digital cameras on a mount placed on a tripod at a certain distance from the object subjected to test conditions.

In order to study the response of sandwich in terms of energy absorbed before damage (pre-crash), Tagarielli et al. (2008) investigated the effects of stress–strain response of two PVC foams (Divinycell H100 and H250) and end grain balsa wood on uniaxial compression. For example, an increase in the strain rate, from quasi-static rates of 10^{-4}/s to the order of 10^{-3}/s, allows doubling the compressive yield strength of both the materials, $90\,kg/m^3$ for balsa wood and $250\,kg/m^3$ for PVC foam [Tagarielli et al. 2008]. Balsa wood demonstrates an impressive compressive strength-to-weight ratio due to its low density. Compared to the $100\,kg/m^3$ PVC foam, for the same strain rate increase, low-density PVC shows an increase of 30% in the uniaxial compressive strength. This explains the trend of the use of high-density foam core compared to low-density balsa of $90\,kg/m^3$ that offers the same compressive yielding strength.

Marine structures are also subjected to repeated water impacts, also called "slamming" loads. Baral et al. (2010) presented the simulation results of the water

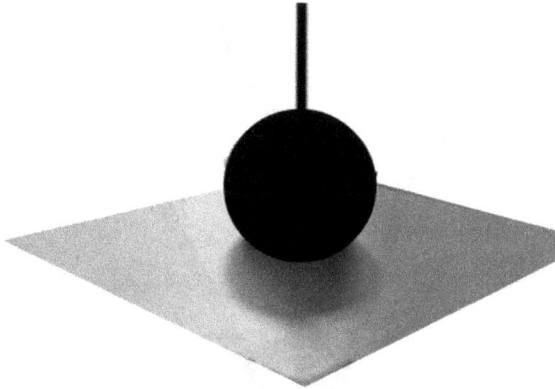

FIGURE 16.19   Drop ball test.

impact (slamming) loading of sandwich boat structures. A weighted ball is dropped from gradually increasing heights onto rigidly supported sandwich panels until damage is detected (see Figure 16.19). The core used is the honeycomb core, composed of impregnated Nomex paper. The material shows excellent specific transverse compression and shear behavior. The results from the tests indicated that honeycomb core sandwich panels, the most widely used material for racing yacht hulls, start to damage due to core crushing at impact energies of around 550 J. Sandwich panels of the same area/weight and the same carbon/epoxy facings, but using a novel foam core reinforced in the thickness direction with pultruded carbon fiber pins [Baral et al. 2010], as shown in Figure 16.20, do not show signs of visible damage until the impact energy reaches the value of 1,200 J, which represents the doubling of the energy absorbed before damage occurrence. Thus, the novel foam core reinforced in the thickness direction with pultruded carbon fiber pins offers significantly improved resistance to wave impact and is a good candidate sandwich material for high vessel speed hulls.

As marine materials have progressed from traditional hand layup glass-reinforced composites to higher-performance carbon fiber-reinforced materials, the distinction between marine and non-marine composites has become less well defined. Many studies don't specifically cite marine applications. Only the type of experiments (compression, bending rig by the drop test, etc.) defines the application desired of the sandwich specimens.

16.8   CONCLUDING REMARKS

Sandwich structural components made of composite materials are used in a wide range of applications in industries (aerospace, automotive, transportation, defense, etc.). These components can be used as energy-absorbing devices, which drives many researchers to study experimentally and numerically the

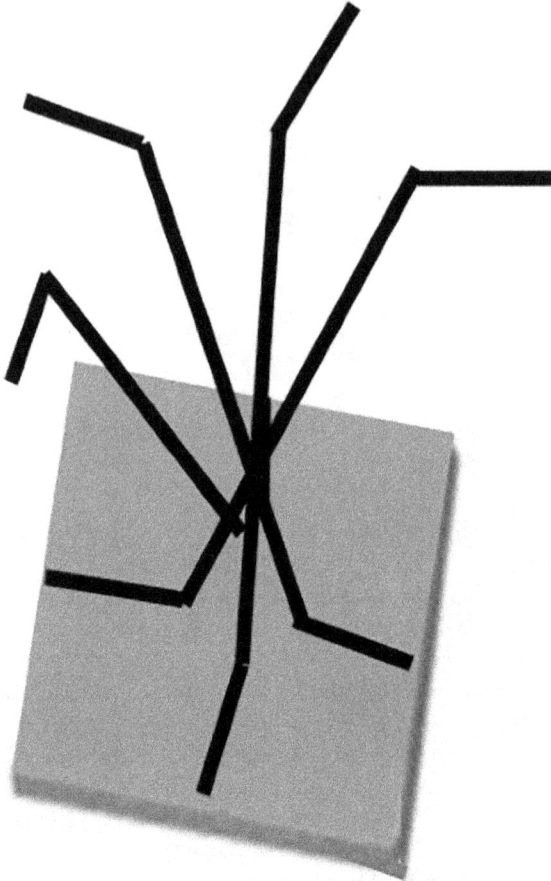

FIGURE 16.20 Foam core with pultruded carbon fiber pins.

crashworthiness of CSS. Characteristics of CSS for crashworthiness applications can be summarized in the following concluding remarks:

- The study of metrics used in crashworthiness studies shows that the specific energy absorption is the essential parameter for crashworthiness applications.
- The energy absorption of CSS is characterized by four different crushing modes, which are transverse shearing, brittle fracturing, lamina bending and local buckling. These modes represent dissipation energy mechanisms.
- The two most used applications to study the crashworthiness of a sandwich specimen are compression loading and three-point flexural bending.
- Axial compression results in three crushing modes. The stable and progressive one is the more energy absorption efficient (mode III).

- Three-point flexural bending shows two different modes. The second mode is progressive and stable and is considered as the efficient energy-absorbing mode.
- In both applications, the crushing mode is controlled by the nature of the foam used. The energy absorption and the peak load increase with the increase in the shear modulus values.
- Under impact loading, composite sandwich panels delaminate at or near the core–face sheet interface, which causes a sudden loss in structural integrity and then catastrophic consequences.
- The face sheets can be tufted together throughout the entire sandwich assembly to improve the energy absorption of composite sandwich materials during crushing.
- Marine materials showed a trend toward high-density foam materials (such as PVC 250).
- Natural cores are pertinent candidates as they offer as good strength yielding as high-density PVC. Among the trending cores, balsa wood is crashworthy with a good strength-to-weight ratio (density 90 km/m^3).
- Core reinforced in the thickness direction with pultruded carbon fiber pins also showed good energy absorption compared to traditional honeycomb core used in racing yacht.

REFERENCES

Abosbaia, AAS, Mahdi, E, Hamouda, AMS, Sahari, BB, (2003). Quasi-static axial crushing of segmented composite tubes. *Composite Structures*, 60, 327–343.

Aly, NM, Saad, MA, Sherazy, EH, Kobesy, OM, Almetwally, AA, (2012). Impact properties of woven reinforced sandwich composite panels for automotive applications. *Journal of Industrial Textiles*, 42, 204–218.

Baral, N, Cartié, DDR, Partridge, IK, Baley, C, Davies, P, (2010). Improved impact performance of marine sandwich panels using through-thickness reinforcement: Experimental results. *Composites Part B: Engineering*, 41, 117–123.

Bulla, PH, Edgrenb, F, (2004).Compressive strength after impact of CFRP-foam core sandwich panels in marine applications. *Composites Part B: Engineering*, 35, 535–541.

Blok, LG, Kratz, J, Lukaszewicz, D, Hesse, S, Ward, C, Kassapoglou, C, (2017). Improvement of the in-plane crushing response of CFRP sandwich panels by through thickness reinforcements. *Composite Structures*, 161, 15–22.

Dell'Anno, G, Treiber, JWG, Partridge, IK, (2016). Manufacturing of composite parts reinforced through-thickness by tufting. *Robotics and Computer-Integrated Manufacturing*, 37, 262–272.

Donga, A, (2011).*Application of Sandwich Beam in Automobile Front Bumper for Frontal Crash Analysis*. PhD Dissertation, Wichita State University.

Elgalai, AM, Mahdi, E, Hamouda, AMS, Sahari, BB, (2004). Crushing response of composite corrugated tubes to quasi-static axial loading. *Composite Structures*, 64, 665–671.

Found, MS, Robinson, AM, Carruthers, JJ, (1997). The influence of FRP inserts on the energy absorption of a foam-cored sandwich panel. *Composite Structures*, 38, 373–381.

Gibson, LJ, Ashby, MF, (1997). *Cellular Solids: Structure and Properties*, 2nd ed., Cambridge University Press, Cambridge.

Gary, LF, Robert, MJ, (1989). *Energy-Absorption Capability of Composite Tubes and Beam*, NASA Technical Memorandum, Virginia Polytechnic Inst and State University Blacksburg, 101634.

Goldsmith, W, Sackman, GL, (1992). An experimental study of energy absorption in impact on sandwich plates. *International Journal of Impact Engineering*, 12, 241–262.

Ha, NS, Lu, G, (2020). Thin-walled corrugated structures: A review of crashworthiness designs and energy absorption characteristics. *Thin-Walled Structures*, 157, 106995.

Haug, E, Fort, O, Trameçon, A, Watanabe, M, Nakada, I, (2018). *Numerical Crashworthiness Simulation of Automotive Structures and Components Made of Continuous Fiber Reinforced Composite and Sandwich Assemblies*. SAE International by University of British Columbia in United States, pp. 245–58, 1991,910152.

Han, Y, Yang, J, Mizuno, K, Matsui, Y, (2012). Effects of vehicle impact velocity, vehicle front-end shapes on pedestrian injury risk. *Traffic Injury Prevention*, 13, 507–518.

Henao, A, Carrera, M, Miravete, A, Castejón, L, (2010).Mechanical performance of through thickness tufted sandwich structures. *Composite Structures*, 92, 2052–2059.

Ji, G, Ouyang, Z, Li, G, Pang, SS, (2011). Impact/debonding tolerant sandwich panel with aluminum tube reinforced foam core. ASME 2011 International Mechanical Engineering Congress and Expositionat, Denver CO, IMECE2011-65724, 281–287.

Ji, G, Li, G, Pang, SS, (2012). Shape memory polymer with aluminum tube reinforced impact resistant sandwich core. ASME 2011 International Mechanical Engineering Congress and Exposition, IMECE2012-86141, 97–101.

Jing, L, b, Su, X, Chen, D, Yang, F, Zhao, L, (2019). Experimental and numerical study of sandwich beams with layered-gradient foam cores under low-velocity impact. *Thin-Walled Structures*, 135, 227–244.

Jurewicz, C, Sobhani, A, Woolley, J, Dutschke, J, Corbenc, B, (2016).Exploration of vehicle impact speed - injury severity relationships for application in safer road design. *Transportation Research Procedia*, 14, 4247–425.

Kalhori, H, Ye, L, Li, Z, Li, B, (2015).Identification of location and magnitude of impact force on a composite sandwich structure with lattice truss core. ASME 2015 International Mechanical Engineering Congress and Exposition at Huston, Texas, IMECE2015-51411, V001T01A023, 7 p.

Khan, MA, Syed, AK, Ijaz, H, Shah, RMBR. (2017). Experimental and numerical analysis of flexural and impact behaviour of glass/pp sandwich panel for automotive structural applications. *Advanced Composite Materials*, 1568–5519.

Khalid, AA, Sahari, BB, Khalid, YA, (2002).Performance of composite cones under axial compression loading. *Composites Science and Technology*, 62, 17–27.

Lukaszewicz, D, Blok, LG, Kratz, J, Ward, C, Kassapoglou, C, (2016). Application of fibre reinforced plastic sandwich structures for automotive crashworthiness applications. Proceedings of the International Conference on Automotive Composites, ICAutoC 2016, Lisboa.

Lukaszewicz, D, Blok, L, Kratz, J, Ward, C, Kassapoglou, C, (2017). Experimental and numerical investigation of full scale impact test on fibre-reinforced plastic sandwich structure for automotive crashworthiness. *International Journal of Automotive Composites*, 3, 339–353.

Mahdi, E, Hamouda, AMS, Sahari, BB, Khalid, YA (2001). An experimental investigation into crushing behavior of filament-wound laminated cone–cone intersection composite shell. *Composite Structures*, 51, 211–219.

Mahdi, E, Hamouda, AMS, Sahari, BB, Khalid, YA, (2003). Effect of hybridization on crushing behavior of carbon/glass fibre/epoxy circular cylindrical shells. *Journal of Materials Processing Technology*, 132, 49–57.

Mahdi, E, Hamouda, AMS, Mokhtar, AS, Majid, DL, (2005). Many aspects to improve damage tolerance of collapsible composite energy absorber devices. *Composite Structures*, 67, 175–187.

Mamalis, AG, Manolakos, DE, Ioannidis, MB, Papapostolou, DP, (2004). Crashworthy characteristics of axially statically compressed thin walled square CFRP composite tubes: Experimental. *Composite Structures*, 63, 347–360.

Mamalis, AG, Manolakos, DE, Ioannidis, MB, Papapostolou, DP, (2005). On the crushing response of composite sandwich panels subjected to edgewise compression: Experimental. *Composite Structures*, 71, 246–257.

Mamalis, AG, Manolakos, DE, Ioannidis, MB, Papapostolou, DP, Kostazos, PK, Konstantinidis, DG, (2002). On the compression of hybrid sandwich composite panels reinforced with internal tube inserts: Experimental. *Composite Structures*, 56, 191–199.

Mamalis, AG, Spentzas, KN, Manolakos, DE, Ioannidis, MB, Papapostolou, DP, (2008). Experimental investigation of the collapse modes and the main crushing characteristics of composite sandwich panels subjected to flexural loading. *International Journal of Crashworthiness*, 13, 349–362.

Mamalis, AG, Spentzas, KN, Papapostolou, DP, Pantelelis, N, (2013). Finite element investigation of the influence of material properties on the crushing characteristics of in-plane loaded composite sandwich panels. *Thin-Walled Structures*, 63, 163–174.

Mulla Salim, H, Yadv Sanjay, D, Shinde, D, Deshpande, D, (2013). Importance of federal motor vehicle safety standards 207/210 in occupant safety - A case study. *Procedia Engineering*, 64, 1099–1108.

Mujeeb, M, Pasupuleti, VDK, Dongre, A, (2020). Effect of Vehicle Impact on Reinforced Concrete Structures. In: Subramaniam, K., Khan, M. (eds) *Advances in Structural Engineering. Lecture Notes in Civil Engineering*, 74, 195–204.

Nguyen, MQ, Jacombs, SS, Thomson, RS, Hachenberg, D, Scott, ML, (2005). Simulation of impact on sandwich structures. *Composite Structures*, 67, 217–227.

Ramakrishna, S, Hamada, H, (1998). Energy absorption characteristics of crashworthy structural composite materials. *Key Engineering Materials*, 141–143, 585–620.

Sun, G, Xu, F, Li, G, Li, Q, (2014). Crashing analysis and multiobjective optimization for thin-walled structures with functionally graded thickness. *International Journal of Impact Engineering*, 64, 62–74.

Sun, G, Wang, Z, Yu, H, Gong, Z, Li, Q, (2019). Experimental and numerical investigation into the crashworthiness of metal-foam-composite hybrid structures. *Composite Structures*, 209, 535–547.

Sen, X, Jikuang, Y, Zhihua, Z, (2013). Research and optimization of crashworthiness in small overlap head-on collision. Fifth International Conference on Measuring Technology and Mechatronics Automation (ICMTMA), Hong Kong, China, 854–857.

Simms, C, Wood, D, (2009). The relationship between vehicle impact speed and pedestrian and cyclist projection distance. In: Pedestrian and Cyclist Impact. *Solid Mechanics and Its Applications*, 166, 51–74.

Srinivas, GR, Deb, A, Sanketh, R, Gupta, NK, (2017). An enhanced methodology for light weighting a vehicle design considering front crashworthiness and pedestrian impact safety requirements. *Procedia Engineering*, 173, 623–630.

Tarlochan, F, Hamouda, AMS, Mahdi, E, Sahari, BB, (2005). On the compression behavior of thin walled polymer composite sandwich panels. The 3rd International Conference on Structural Stability and Dynamics, Florida, 19–22 June, pp. 71–72.

Tarlochan, F, Hamouda, AMS, Mahdi, E, Sahari, BB, (2007). Composite sandwich structures for crashworthiness applications. *Proceedings of the Institution of Mechanical Engineers Part L Journal of Materials Design and Applications*, 221, 121–130.

Tagarielli, VL, Deshpande, VS, Fleck, NA, (2008). The high strain rate response of PVC foams and end-grain balsa wood. *Composites Part B: Engineering*, 39, 83–91.

Timesli, A, (2020a). An efficient approach for prediction of the nonlocal critical buckling load of double-walled carbon nanotubes using the nonlocal Donnell shell theory. *SN Applied Sciences*, 2, 407.

Timesli, A, (2020b). Buckling analysis of double walled carbon nanotubes embedded in Kerr elastic medium under axial compression using the nonlocal Donnell shell theory. *Advances in Nano Research*, 9, 69–82.

Timesli, A, (2020c).Prediction of the critical buckling load of SWCNT reinforced concrete cylindrical shell embedded in an elastic foundation. *Computers and Concrete*, 26, 53–62.

Warrior, NA, Turner, TA, Robitaille, F, Rudd, CD, (2003). Effect of resin properties and processing parameters on crash energy absorbing composite structures made by RTM. *Composites Part A: Applied Science and Manufacturing*, 34, 543–550.

Wang, Z, Li, Z, Xiong, W, (2019). Numerical study on three-point bending behavior of honeycomb sandwich with ceramic tile. *Composites Part B: Engineering*, 167, 63–70.

Wekezer, JW, Oskard, MS, Logan, RW, Zywicz, E, (1993). Vehicle Impact Simulation. *Journal of Transportation Engineering*, 119, 598–617.

Zhu, X, Xiong, C, Yin, J, Yin, D, Deng, H, (2019). Bending experiment and mechanical properties analysis of composite sandwich laminated box beams. *Materials*, 12, 2959.

17 Role of 3D Printing in the Fabrication of Composite Sandwich Structures

Athul Joseph
Indian Institute of Science

Vinyas Mahesh
Indian Institute of Science
Nitte Meenakshi Institute of Technology
National Institute of Technology Silchar

Vishwas Mahesh
Siddaganga Institute of Technology
Indian Institute of Science

Dineshkumar Harursampath
Indian Institute of Science

Vasu Mallesha
National Institute of Technology

CONTENTS

17.1 Introduction 350
17.2 3D Printed Sandwich Composites for Aeronautical and Automotive
     Applications 351
17.3 3D Printed Sandwich Composites in Biomedical Applications 352
17.4 3D Printed Sandwich Composites for Energy Storage and Soft
     Robotics 356
17.5 3D Printed Sandwich Composites for Space Structures....................... 358
17.6 Sports Applications... 359
17.7 3D Printed Sandwich Composites for Military Applications................ 360
17.8 3D Printed Sandwich Composites for Construction Applications 362

DOI: 10.1201/9781003143031-17

17.9　　Challenges in 3D Printing Composite Sandwich Structures................. 363
17.10　Conclusions and Future Prospects.. 365
Acknowledgement 366
Bibliography 367

17.1　INTRODUCTION

The advantages of composite structures were an untold truth until when they were revealed in the engineering front even when they were present since ever. Composites offer a plethora of advanced and improved characteristics that enable better engineering prospects for maximum benefit. The introduction of sandwich composites on this front led to the expansion of the application spectrum of composites with enhanced behaviour. When this effort is combined with additive manufacturing (AM) techniques, sandwich composites offer great advantages. Recent studies have greatly focused on producing additively manufactured novel sandwich composite structures, which is evident from Figure 17.1, which describes an increasing trend in this topic.

Recently, fibre-reinforced composites have extensively been researched to serve various engineering applications (Eichenhofer et al. 2018; Austermann et al. 2019). In this regard, continuous lattice fabrication (CLF) is a novel AM solution that directs fibres in all special coordinates as opposed to the highly directional routes adopted by fused deposition modelling (FDM) and other similar direct printing methods (Goh, Agarwala, and Yeong 2019; Eichenhofer, Wong,

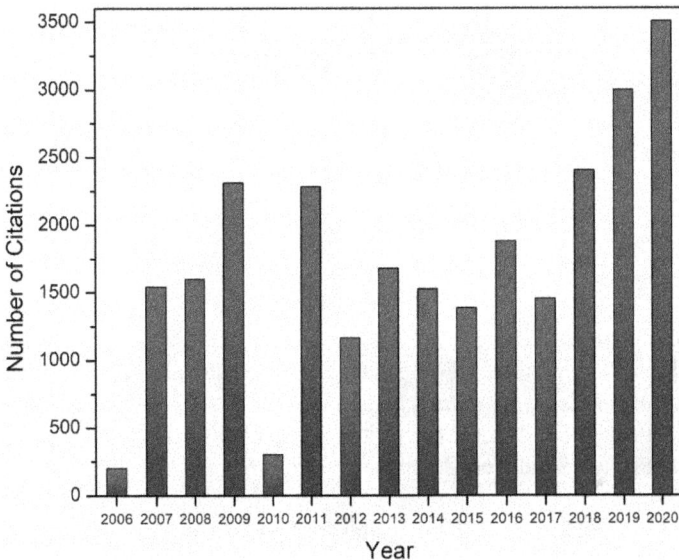

FIGURE 17.1 Increasing research interest on additively manufactured sandwich composite structures. ("Digital Science. (2018) Dimensions [Software] Available from https://App.Dimensions.Ai. Accessed on (20-11-2020), under Licence Agreement," 2018.)

and Ermanni 2017; Picha, Spackman, and Samuel 2016). Attempts have also been made to integrate polymer matrices with metallic inclusions for novel applications such as electroluminescent devices, sensors, actuators, robotic systems and several other applications through FDM (Brubaker et al. 2019; Carricoa et al. 2015). However, a much more sophisticated and rapid method to fabricate sensors and actuators can be achieved by drop-on-demand high-speed 3D printing (Davoodi et al. 2020). Porous 3D printed thermoplastics with metal–organic frameworks for higher chemical activity were realized using a regular 3D printer for various structural and mechanical applications (Evans et al. 2018; Mu et al. 2017). Similarly, bio-inspired structures made from plastics, metal-reinforced thermoplastics and elastomers showed promising applicability in soft robotics and structural applications (Korde, Shaikh, and Kandasubramanian 2018; Hamidi and Tadesse 2019).

Furthermore, 3D printing of sandwich composite panels has been used in more advanced applications such as tooling, electronic displays, liquid crystal displays and energy storage applications (Turk et al. 2018; Z. Wang et al. 2018; Yuan et al. 2017; Zhu et al. 2017). Although AM offers great rapid prototyping solutions, the defects observed in the produced parts are a source of worry for its complete adaptation (Echeta et al. 2020). While manufacturing challenges are ever present, the properties of the structures prepared by AM techniques can be enhanced by post-processing methods such as thermal ageing and surface engineering (Davoudinejad et al. 2020). Therefore, the upcoming sections provide a meticulous review of the current applications of additively manufactured sandwich composite structures while also addressing the challenges faced during their production.

17.2   3D PRINTED SANDWICH COMPOSITES FOR AERONAUTICAL AND AUTOMOTIVE APPLICATIONS

Sandwich composites have found significant applications in the field of aeronautical and automobile engineering. Along with significant weight reductions, the use of these structures has improved the performance parameters of the parent structure with significant economic advantages. One of the prime uses of these composites is their energy absorption capabilities. Several additively manufactured sandwich cores such as corrugated tubes, honeycomb structures and cellular structures have enabled better crashworthiness and enhanced energy absorption capabilities as shown in many studies (Alkhatib et al. 2019; Anandan et al. 2019; Brenne, Niendorf, and Maier 2013; Chantarapanich et al. 2014). A fine example of this is the AlSi10Mg sandwich cores prepared by direct metal laser sintering (DMLS) (Alkhatib et al. 2019). Similarly, 304L stainless steel honeycomb structures were fabricated using selective laser melting (SLM) (Anandan et al. 2019). These composites showed great compressive behaviour with the increase in the thickness of the cell wall under out-of-plane tests. Tensile and flexural studies of sandwich of carbon fibre-reinforced Onyx structures showed improved strengths with the increase in the number of fibre layers and fibre rings in each layer in the sandwich structure (Yu et al. 2019). The programmable negative Poisson's ratio (NPR) achieved through AM techniques has enabled the usage of auxetic sandwich composites in aeronautical applications for similar purposes (Ling et al. 2020).

In addition to superior mechanical properties, additively manufactured sandwich composite structures exhibit appreciable thermal capabilities. The composites exhibited better creep resistance, heat transfer and insulation, which endorses their usage for the intended application (Zhang et al. 2020; Yu et al. 2019; Sun et al. 2020; Mohammadizadeh et al. 2018; Mohamed et al. 2019). Forced convectional heat transfer studies on AlSi10Mg sandwich cylinders with different internal structures reported improved heat transfer and reduced flow resistance for varying pumping powers and Reynolds numbers (Sun et al. 2020). Dynamic stress influences realized through creep tests were also analysed for Kevlar fibre-reinforced nylon composites (Mohammadizadeh et al. 2018). Different fibre compositions such as carbon fibres and glass fibres with nylon were tested as well. It is seen that at a temperature of about 100°C, the strains observed for nylon–Kevlar composites were the least. At the same time, subjecting sandwich composites to thermal fields as a part of a treatment process has significantly improved their mechanical properties. Studies by Mohamed et al. (2019) showed that heat treatment of slitted tubes made of AlSi10Mg improved their crushing force efficiency and energy absorption parameters.

However, the strength and energy absorption capabilities of such structures are greatly influenced by numerous parameters such as microstructure, local strains, printing parameters and lattice geometry (Wang et al. 2020; Brenne, Niendorf, and Maier 2013; Cuan-Urquizo, Yang, and Bhaskar 2015; Davami et al. 2019; Friedrich and Almajid 2013). In addition to common 3D printing such as FDM and SLS, several other techniques such as DMLS, SLM, stereolithography (SLA) and direct laser writing (DLW) have also been used (Harris, Winter, and McShane 2017; Gao et al. 2020; Zhang et al. 2020; Ullah et al. 2014). A majority of these properties are used to design and fabricate components such as bumpers, helmets, wing structures for unmanned aerial vehicles (UAVs) and engine thermal barriers where energy absorption and strength are of prime consideration.

17.3 3D PRINTED SANDWICH COMPOSITES IN BIOMEDICAL APPLICATIONS

Sandwich composites used in biomedical applications mostly focus on five major factors: biocompatibility, biochemical activity, lightweight, structural feasibility and mechanical strength. Metal- and ceramic-based biocomposites find widespread biomedical and tissue engineering applications (Jakus 2019; Haleem et al. 2020). In a study by Antolak-Dudka et al. (2019), different Ti6Al4V hexagonal honeycomb structures manufactured by Laser Engineered Net Shaping (LENS™) Technology were subjected to static and dynamic deformation tests. It is seen that the cell size of the individual unit cells played a predominant role in gauging their energy absorption capabilities, as illustrated in Figure 17.2. Such structures were

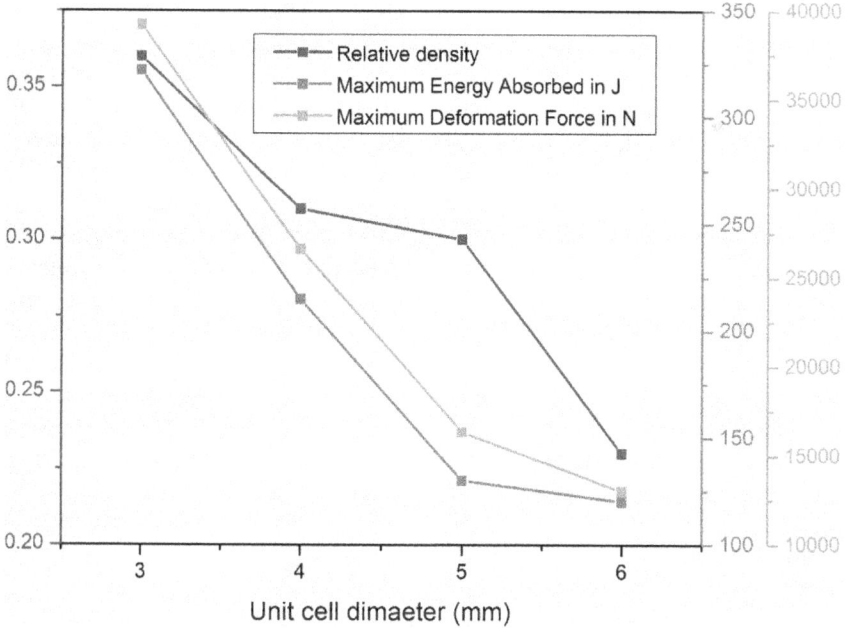

FIGURE 17.2 Comparison of the maximum energy absorbed, maximum compressive force and relative densities of the structures for varying unit cell diameter (Antolak-Dudka et al. 2019).

beneficial to manufacture biomedical implants. Similarly, orthopaedic scaffolds for bone tissue engineering were realized using 3D printing technologies (Haleem et al. 2020). These structures are instrumental in aiding with the desired mechanical strength and integrity the patients in need of them. While the mechanical strength of the scaffold is a prime factor, the chemical activity and bioactivity of the scaffolds should be good enough for cell re-growth. Additionally, these structures need to be highly porous and precise in their size and shape to facilitate cellular penetration. Electroactive scaffolds were realized by 3D printing 30% Pluronic F127 with varying compositions of aniline tetramer-grafted polyethylenimine (AT-PEI) (Dong et al. 2017). The excellent electroactivity of AT-PEI nanoparticles aided in higher conductivity observed in the final composites. All the composites prepared using this method possessed a conductivity value that was greater than 2×10^{-3} S/cm. Hence, these scaffolds are apt candidates for the electrical stimulation of tissue regeneration, such as muscle and cardiac nerve tissue repair. Bioprinting of advanced biomaterials such as Hyperelastic Bone® was capable of inducing appreciable regeneration (Jakus 2019). The Hyperelastic Bone sandwich composites tested in higher-order primates with full-thickness cranial defects showed complete bone regeneration and integration in 15 months. The material is also highly cost-effective, manufacturing-friendly and economically feasible.

However, a contrasting deduction may be observed for auxetic lattices manufactured through SLM (Arjunan et al. 2020). In addition to the impressive properties of these structures, optimization routes were seen to provide better weight
reductions for femoral nails with significant material savings (Cucinotta et al.
2019). 3D printing and electrospinning techniques were used to manufacture
sandwiched composite hydrogels for cartilage and bone tissue engineering (De
Mori et al. 2018). Similarly, it is seen that porous structures realized through
AM technologies possessed load-bearing capabilities that are apt for real-world
biomedical applications (Wang et al. 2016; Zhang et al. 2019). Additively manufactured porous Ti6Al4V structures were studied to gauge the effect of pore density and pore sizes on the mechanical properties of the structure (Zhang et al.
2019). The study also laid special emphasis on the functionally graded design
which possessed different pore sizes in each layer. It is seen that the reduction in
pore sizes could cater to higher strength and modulus. However, while employing a step-wise functionally graded configuration, the improvement was not as
large as those with pore sizes <600 μm. Figure 17.3 illustrates the variation of
the ultimate strength and tensile modulus of these composites when subjected to
uniaxial tests.

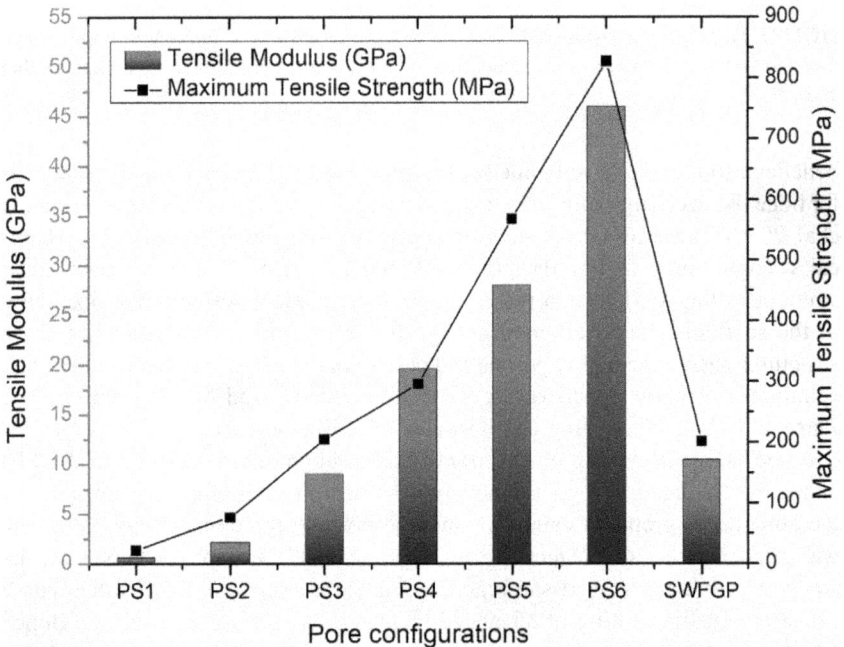

FIGURE 17.3 Variation of ultimate strength and tensile modulus of additively manufactured Ti6Al4V for different pore sizes. PS1–PS6 refers to pore sizes 1,331, 1,058, 808,
651, 490 and 351 μm, respectively, and SWFGP is the step-wise functionally graded pore
configuration (Zhang et al. 2019).

Topological optimizations are a highly effective strategy to obtain optimal internal architectures for porous metallic composites which satisfy the requirements of multifunctionality and biomimicry (Wang et al. 2016). This optimization route is advantageous in AM-based production of porous implants. Further enhancements of the mechanical and biological properties of such porous composites could be achieved by heat treatment and surface modification techniques. Similarly, porous orthopaedic implants fabricated using SLM included functionally graded structures with bone ingrowth surfaces (Mullen et al. 2009). The properties observed for these structures matched with those of regular human bones. The structures have highly tuneable properties that enable easy adaption for various requirements.

The superior mechanical properties, lightweight and biocompatible nature of additively manufactured sandwich composites make them ideal candidates in the production of medical implants, musculoskeletons, bone ingrowth structures and many others (Mullen et al. 2009; Kussmaul et al. 2019; Jeyachandran et al. 2020). Bioactive glass (BAG) was fused with high-density polyethylene (HDPE) to obtain 3D printable structures for bone regeneration applications (Jeyachandran et al. 2020). It is seen that the addition of bioactive glass did not drastically deteriorate the mechanical properties of the structure as compared to pure nylon filaments and improved the overall modulus of the structure, as shown in Figure 17.4. Therefore, as seen from the above examples, the use of additively manufactured

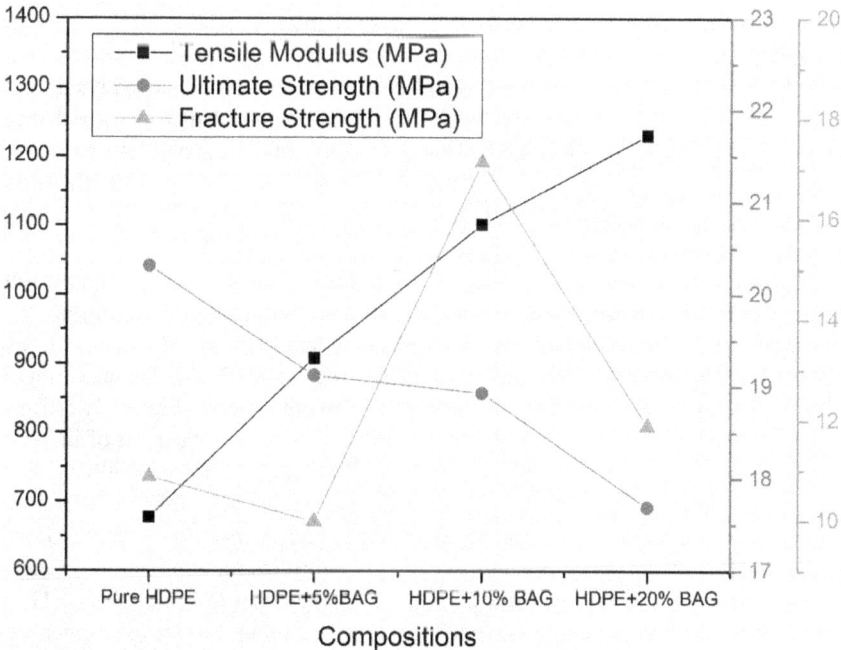

FIGURE 17.4 Comparison of key mechanical parameters for different compositions of BAG in HDPE prints (Jeyachandran et al. 2020).

sandwich composite structures in biomedical applications is of great significance with their superior performance, easier adaptation and excellent comfort. Besides the superior properties, biomedical structures realized through AM are highly cost-efficient in their production through the fabrication of an almost single structure (Kussmaul et al. 2019). The need for multiple attachments and complex joining techniques are eliminated, which save time, effort and finances.

17.4 3D PRINTED SANDWICH COMPOSITES FOR ENERGY STORAGE AND SOFT ROBOTICS

3D printed composites have an abundance of applications in electronics devices right from minor sensors to large energy storage devices. The versatility and quick production offered by 3D printing are well exploited in electronic applications. One of the recurrent applications of this technology is in the development of energy storage devices. This particular application can be realized through many 3D printing techniques such as direct ink writing (DIW), FDM, SLM and SLS (Zhang et al. 2017; Chang et al. 2019; Foster et al. 2017). In a study presented by Zhang et al. (2017), an array of AM techniques for the production of electrochemical energy storage devices such as supercapacitors and batteries have been discussed. On comparing methods such as inkjet printing and direct writing, the application of these techniques depends on the properties and qualities of the print that is needed (Chang et al. 2019). The former offers multi-material capabilities and higher resolution, while the latter is characterized by higher active material loading and lower clogging risks. Similarly, SLA offers better resolution over FDM, while FDM parts are cheaper to produce. Several high-performance materials such as graphene and carbon nanotubes have been embedded in 3D printed sandwich matrices to enhance the electrochemical energy storage of these components (Tian et al. 2017; Gulzar, Glynn, and O'Dwyer 2020; Zhang et al. 2017; Thaler, Aliheidari, and Ameli 2019). Energy storage devices made by 3D printing sandwiched graphene/polylactic acid (PLA) composites were reported in the work of Foster et al. (2017).

Acrylonitrile butadiene styrene (ABS)/carbon nanotube (CNT) composites offer appreciable mechanical, electrical and piezoresistive properties that can serve various electrical and electromechanical applications such as sensors, transducers and actuators (Thaler, Aliheidari, and Ameli 2019). The addition of CNTs is seen to increase the conductivity of the composites. Figure 17.5 illustrates the comparison between the maximum conductivity observed of in-layer printed composites and compression-moulded composites at a broadband frequency of 10^6 Hz. These structures were seen as viable additions to freestanding lithium-ion anodes and solid-state graphene supercapacitors. Carbon-based materials have reported great energy-related capabilities for applications in electronic circuits, and high-temperature thermal energy uses (Fu et al. 2017). They have also exhibited greater cyclic performances with minimal loss in amperage.

Another major application of 3D printing technology is in the manufacture of sandwich structures for sensors and actuators of various kinds (Ghazanfari et al. 2016; Liu et al. 2018; Mousavi et al. 2018, 2020; Wang et al. 2019). Ceramic

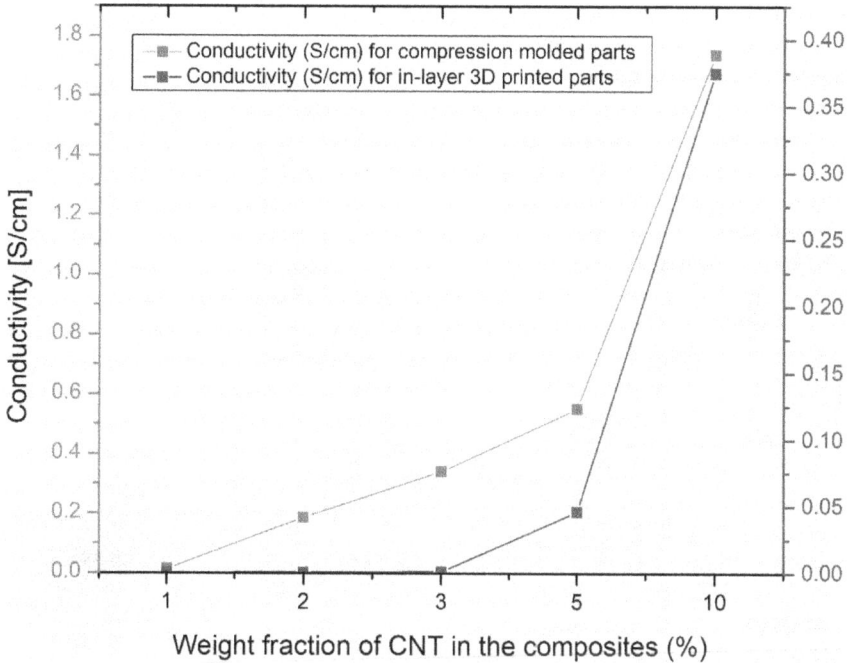

FIGURE 17.5 Variation of conductivity of compression-moulded and in-layer 3D printed ABS/CNT composites for the highest broadband frequency (10^6 Hz) with respect to varying CNT content (Thaler, Aliheidari, and Ameli 2019).

composites embedded with sapphire optical fibres were manufactured by an on-demand extrusion-based AM process (Ghazanfari et al. 2016). Two classes of fibre groups of 125 and 250 μm were used to evaluate the feasibility in manufacturing these components. It is seen that as the fibre size and relative density increase, it is tougher to embed them in ceramic matrices. However, the fibres with lower dimensions were successfully embedded and can be used in functional strain measurement of parts subjected to thermal and mechanical loads. Another fine example for the application of additively manufactured composite sandwich structures is in the manufacture of tactile sensors for wearable electronics and soft robotics (Liu et al. 2018). However, the materials used for these applications need to be tuned appropriately to achieve maximum reproducibility. A more effective strategy to realize completely functional tactile sensors is the adoption of multi-material 3D printing processes wherein one method can deposit a variety of materials. Soft robots with multidirectional proprioception were realized using constriction-resistive sensors manufactured by direct 3D printing (Mousavi et al. 2020). The study reported the highest multidirectional sensitivity ever measured where CNT–PLA layers were sandwiched with thermoplastic polyurethane (TPU). A similar attempt was made by Mousavi et al. (2018) for the manufacture of ultrasensitive self-healing tactile sensors for soft robotic applications made by graphene–PLA and TPU filaments.

Vibration studies on 3D printed sandwich piezoelectric transducers showed great competency in symmetrical and anti-symmetrical conditions with great agreements to the theoretical estimations (Wang et al. 2019). These composites can be used for effective grappling and moving applications in soft robotics. Inkjet printing has enabled the inclusion of piezoelectric entities in sandwich polymer actuators for lab-on-a-chip (LOC) systems for enhanced functionality (Pabst et al. 2013). Sandwich structures of the piezoelectric layer in between two silver electrodes were printed on to a polyethylene terephthalate (PET) substrate. On the application of a voltage across the piezoelectric layer, a bending deflection is observed as a result of the reverse piezoelectric effect. This translates to a flow rate of 100 µL/min, which is phenomenal for LOC applications. The method also eliminates the need for passive actuating membranes while offering the benefit of a low-cost printed device.

In addition to the above applications, AM techniques have been used to fabricate sandwich electrodes for hard disk drives with a piezoelectric film (Xu et al. 2020). Lead zirconate titanate (PZT) inks were used in the fabrication of hard disk drive actuators using inkjet printing on a flexible substrate. The vibration tests and finite element simulations demonstrate appreciable resonant response. Inter-digitated silver electrodes (IDEs) were possible to manufacture over simple sandwich electrodes through this method. Moreover, a more reliable manufacturing route would be to print IDEs on a flexible PET substrate and finish it off with a PZT-silane film. 3D printed polymeric mesh structures embedded in epoxy-carbon fibre structures provided appreciable shielding against electromagnetic radiation (Mishra et al. 2019). Hence, as seen from the above description, 3D printing technology can delve much deeper in electronic applications, especially in fields such as soft robotics. The excellent geometric control and customizability offered by various 3D printing techniques were seen to greatly improve the charge storage mechanisms and electrochemical performances (Tian et al. 2017). However, the limitations concerning the number of 3D printable materials in the current scenario pose a huge barrier to reach state-of-the-art power and energy densities.

17.5 3D PRINTED SANDWICH COMPOSITES FOR SPACE STRUCTURES

Outer space devices require great effort and consideration in their design and fabrication, especially for a new method such as 3D printing. The composites selected for these applications require to be highly multifunctional. Additively manufactured sandwich composites have found bespoke applications in X-ray optics for extreme outer space applications as demonstrated by Atkins et al. (2017). The study emphasized on assessing the quality of the mirror surfaces through stringent experimental routes. Also, topology optimization techniques have been employed to investigate new lightweight techniques. Similarly, AM of mirrors for space-borne applications such as telescopic and spectrometric optical systems is becoming increasingly popular due to the significant weight reduction it offers without hampering the mechanical stability of the system (Hilpert et al. 2019; Atkins et al. 2018). About five different AM mirror structures were produced by Atkins et al. (2017). These mirrors were made by AlSi10Mg, nickel–phosphorus (NiP)-coated AlSi10Mg and Ti64, where each variant has one pair of hand-polished ones and another pair of diamond-turned ones. Similarly Eberle et al. (2018) developed additively manufactured AlSi40 mirrors that were suited for cryogenic space applications. The study emphasized on the importance of AM in the manufacture of such structures and offered a step-wise approach to design iterations which reduced the mass by 30% and the wavefront error (WFE) by 20%.

AM has also helped to reform satellite structures and other associated systems. Multifunctional satellite buses made using triply periodic minimal surface (TPMS) sandwich beam structures showed superior performance characteristics as opposed to those realized with traditional materials and methods (Macchia 2019). It affirms the ability of AM alone to produce such metamaterials, which ascertains its importance in the design of space systems. One of the primary requirements for satellite and space vehicle structures is weight reduction. Several studies (Schubert, Perfetto, Dafnis, Mayer, et al. 2018; Venkatesan et al. 2020; Schubert, Perfetto, Dafnis, Atzrodt, et al. 2018; Ferrari et al. 2016; Ferrari 2015) have shown that AM can not only produce fine structures for the intended applications, but can also improve the properties and performance of those structures due to its ability to handle intricate geometries and multifunctional materials.

Additively manufactured sandwich composites presented in the works of (Schubert, Perfetto, Dafnis, Atzrodt, et al. 2018) possessed both passive and active functionality embedded on to a single structure. The passive functions included heat transfer, radiation shielding and impact protection, while the active functions included vibration reduction and energy and data transfer. The results show good multifunctional prospects at the coupon-level tests. Hypervelocity impact tests on similar composites for space debris protection applications showed good impact resistance along with significant weight efficiencies (Schubert and Dafnis 2019). The study also suggested the use of cellular aluminium structures integrated with intermediate layers of high-performance fabrics for better results. Ceramic matrix composite cellular structures were seen as promising candidates for thermal protection systems (TPS) in space vehicles (Ferrari et al. 2016). It was the first time that a complete ceramic matrix composite prototype was tested in earth re-entry conditions. The structures could bear good amounts of thermomechanical loads with significant weight reductions and efficient heat removal by passing a gas through the porous structure. Furthermore, these sandwich structures have shown promising efficiency to carry different sets of electronics and communication systems within them (Echsel, Springer, and Hümbert 2020).

17.6 SPORTS APPLICATIONS

3D printed sandwich composites have found an increasing interest in the production of sports products and equipment. One of the main advantages of these composites is their impact absorption capabilities, which is beneficial in producing personal protection equipment such as helmets, shin guards and crash pads (Xiao and Song 2018; Schaedler and Carter 2016; Aguirre et al. 2020; Tafazoli and Nouri 2020; Nallavan 2020; Subic et al. 2017; Buchanan et al. 2017). In this respect, architected cellular materials have found an increasing interest (Schaedler and Carter 2016). These structures provide extended impulse durations by reducing the incoming pressure below the damage threshold. They also provide reductions in the volume and are extremely light in weight. Polystyrene and polypropylene foams are currently used in sporting gear and helmets. However, modern 3D printed architected materials such as polyurethane with twin hemispheres, hollow nickel microlattice with vertical posts and solid aluminium microlattice with octahedral architecture are gaining popularity.

Auxetic materials, particularly auxetic foams, possess excellent energy absorption capabilities that can be exploited for sports-related applications (Duncan 2019). Auxetic foam structures with advanced mechanical and impact performance were used to produce personal protection equipment (PPE) such as shin pads, cricket helmets, cricket pads, crash helmets and car bumpers for sports personnel. Cylinder-ligament honeycomb auxetic structures show great indentation resistance and resilience to be a suitable candidate for sports helmet applications (Sanami et al. 2014). Auxetic fabrics have a wide range of uses in PPE to boost comfort and efficiency with easier adaption to curves combined with good indentation resistance and energy absorption (Nallavan 2020). AM processes optimized through FE simulations can produce sports PPE of the highest quality (Shepherd et al. 2020). The adoption of FE techniques results in better optimizations that eliminate the need for continuous fabrication and testing cycles.

Bio-inspired core designs were utilized to additively manufacture on-water sports boards with better bending resistance compared to natural core designs (Soltani, Reza Noroozi 2020). The study incorporated a wide variety of core designs to offer a comparative analysis of the strength capabilities of each of the designs. Figure 17.6 depicts the comparison of the maximum reaction force observed at a maximum deflection of 4 mm for different core designs. Bio-inspired designs were also adopted for improved energy absorption of sandwich structures for sporting devices (Aguirre et al. 2020). The biomimicking design approach showed a quadratic relationship between the stiffness and the impact force for porous bone mimics. These properties are uplifted when it is coupled with 4D printing where shape-memory properties can provide excellent shape retention when subjected to flexural, impact and compressive loads (Mehrpouya et al. 2020).

17.7 3D PRINTED SANDWICH COMPOSITES FOR MILITARY APPLICATIONS

The defence industry has continuously expressed interest in better protective equipment for its personnel to withstand unfortunate events and lethal conflicts. A variety of requirements such as high mechanical strength and integrity, lower density, good thermal resistance and fire retardancy must be fulfilled to qualify

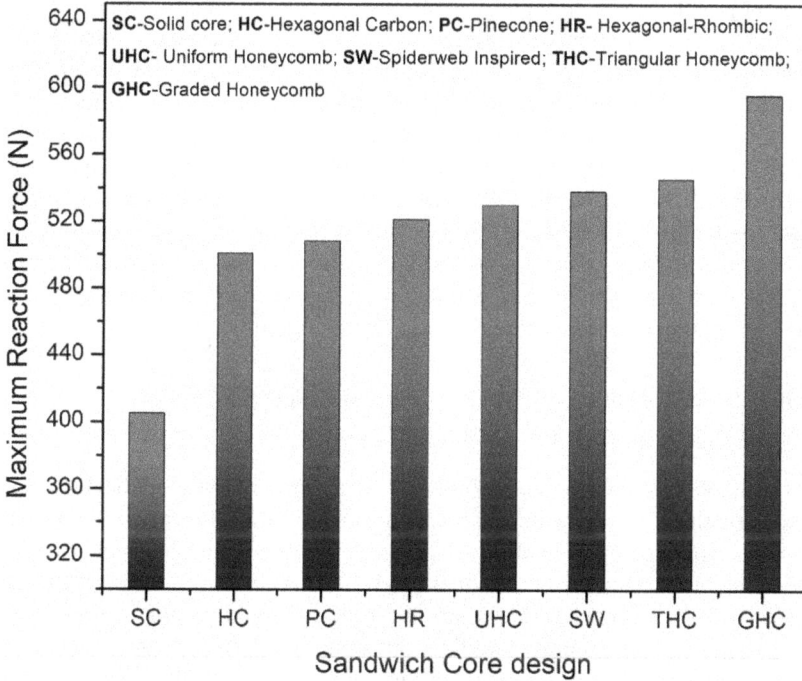

SC-Solid core; HC-Hexagonal Carbon; PC-Pinecone; HR- Hexagonal-Rhombic; UHC- Uniform Honeycomb; SW-Spiderweb Inspired; THC-Triangular Honeycomb; GHC-Graded Honeycomb

FIGURE 17.6 Variation in the reaction force for different core structures for the highest deflection measured (Soltani, Reza Noroozi 2020).

for battleground operations. 3D printing has provided an exemplary platform for this initiative with feasible and effective manufacturing solutions. A study by Ngo et al. (2018) showed that sandwiched honeycomb panels can be extensively used to fabricate armour bodies as a result of their superior energy absorption capabilities and lower density. Additionally, the properties of such structures can be enhanced significantly by introducing nanomaterials in their matrices. Nanoparticles such as graphene and CNTs were also used in multiphase shear-thickening fluids for advanced military applications. Additively manufactured polymer composite materials can be used to manufacture parts of military land vehicles, military aircraft, warships and other structural elements (Koniuszewska and Kaczmar 2016). Other structures that can be realized from these materials include masts, antennas and radars. In addition to mechanical strength, certain 3D printed sandwich materials possess flame-retardancy and temperature-resistance properties that may be beneficial on the military front.

Sandwiched honeycomb structures were of great use in military transportation applications (Koštial et al. 2020). The properties of the structure were highly dependent on the design and manufacture of the core. It is seen that these structures offered high flexibility and strength even when the honeycomb density was as low as 64 kg/m^3. Further, the properties varied along the length and width of the honeycomb, where strength was observed along the length of the structure. On comparing aluminium and Nomex honeycomb structures, the former showed lower impact properties of all of the values of core thicknesses. However, when coated with different materials, the impact properties see a positive influence for both the structures. The use of smart materials to obtain pre-programmed recoverable materials has given rise to a new class of manufacturing technique called "4D printing" (Mitchell et al. 2018). These self-transforming structures have great prospects for application on the military front including composite hinges, reflector antennas and truss booms. Additively manufactured sandwich metallic lattice structures are also used in the manufacture of heat exchangers, de-icing systems and other non-conventional applications (Ngo et al. 2018).

17.8   3D PRINTED SANDWICH COMPOSITES FOR CONSTRUCTION APPLICATIONS

AM technologies have increasing popularity in more intensive applications such as construction. This high-precision, low-demand venture is seen as a viable technique for concrete and metallic 3D printing ventures (Buchanan and Gardner 2019). Out of the numerous techniques used in AM, powder bed fusion (PBF) and directed energy deposition (DED) are the most suitable for construction. These processes aid in the accurate building, but come at the cost of time and dimensional limitations. Functionally graded materials and controlled microstructure can enhance the mechanical properties of the structures. Rather than building entire structures with AM, it is more feasible to include AM composites to provide hybrid solutions in the field of construction. Several meta-functional composites have been used in bridge bearings, protective structures, beams and other structural components (Türk et al. 2017; Huynh, Rotella, and Sangid 2016; Li et al. 2020; Sengsri and Kaewunruen 2020).

In addition to metallic structures, polymer composites made of HDPE are prospective candidates in construction due to their enhanced compressive behaviour over tensile strength (Hinchcliffe, Hess, and Srubar 2016). PLA does provide good mechanical strength, but its poor thermal properties limit it. In addition to better properties, these structures are lower in density and easier to produce, which results in minimized expenditure. Additionally, the inclusion of fibres, such as carbon and aramid, can further improve the load-bearing capabilities of these structures. Curved layer fused deposition modelling (CLFDM) of polymer composites has produced structures with better surface quality and surface strength in very short intervals of time by minimizing the staircase effect seen in flat-printed specimens (Lim et al. 2016). Glass fibre-reinforced composites used in building facade have gained increasing popularity over the years. However, one of the primary reasons for their hindered usage is their fire-retardant capabilities. A study by Nguyen et al. (2014) proposed a computation-assisted experimental approach to improve the fire-retardant properties of these structures. When aluminium hydroxide hydrate (ATH) was introduced into the matrix, the flame-retardant properties such as total heat

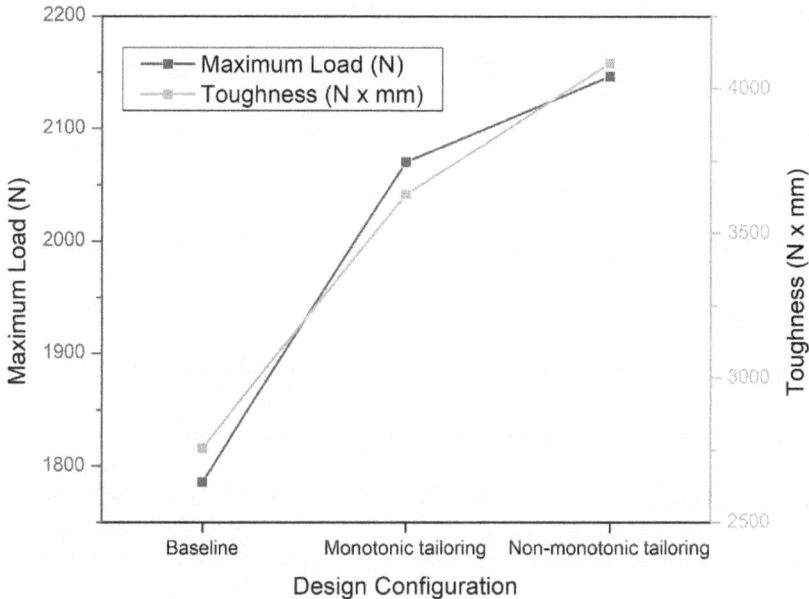

FIGURE 17.7 Maximum load and toughness of tailored sandwich composites prepared by multi-material jetting (Ubaid, Wardle, and Kumar 2018).

release (THR) and fire growth rate (FIGRA) improved. The addition of a gel coating also improves these parameters. However, efficient control must be exercised for optimum results.

Fatigue studies of microtrusses produced by additively manufactured IN718 nickel–chromium alloys showed improved strength parameters and fatigue life (Huynh, Rotella, and Sangid 2016). This makes the structure extremely durable for load-bearing structural applications. Fracture studies on Ti6Al4V sandwich structures indicate that they can serve as suitable structural members of various structural applications (Daynes et al. 2020). Novel AM techniques such as multi-material jetting improved the strength and performance of multilayered sandwich composites (Ubaid, Wardle, and Kumar 2018). A notable increase in the strength and toughness of the composites was observed for monotonic and non-monotonic spatial tailoring of adherend layers prepared by the same method. Figure 17.7 demonstrates this variation.

17.9 CHALLENGES IN 3D PRINTING COMPOSITE SANDWICH STRUCTURES

This section of the chapter discusses various challenges, limitations and drawbacks of AM techniques for sandwich composite structures. One of the major drawbacks of AM is the manufacturing cost (Buchanan and Gardner 2019). While it is believed that AM can offer substantial material and economic savings in the future, its current adoption is an expensive venture. Characterized by low-volume production runs, AM has huge prospects of cost savings in industries such as construction when compared to conventional methods of manufacture. Automation in AM has reduced significant labour expenditures. However, it is not entirely eliminated as several pre-processing, maintenance, optimization and post-processing techniques demand skilled labour.

Another important factor is the time consumption. AM is a relatively slow, layer-by-layer manufacturing technique. This consumes much time to get quality prints with close dimensional tolerances. Moreover, the intricacy of the involved geometry, properties of the material used, calibration and other parameters strictly affect the printing speed. Therefore, there is an important need to increase the material formation/deposition rate for faster printing. However, such alterations may affect the geometrical dimensions and print quality, which may result in poor parts. Unlike subtractive techniques of manufacturing, AM parts require support structures to resist self-weight and other internal and external stresses as the strength of the components increases only as the printing proceeds. This is particularly seen in the manufacture of sandwiched panels with different lattice cores.

One of the major challenges in AM is the limited availability of materials that are apt for 3D printing (Quan et al. 2015). While thermoplastics, metals and a few ceramics have been adopted, load-bearing entities such as thermosets have not been widely adopted for AM. Limitations with respect to the printing resolutions available hinder the printing accuracy and surface smoothness. Currently, wall thicknesses and layer thicknesses only in the millimetre and sub-millimetre scales are possible. Another challenge that exists on this front is the in situ monitoring of the fabrication of sandwich structures (Raghavan et al. 2013; Manvatkar, De, and Debroy 2014; Park and Ahn 2014). Since AM follows a direct deposition approach, it is important to exercise great control and monitoring of the starting materials, processing parameters and product quality so that there is sufficient layer adhesion which is important to achieve optimal properties for sandwich composites in particular. Another major drawback is the lower volume fraction of additives in reinforced polymer composites due to issues such as nozzle clogging and high printing temperature requirements (Spadaccini 2019).

Although these limitations stop serious stopper on the application spectrum of AM, a good amount of these shortcomings can be compensated for with well-thought-out optimization techniques such as topology optimization, design optimization and load-dependent optimizations. Topological optimization is seen to aid in the production of highly efficient femoral nails with excellent utilization factors (Cucinotta et al. 2019). The adopted optimization route is seen to have produced significant mass savings with almost no changes in the stress on the tissues of the femur. A similar optimization was adopted for AM of AlSi40 mirrors for cryogenic space applications (Eberle et al. 2018). The topology optimization was beneficial in producing about 30% mass reduction and 20% wavefront error reduction. Lattice design optimizations were used to enhance the properties of multi-material sandwich lattice composites (Stankovic et al. 2015). Similar to the previous optimization route, there are significant mass savings for materials with minimal densities. A combination of this route with topological optimizations can result in excellent property enhancements as well.

Optimizations are crucial when it comes to biomedical applications. In a study conducted by El-Sayed et al. (2020), a design optimization route was adopted to extract optimal functionality from Ti6Al4V diamond lattices manufactured through PBF. The technique was highly beneficial in regulating the strut length, strut diameter and orientation angle to produce structures with adequate structural stability and porosity for enhanced bioactivity. Load-dependent optimization techniques facilitate new possibilities when it comes to AM sandwich composite structures (Riss, Schilp, and Reinhart 2014). These optimization routes were applied for honeycomb structures on a freeform surface. The structural feasibility and AM technology provide excellent optimization opportunities for significant weight reduction as opposed to conventional processes. Functionally graded Ti6Al4V lattice structures followed a multiscale design optimization route to reduce the mass by about 44%, which was about 22% less than the remodelled topology optimization carried out in the study (Trudel et al. 2019). However, an ordered number of optimization runs must be carried out to obtain the global maxima and minima without losing the robustness of the structure. Similarly, chemical and physical optimization can improve the ionic transport capabilities in electrochemical energy storage devices (Tian et al. 2017). Another effective optimization strategy would be the introduction of nanomaterials in such devices. These entities have high surface areas that facilitate effective ionic transport. Electric transport optimization can be achieved through the inclusion of highly conductive additives or matrices. Such materials can be realized through multi-material 3D printing. Additionally, while scaling up these devices, increasing the layer adhesion/layer bonding can prove to be an effective optimization strategy. Hence, as seen from the above examples, optimization methods are application dependent and are often employed with an objective. However, each optimization run must be followed with careful scrutiny of the observed parameters so that there are no excessive runs which diminish the robustness of the structure.

17.10 CONCLUSIONS AND FUTURE PROSPECTS

From the above studies, it is clear that 3D printing has played a pivotal role in multiple fields of application, reinventing and redesigning our approaches to prototyping and production. Since over 30 years of the technology's initiation with almost no noticeable revenue, 3D printing today acquires about 28% of the world's total global product and revenue (Picco et al. 2014). From heavy structural components to sophisticated electronics and biomedical implants, sandwich composites prepared through numerous AM techniques have helped in transforming a new generation of manufacturing of common and bespoke goods. However, the physical limitations in the adopted technologies have hindered their prospective widespread use.

As seen in this study, the applications of these composites lie across a vast spectrum. When AM composites are needed for their structural strength in aerospace and automotive applications, they have good biocompatibility and electrical parameters for biomedical and soft robotic applications. This calls in for multifunctional capabilities to cater to the advanced requirements of these applications. Sandwich composites have expressed increasing relevance in more civil applications including sports, transport and construction through various devices, equipment and structures. In construction applications, AM sandwich structures play a more supportive role in the production of hybrid structures. Lower values of properties often characterize 3D printed parts compared to their conventionally manufactured counterparts. However, certain post-processing techniques and optimization methods have aided in producing parts with good mechanical integrity and appreciable thermal properties, as seen in the literature. The highly anisotropic nature of a majority of 3D printing techniques is at times disadvantageous in multi-stress fields. However, on comparing the advantages offered by AM, such as versatility, manufacturing flexibility, weight reduction, property enhancement, intricate part production and economic gain, the disadvantages are easily outweighed. Moreover, these difficulties can be overcome through object-oriented research to address and mitigate these insufficiencies with enhanced technological features and easier operation. Several optimization techniques such as topological, design, load-dependent and multi-scale optimization techniques have improved the properties of these structures with significant weight reductions. However, care must be taken to properly analyse these optimization runs so that they do not shun the robustness of the structure.

When compared to other production techniques, AM has a vibrant feature with the capability to seep into any field. The prospects of low-volume manufacturing and significant economic and material savings make it the technology of the future. Moreover, recent research focuses on maximizing the capabilities of AM by improving the manufacturing-related parameters of the current techniques and the introduction of new methods of additive manufacturing. The recent use of multifunctional and advanced materials exemplifies this prospect into the go-to production process in the near future. Therefore, it is very critical to create, analyse, scrutinize and calibrate AM and AM-related materials and structures to reap maximum benefits in the shortest interval.

ACKNOWLEDGEMENT

The author acknowledges the support of Indian Institute of Science, Bangalore, through C.V. Raman Post-doctoral Fellowship R(IA)/CVR-PDF/2019/1630, under Institution of Eminence scheme.

BIBLIOGRAPHY

Aguirre, Trevor G., Luca Fuller, Aniket Ingrole, Tim W. Seek, Benjamin B. Wheatley, Brett D. Steineman, Tammy L. Haut Donahue, and Seth W. Donahue. 2020. "Bioinspired material architectures from bighorn sheep horncore velar bone for impact loading applications." *Scientific Reports* 10 (1). Nature Publishing Group UK: 1–14. doi:10.1038/s41598-020-76021-5.

Alkhatib, Sami E., Mohammad S. Matar, Faris Tarlochan, Othman Laban, Ahmed S. Mohamed, and Nouman Alqwasmi. 2019. "Deformation modes and crashworthiness energy absorption of sinusoidally corrugated tubes manufactured by direct metal laser sintering." *Engineering Structures* 201 (December). Elsevier: 109838. doi:10.1016/j.engstruct.2019.109838.

Anandan, Sudharshan, Rafid M. Hussein, Myranda Spratt, Joseph Newkirk, K. Chandrashekhara, Heath Misak, and Michael Walker. 2019. "Failure in metal honeycombs manufactured by selective laser melting of 304 L stainless steel under compression." *Virtual and Physical Prototyping* 14 (2). Taylor & Francis: 114–22. doi:1 0.1080/17452759.2018.1531336.

Antolak-Dudka, Anna, Paweł Płatek, Tomasz Durejko, Paweł Baranowski, Jerzy Małachowski, Marcin Sarzyński, and Tomasz Czujko. 2019. "Static and dynamic loading behavior of Ti6Al4V honeycomb structures manufactured by laser engineered net shaping (LENSTM) technology." *Materials* 12 (8). doi:10.3390/ma12081225.

Arjunan, Arun, Manpreet Singh, Ahmad Baroutaji, and Chang Wang. 2020. "Additively Manufactured AlSi10Mg inherently stable thin and thick-walled lattice with negative Poisson's ratio." *Composite Structures* 247. Elsevier Ltd: 112469. doi:10.1016/j. compstruct.2020.112469.

Atkins, Carolyn, Charlotte Feldman, David Brooks, Stephen Watson, William Cochrane, Mélanie Roulet, Emmanuel Hugot, et al. 2018. "Topological design of lightweight additively manufactured mirrors for space." In Proc. SPIE 10706, Advances in Optical and Mechanical Technologies for Telescopes and Instrumentation III. doi:10.1117/12.2313353.

Atkins, Carolyn, Charlotte H. Feldman, David Brooks, Richard Willingale, Peter Doel, Melanie Roulet, Stephen Watson, William Cochrane, and Emmanuel Hugot. 2017. "Additive manufactured X-ray optics for astronomy." In Proc. SPIE 10399, Optics for EUV, X-Ray, and Gamma-Ray Astronomy VIII. doi:10.1117/12.2274011.

Austermann, Johannes, Alec J. Redmann, Vera Dahmen, Adam L. Quintanilla, Sue J. Mecham, and Tim A. Osswald. 2019. "Fiber-reinforced composite sandwich structures by co-curing with additive manufactured epoxy lattices." *Journal of Composites Science* 3 (2): 53. doi:10.3390/jcs3020053.

Brenne, F., T. Niendorf, and H. J. Maier. 2013. "Additively manufactured cellular structures: Impact of microstructure and local strains on the monotonic and cyclic behavior under uniaxial and bending load." *Journal of Materials Processing Technology* 213 (9). Elsevier B.V.: 1558–64. doi:10.1016/j.jmatprotec.2013.03.013.

Brubaker, Cole D., Kailey N. Newcome, G. Kane Jennings, and Douglas E. Adams. 2019. "3D-printed alternating current electroluminescent devices." *Journal of Materials Chemistry C* 7 (19): 5573–78. doi:10.1039/c9tc00619b.

Buchanan, C., and L. Gardner. 2019. "Metal 3D printing in construction: A review of methods, research, applications, opportunities and challenges." *Engineering Structures* 180 (2): 332–48.

Buchanan, Craig, Ville Pekka Matilainen, Antti Salminen, and Leroy Gardner. 2017. "Structural performance of additive manufactured metallic material and cross-sections." *Journal of Constructional Steel Research* 136: 35–48. doi:10.1016/j.jcsr.2017.05.002.

Carricoa, J. D., Nicklaus W. Traedena, Matteo Aurelib, and Kam K. Leang. 2015. "Fused filament 3D printing of ionic polymer-metal composites (IPMCs)." *Smart Materials and Structures* 24: 125021.

Chang, Peng, Hui Mei, Shixiang Zhou, Konstantinos G. Dassios, and Laifei Cheng. 2019. "3D printed electrochemical energy storage devices." *Journal of Materials Chemistry A* 7 (9). Royal Society of Chemistry: 4230–58. doi:10.1039/c8ta11860d.

Chantarapanich, Nattapon, Apinya Laohaprapanon, Sirikul Wisutmethangoon, Pongnarin Jiamwatthanachai, Prasert Chalermkarnnon, Sedthawatt Sucharitpwatskul, Puttisak Puttawibul, and Kriskrai Sitthiseripratip. 2014. "Fabrication of three-dimensional honeycomb structure for aeronautical applications using selective laser melting: A preliminary investigation." *Rapid Prototyping Journal* 20 (6): 551–58. doi:10.1108/RPJ-08-2011-0086.

Cuan-Urquizo, E., S. Yang, and A. Bhaskar. 2015. "Mechanical characterization of additively manufactured material having lattice microstructure." *IOP Conference Series: Materials Science and Engineering* 74 (1). doi:10.1088/1757-899X/74/1/012004.

Cucinotta, F., E. Guglielmino, G. Longo, G. Risitano, D. Santonocito, and F. Sfravara. 2019. "Topology optimization additive manufacturing-oriented for a biomedical application." *Lecture Notes in Mechanical Engineering* 1: 184–93 Springer International Publishing. doi:10.1007/978-3-030–12346-8_18.

Davami, Keivan, Mehrdad Mohsenizadeh, Michael Munther, Tyler Palma, Ali Beheshti, and Kasra Momeni. 2019. "Dynamic energy absorption characteristics of additively manufactured shape-recovering lattice structures." *Materials Research Express* 6 (4). IOP Publishing. doi:10.1088/2053-1591/aaf78c.

Davoodi, Elham, Haniyeh Fayazfar, Farzad Liravi, Elahe Jabari, and Ehsan Toyserkani. 2020. "Drop-on-demand high-speed 3D printing of flexible milled carbon fiber/silicone composite sensors for wearable biomonitoring devices." *Additive Manufacturing* 32 (December 2019). Elsevier: 101016. doi:10.1016/j.addma.2019.101016.

Davoudinejad, Ali, Mohammad Reza Khosravani, David Bue Pedersen, and Guido Tosello. 2020. "Influence of thermal ageing on the fracture and lifetime of additively manufactured mold inserts." *Engineering Failure Analysis* 115. Elsevier Ltd: 104694. doi:10.1016/j.engfailanal.2020.104694.

Daynes, Stephen, Joseph Lifton, Wen Feng Lu, Jun Wei, and Stefanie Feih. 2020. "Fracture toughness characteristics of additively manufactured Ti-6Al-4V lattices." *European Journal of Mechanics - A/Solids*. Elsevier Masson SAS: 104170. doi:10.1016/j.euromechsol.2020.104170.

Digital Science. 2018. Dimensions [Software] Available from Https://App.Dimensions.Ai. Accessed on (20-11-2020), under Licence Agreement.

Dong, Shi Lei, Lu Han, Cai Xia Du, Xiao Yu Wang, Lu Hai Li, and Yen Wei. 2017. "3D printing of aniline tetramer-grafted-polyethylenimine and pluronic F127 composites for electroactive scaffolds." *Macromolecular Rapid Communications* 38 (4): 1–7. doi:10.1002/marc.201600551.

Duncan, Oliver. 2019. *Auxetic Foams for Sports Applications*. Sheffield Hallam University. http://shura.shu.ac.uk/information.html%0ASheffield.

Eberle, Sebastian, Arnd Reutlinger, Bailey Curzadd, Michael Mueller, Mirko Riede, Christoph Wilsnack, Ana Brandão, et al. 2018. "Additive manufacturing of an AlSi40 mirror coated with electroless nickel for cryogenic space applications." In *International Conference on Space Optics-ICSO 2018*, 11180: 40. doi:10.1117/12.2535960.

Echeta, Ifeanyichukwu, Xiaobing Feng, Ben Dutton, Richard Leach, and Samanta Piano. 2020. "Review of defects in lattice structures manufactured by powder bed fusion." *International Journal of Advanced Manufacturing Technology* 106 (5–6): 2649–68. doi:10.1007/s00170-019-04753-4.

Echsel, M., P. Springer, and S. Hümbert. 2020. "Production and planned in-orbit qualification of a function-integrated, additive manufactured satellite sandwich structure with embedded automotive electronics." *CEAS Space Journal*, no. 0123456789. Springer Vienna. doi:10.1007/s12567-020-00328-2.

Eichenhofer, Martin, Daniel Schupp, Michael Niedermeier, Daniel Huelsbusch, Frank Walther, and Patrick Striemann. 2018. "Compression testing of additively manufactured continuous carbon fiber-reinforced sandwich structures." *Materialpruefung/ Materials Testing* 60 (9): 801–8. doi:10.3139/120.111216.

Eichenhofer, Martin, Joanna C.H. Wong, and Paolo Ermanni. 2017. "Continuous lattice fabrication of ultra-lightweight composite structures." *Additive Manufacturing* 18. Elsevier B.V.: 48–57. doi:10.1016/j.addma.2017.08.013.

El-Sayed, Mahmoud Ahmed, Khamis Essa, Mootaz Ghazy, and Hany Hassanin. 2020. "Design optimization of additively manufactured titanium lattice structures for biomedical implants." *International Journal of Advanced Manufacturing Technology* 110 (9–10): 2257–68. doi:10.1007/s00170-020-05982-8.

Evans, Kent A., Zachary C. Kennedy, Bruce W. Arey, Josef F. Christ, Herbert T. Schaef, Satish K. Nune, and Rebecca L. Erikson. 2018. "Chemically active, porous 3D-printed thermoplastic composites." *ACS Applied Materials and Interfaces* 10 (17): 15112–21. doi:10.1021/acsami.7b17565.

Ferrari, Michael. 2015. "*Structurally Optimized and Additively Manufactured Inserts for Sandwich Panels of Spacecraft Structures.*" ETH Zürich, Switzerland, Master Thesis 107.

Ferrari, Luca, Maurizio Barbato, Burkard Esser, Ivaylo Petkov, Markus Kuhn, Sandro Gianella, Jorge Barcena, et al. 2016. "Sandwich structured ceramic matrix composites with periodic cellular ceramic cores: An active cooled thermal protection for space vehicles." *Composite Structures* 154: 61–68. doi:10.1016/j.compstruct. 2016.07.043.

Foster, Christopher W., Michael P. Down, Yan Zhang, Xiaobo Ji, Samuel J. Rowley-Neale, Graham C. Smith, Peter J. Kelly, and Craig E. Banks. 2017. "3D printed graphene based energy storage devices." *Scientific Reports* 7 (March): 1–11. doi:10.1038/srep42233.

Friedrich, Klaus, and Abdulhakim A. Almajid. 2013. "Manufacturing aspects of advanced polymer composites for automotive applications." *Applied Composite Materials* 20 (2): 107–28. doi:10.1007/s10443-012-9258-7.

Fu, Kun, Yonggang Yao, Jiaqi Dai, and Liangbing Hu. 2017. "Progress in 3D printing of carbon materials for energy-related applications." *Advanced Materials* 29 (9). doi:10.1002/adma.201603486.

Gao, Shuyue, Cao Wang, Bohang Xing, Minhao Shen, Weiming Zhao, and Zhe Zhao. 2020. "Experimental investigation on bending behaviour of ZrO2 honeycomb sandwich structures prepared by DLP stereolithography." *Thin-Walled Structures* 157 (September). Elsevier Ltd: 107099. doi:10.1016/j.tws.2020.107099.

Ghazanfari, Amir, Wenbin Li, Ming C. Leu, Yiyang Zhuang, and Jie Huang. 2016. "Advanced ceramic components with embedded sapphire optical fiber sensors for high temperature applications." *Materials and Design* 112. Elsevier B.V.: 197–206. doi:10.1016/j.matdes.2016.09.074.

Goh, Guo Liang, Shweta Agarwala, and Wai Yee Yeong. 2019. "Directed and on-demand alignment of carbon nanotube: A review toward 3D printing of electronics." *Advanced Materials Interfaces* 6 (4). doi:10.1002/admi.201801318.

Gulzar, Umair, Colm Glynn, and Colm O'Dwyer. 2020. "Additive manufacturing for energy storage: Methods, designs and material selection for customizable 3D printed batteries and supercapacitors." *Current Opinion in Electrochemistry* 20: 46–53. doi:10.1016/j.coelec.2020.02.009.

Haleem, Abid, Mohd Javaid, Rizwan Hasan Khan, and Rajiv Suman. 2020. "3D printing applications in bone tissue engineering." *Journal of Clinical Orthopaedics and Trauma* 11. Delhi Orthopedic Association: S118–24. doi:10.1016/j.jcot.2019. 12.002.

Hamidi, Armita, and Yonas Tadesse. 2019. "Single step 3D printing of bioinspired structures via metal reinforced thermoplastic and highly stretchable elastomer." *Composite Structures* 210. Elsevier Ltd: 250–61. doi:10.1016/j.compstruct.2018.11.019.

Harris, J. A., R. E. Winter, and G. J. McShane. 2017. "Impact response of additively manufactured metallic hybrid lattice materials." *International Journal of Impact Engineering* 104. Elsevier Ltd: 177–91. doi:10.1016/j.ijimpeng.2017.02.007.

Hilpert, Enrico, Johannes Hartung, Henrik von Lukowicz, Tobias Herffurth, and Nils Heidler. 2019. "Design, additive manufacturing, processing, and characterization of metal mirror made of aluminum silicon alloy for space applications." *Optical Engineering* 58 (09): 092613. doi:10.1117/1.oe.58.9.092613.

Hinchcliffe, Sean A., Kristen M. Hess, and Wil V. Srubar. 2016. "Experimental and theoretical investigation of prestressed natural fiber-reinforced polylactic acid (PLA) composite materials." *Composites Part B: Engineering* 95. Elsevier Ltd: 346–54. doi:10.1016/j.compositesb.2016.03.089.

Huynh, Lena, John Rotella, and Michael D. Sangid. 2016. "Fatigue behavior of IN718 microtrusses produced via additive manufacturing." *Materials and Design* 105. Elsevier B.V.: 278–89. doi:10.1016/j.matdes.2016.05.032.

Jakus, Adam E. 2019. "Tissue engineering and additively manufactured ceramic-based biomaterials" *American Ceramic Society Bulletin* 98(4): 21–27.

Jeyachandran, Praveen, Srikanth Bontha, Subhadip Bodhak, Vamsi Krishna Balla, Biswanath Kundu, and Mrityunjay Doddamani. 2020. "Mechanical behaviour of additively manufactured bioactive glass/high density polyethylene composites." *Journal of the Mechanical Behavior of Biomedical Materials* 108 (April). Elsevier Ltd: 103830. doi:10.1016/j.jmbbm.2020.103830.

Koniuszewska, Anna G., and Jacek W. Kaczmar. 2016. "Application of polymer based composite materials in transportation." *Progress in Rubber, Plastics and Recycling Technology* 32 (1): 1–23. doi:10.1177/147776061603200101.

Korde, Jay M., Mohsin Shaikh, and Balasubramanian Kandasubramanian. 2018. "Bionic prototyping of honeycomb patterned polymer composite and its engineering application." *Polymer - Plastics Technology and Engineering* 57 (17). Taylor & Francis: 1828–44. doi:10.1080/03602559.2018.1434667.

Koštial, Pavel, Zora Koštialová Jančíková, Ivan Ružiak, and Milada Gajtanska. 2020. "Case study of chosen sandwich-structured composite materials for means of transport." *Coatings* 10 (8). doi:10.3390/COATINGS10080750.

Kussmaul, Ralph, Manuel Biedermann, Georgios A. Pappas, Jónas Grétar Jónasson, Peter Winiger, Markus Zogg, Daniel-Alexander Türk, Mirko Meboldt, and Paolo Ermanni. 2019. "Individualized lightweight structures for biomedical applications using additive manufacturing and carbon fiber patched composites." *Design Science* 5: 1–24. doi:10.1017/dsj.2019.19.

Li, Shi, Menglei Hu, Lijun Xiao, and Weidong Song. 2020. "Compressive properties and collapse behavior of additively-manufactured layered-hybrid lattice structures under static and dynamic loadings." *Thin-Walled Structures* 157 (September). Elsevier Ltd: 107153. doi:10.1016/j.tws.2020.107153.

Lim, Sungwoo, Richard A. Buswell, Philip J. Valentine, Daniel Piker, Simon A. Austin, and Xavier De Kestelier. 2016. "Modelling curved-layered printing paths for fabricating large-scale construction components." *Additive Manufacturing* 12: 216–30. doi:10.1016/j.addma.2016.06.004.

Ling, Bin, Kai Wei, Zhonggang Wang, Xujing Yang, Zhaoliang Qu, and Daining Fang. 2020. "Experimentally program large magnitude of Poisson's ratio in additively manufactured mechanical metamaterials." *International Journal of Mechanical Sciences* 173 (December 2019). Elsevier Ltd. doi:10.1016/j.ijmecsci.2020.105466.

Liu, Changyong, Ninggui Huang, Feng Xu, Junda Tong, Zhangwei Chen, Xuchun Gui, Yuelong Fu, and Changshi Lao. 2018. "3D printing technologies for flexible tactile sensors toward wearable electronics and electronic skin." *Polymers* 10 (6): 1–31. doi:10.3390/polym10060629.

Macchia, Michael A. 2019. "Application of Metamaterials for Multifunctional Satellite Bus Enabled via Additive Manufacturing.", Air Force Institute of Technology, Wright-Patterson Institute of Technology, Ohio, Thesis, 2239

Manvatkar, V., A. De, and T. Debroy. 2014. "Heat transfer and material flow during laser assisted multi-layer additive manufacturing." *Journal of Applied Physics* 116 (12). doi:10.1063/1.4896751.

Mehrpouya, Mehrshad, Annamaria Gisario, Ali Azizi, and Massimiliano Barletta. 2020. "Investigation on shape recovery of 3D printed honeycomb sandwich structure." *Polymers for Advanced Technologies* 31 (12): 3361–65. doi:10.1002/pat.5020.

Mishra, Saswat, Prajakta Katti, S. Kumar, and Suryasarathi Bose. 2019. "Macroporous epoxy-carbon fiber structures with a sacrificial 3D printed polymeric mesh suppresses electromagnetic radiation." *Chemical Engineering Journal* 357: 384–94. doi:10.1016/j.cej.2018.09.119.

Mitchell, A., U. Lafont, M. Hołyńska, and C. Semprimoschnig. 2018. "Additive manufacturing — A review of 4D printing and future applications." *Additive Manufacturing* 24. Elsevier B.V.: 606–26. doi:10.1016/j.addma.2018.10.038.

Mohamed, Ahmed S., Othman Laban, Faris Tarlochan, Sami E. Al Khatib, Mohammed S. Matar, and Elsadig Mahdi. 2019. "Experimental analysis of additively manufactured thin-walled heat-treated circular tubes with slits using AlSi10Mg alloy by quasi-static axial crushing test." *Thin-Walled Structures* 138 (January). Elsevier Ltd: 404–14. doi:10.1016/j.tws.2019.02.022.

Mohammadizadeh, M., I. Fidan, M. Allen, and A. Imeri. 2018. "Creep behavior analysis of additively manufactured fiber-reinforced components." *International Journal of Advanced Manufacturing Technology* 99 (5–8): 1225–34. doi:10.1007/s00170-018-2539-z.

Mori, Arianna De, Marta Peña Fernández, Gordon Blunn, Gianluca Tozzi, and Marta Roldo. 2018. "3D printing and electrospinning of composite hydrogels for cartilage and bone tissue engineering." *Polymers* 10 (3): 1–26. doi:10.3390/polym10030285.

Mousavi, Saeb, David Howard, Shuying Wu, and Chun Wang. 2018. "An ultrasensitive 3D printed tactile sensor for soft robotics." In ICRA *2018*: Workshop on Soft Robotics (IEEE), Brisbane, Queensland, , 1–1.

Mousavi, Saeb, David Howard, Fenghua Zhang, Jinsong Leng, and Chun H. Wang. 2020. "Direct 3D printing of highly anisotropic, flexible, constriction-resistive sensors for multidirectional proprioception in soft robots." *ACS Applied Materials and Interfaces* 12 (13): 15631–43. doi:10.1021/acsami.9b21816.

Mu, X., T. Bertron, C. Dunn, H. Qiao, J. Wu, Z. Zhao, C. Saldana, and H. J. Qi. 2017. "Porous polymeric materials by 3D printing of photocurable resin." *Materials Horizons* 4 (3): 442–49. doi:10.1039/c7mh00084g.

Mullen, Lewis, Robin C. Stamp, Wesley K. Brooks, Eric Jones, and Christopher J. Sutcliffe. 2009. "Selective laser melting: A regular unit cell approach for the manufacture of porous, titanium, bone in-growth constructs, suitable for orthopedic applications." *Journal of Biomedical Materials Research - Part B Applied Biomaterials* 89 (2): 325–34. doi:10.1002/jbm.b.31219.

Nallavan, G. 2020. "Impact of recent developments in fabrication of auxetic materials on safety and protection in sport." *AIP Conference Proceedings* 2271 (September). doi:10.1063/5.0024805.

Ngo, Tuan D., Alireza Kashani, Gabriele Imbalzano, Kate T.Q. Nguyen, and David Hui. 2018. "Additive manufacturing (3D Printing): A review of materials, methods, applications and challenges." *Composites Part B: Engineering* 143. Elsevier Ltd: 172–96. doi:10.1016/j.compositesb.2018.02.012.

Nguyen, Quynh T., Phuong Tran, Tuan D. Ngo, Phong A. Tran, and Priyan Mendis. 2014. "Experimental and computational investigations on fire resistance of GFRP composite for building façade." *Composites Part B: Engineering* 62. Elsevier Ltd: 218–29. doi:10.1016/j.compositesb.2014.02.010.

Pabst, Oliver, Jolke Perelaer, Erik Beckert, Ulrich S. Schubert, Ramona Eberhardt, and Andreas Tünnermann. 2013. "All inkjet-printed piezoelectric polymer actuators: Characterization and applications for micropumps in lab-on-a-chip systems." *Organic Electronics* 14 (12). Elsevier B.V.: 3423–29. doi:10.1016/j.orgel.2013.09.009.

Park, Seoung Hwan, and Doyeol Ahn. 2014. "Internal field effects on electronic and optical properties of ZnO/BeZnO quantum well structures." *Physica B: Condensed Matter* 441. Elsevier: 12–16. doi:10.1016/j.physb.2014.02.008.

Picco, Sergio, Liliana Villegas, Franco Tonelli, Mario Merlo, Javier Rigau, Dario Diaz, and Martin Masuelli. 2014. "Additive manufacturing of al alloys and aluminium matrix composites (AMCs)." In *Light Metal Alloys Applications*, 13. https://www.intechopen.com/books/advanced-biometric-technologies/liveness-detection-in-biometrics.

Picha, Kyle, Clayson Spackman, and Johnson Samuel. 2016. "Droplet spreading characteristics observed during 3D printing of aligned fiber-reinforced soft composites." *Additive Manufacturing* 12. Elsevier B.V.: 121–31. doi:10.1016/j.addma.2016.08.004.

Quan, Zhenzhen, Amanda Wu, Michael Keefe, Xiaohong Qin, Jianyong Yu, Jonghwan Suhr, Joon Hyung Byun, Byung Sun Kim, and Tsu Wei Chou. 2015. "Additive manufacturing of multidirectional preforms for composites: Opportunities and challenges." *Materials Today* 18 (9). Elsevier Ltd.: 503–12. doi:10.1016/j.mattod.2015.05.001.

Raghavan, A., H. L. Wei, T. A. Palmer, and T. DebRoy. 2013. "Heat transfer and fluid flow in additive manufacturing." *Journal of Laser Applications* 25 (5): 052006. doi:10.2351/1.4817788.

Riss, Fabian, Johannes Schilp, and Gunther Reinhart. 2014. "Load-dependent optimization of honeycombs for sandwich components-new possibilities by using additive layer manufacturing." *Physics Procedia* 56 (C). Elsevier B.V.: 327–35. doi:10.1016/j.phpro.2014.08.178.

Sanami, Mohammad, Naveen Ravirala, Kim Alderson, and Andrew Alderson. 2014. "Auxetic materials for sports applications." *Procedia Engineering* 72. Elsevier B.V.: 453–58. doi:10.1016/j.proeng.2014.06.079.

Schaedler, Tobias A., and William B. Carter. 2016. "Architected cellular materials." *Annual Review of Materials Research* 46: 187–210. doi:10.1146/annurev-matsci-070115-031624.

Schubert, Martin, and Anthanasios Dafnis. 2019. "Multifunctional load-bearing aerostructures with integrated space debris protection." *MATEC Web of Conferences* 304: 07003. doi:10.1051/matecconf/201930407003.

Schubert, Martin, Sara Perfetto, Anthanasios Dafnis, Heiko Atzrodt, and Dirk Mayer. 2018. "Multifunctional and lightweight load-bearing composite structures for satellites." *MATEC Web of Conferences EASN-CEAS* 2018 (233): 1–8. doi:10.1051/matecconf/201823300019.

Schubert, Martin, Sara Perfetto, Anthanasios Dafnis, Dirk Mayer, and Heiko Atzrodt. 2018. "Development and design of multifunctional lightweight structures for satellite applications." *Proceedings of the International Astronautical Congress, IAC* 2018-October, Noordwijk (1).

Sengsri, Pasakorn, and Sakdirat Kaewunruen. 2020. "Additive manufacturing meta-functional composites for engineered bridge bearings: A review." *Construction and Building Materials* 262. Elsevier Ltd: 120535. doi:10.1016/j.conbuildmat.2020.120535.

Shepherd, Todd, Keith Winwood, Prabhuraj Venkatraman, Andrew Alderson, and Thomas Allen. 2020. "Validation of a finite element modeling process for auxetic structures under impact." *Physica Status Solidi (B) Basic Research* 1900197: 1–14. doi:10.1002/pssb.201900197.

Soltani, Aref, Reza Noroozi, Mahdi Bodaghi and Reza Hedayati. 2020. "3D printing on-water sports boards." *Polymers* 12(250): 1–18.

Spadaccini, Christopher M. 2019. "Additive manufacturing and processing of architected materials." *MRS Bulletin* 44 (10): 782–88. doi:10.1557/mrs.2019.234.

Stankovic, Tino, Jochen Mueller, Paul Egan, and Kristina Shea. 2015. "Optimization of additively manufactured multi-material lattice structures using generalized optimality criteria." *Proceedings of the ASME Design Engineering Technical Conference* 1A-2015: 1–10. doi:10.1115/DETC2015-47403.

Subic, Editors A, M Scheinowitz, F K Fuss, Todd Shepherd, Heather Driscoll, Keith Winwood, Praburaj Venkatraman, and Tom Allen. 2017. Review of modelling and additive manufacturing of auxetic materials for application. In : *8th Asia-Pacific Congress on Sports Technology* Downloaded from : http://E-S," no. October.

Sun, Shanyouming, Shangsheng Feng, Qiancheng Zhang, and Tian Jian Lu. 2020. "Forced convection in additively manufactured sandwich-walled cylinders with thermo-mechanical multifunctionality." *International Journal of Heat and Mass Transfer* 149. Elsevier Ltd: 119161. doi:10.1016/j.ijheatmasstransfer.2019.119161.

Tafazoli, Mohammad, and Mohammad Damghani Nouri. 2020. "Experimental and numerical study and multi-objective optimization of quasi-static compressive test on three-dimensional printed lattice-core sandwich structures." *International Journal of Crashworthiness* 0 (0). Taylor & Francis: 1–11. doi:10.1080/13588265.2020.1775053.

Thaler, Dominic, Nahal Aliheidari, and Amir Ameli. 2019. "Mechanical, electrical, and piezoresistivity behaviors of additively manufactured acrylonitrile butadiene styrene/carbon nanotube nanocomposites." *Smart Materials and Structures* 28 (084004). https://iopscience.iop.org/article/10.1088/1361-665X/ab256e

Tian, Xiaocong, Jun Jin, Shangqin Yuan, Chee Kai Chua, Shu Beng Tor, and Kun Zhou. 2017. "Emerging 3D-printed electrochemical energy storage devices: A critical review." *Advanced Energy Materials* 7 (17): 1–17. doi:10.1002/aenm.201700127.

Trudel, Eric, Mostafa S.A. Elsayed, Leo Kok, and Emmanuel Provost. 2019. "Multiscale design optimization of additively manufactured aerospace structures employing functionally graded lattice materials." *AIAA Scitech 2019 Forum*, no. January. doi:10.2514/6.2019-0420.

Turk, Daniel Alexander, Andreas Ebnother, Markus Zogg, and Mirko Meboldt. 2018. "Additive manufacturing of structural cores and washout tooling for autoclave curing of hybrid composite structures." *Journal of Manufacturing Science and Engineering, Transactions of the ASME* 140 (10): 1–14. doi:10.1115/1.4040428.

Türk, Daniel Alexander, Ralph Kussmaul, Markus Zogg, Christoph Klahn, Bastian Leutenecker-Twelsiek, and Mirko Meboldt. 2017. "Composites part production with additive manufacturing technologies." *Procedia CIRP* 66 (section 4). The Author(s): 306–11. doi:10.1016/j.procir.2017.03.359.

Ubaid, Jabir, Brian L. Wardle, and S. Kumar. 2018. "Strength and performance enhancement of multilayers by spatial tailoring of adherend compliance and morphology via multimaterial jetting additive manufacturing." *Scientific Reports* 8 (1). Springer US: 1–10. doi:10.1038/s41598-018-31819-2.

Ullah, I., J. Elambasseril, M. Brandt, and S. Feih. 2014. "Performance of bio-inspired kagome truss core structures under compression and shear loading." *Composite Structures* 118 (1): 294–302. doi:10.1016/j.compstruct.2014.07.036.

Venkatesan, K. R., A. Rai, A. Chattopadhyay, T. Stoumbos, and D. Inoyama. 2020. "Finite element-based damage and failure analysis of honeycomb core sandwich composite structures for space applications." *AIAA Scitech 2020 Forum* 1 PartF (January): 1–10. doi:10.2514/6.2020-0249.

Wang, Liang, Viktor Hofmann, Fushi Bai, Jiamei Jin, and Jens Twiefel. 2019. "A novel additive manufactured three-dimensional piezoelectric transducer: Systematic modeling and experimental validation." *Mechanical Systems and Signal Processing* 114. Elsevier Ltd: 346–65. doi:10.1016/j.ymssp.2018.05.025.

Wang, Zhenyu, Xu Liu, Xi Shen, Ne Myo Han, Ying Wu, Qingbin Zheng, Jingjing Jia, Ning Wang, and Jang Kyo Kim. 2018. "An ultralight graphene honeycomb sandwich for stretchable light-emitting displays." *Advanced Functional Materials* 28 (19): 1–9. doi:10.1002/adfm.201707043.

Wang, Kui, Xun Xie, Jin Wang, Andong Zhao, Yong Peng, and Yanni Rao. 2020. "Effects of infill characteristics and strain rate on the deformation and failure properties of additively manufactured polyamide-based composite structures." *Results in Physics* 18 (August). Elsevier: 103346. doi:10.1016/j.rinp.2020.103346.

Wang, Xiaojian, Shanqing Xu, Shiwei Zhou, Wei Xu, Martin Leary, Peter Choong, M. Qian, Milan Brandt, and Yi Min Xie. 2016. "Topological design and additive manufacturing of porous metals for bone scaffolds and orthopaedic implants: A review." *Biomaterials* 83. Elsevier Ltd: 127–41. doi:10.1016/j.biomaterials.2016.01.012.

Xiao, Lijun, and Weidong Song. 2018. "Additively-manufactured functionally graded Ti-6Al-4V lattice structures with high strength under static and dynamic loading: Experiments." *International Journal of Impact Engineering* 111 (October): 255–72. doi:10.1016/j.ijimpeng.2017.09.018.

Xu, Weiwei, Felippe J. Pavinatto, G. Z. Cao, J. Devin MacKenzie, and I. Y. Shen. 2020. "Additively manufactured lead-zirconate-titanate thin films for hard-disk drive applications." *IEEE Transactions on Magnetics* 56(2): 1–6. doi:10.1109/TMAG.2019.2952464.

Yu, Tianyu, Ziyang Zhang, Shutao Song, Yuanli Bai, and Dazhong Wu. 2019. "Tensile and flexural behaviors of additively manufactured continuous carbon fiber-reinforced polymer composites." *Composite Structures* 225 (June). doi:10.1016/j.compstruct.2019.111147.

Yuan, Chao, Devin J. Roach, Conner K. Dunn, Quanyi Mu, Xiao Kuang, Christopher M. Yakacki, T. J. Wang, Kai Yu, and H. Jerry Qi. 2017. "3D printed reversible shape changing soft actuators assisted by liquid crystal elastomers." *Soft Matter* 13 (33): 5558–68. doi:10.1039/c7sm00759k.

Zhang, Xiang Yu, Gang Fang, Sander Leeflang, Amir A. Zadpoor, and Jie Zhou. 2019. "Topological Design, permeability and mechanical behavior of additively manufactured functionally graded porous metallic biomaterials." *Acta Biomaterialia* 84. Acta Materialia Inc.: 437–52. doi:10.1016/j.actbio.2018.12.013.

Zhang, Chao, Shuai Wang, Jian Li, Yi Zhu, Tao Peng, and Huayong Yang. 2020. "Additive manufacturing of products with functional fluid channels: A review." *Additive Manufacturing* 36: 1–50. doi:10.1016/j.addma.2020.101490.

Zhang, Feng, Min Wei, Vilayanur V. Viswanathan, Benjamin Swart, Yuyan Shao, Gang Wu, and Chi Zhou. 2017. "3D printing technologies for electrochemical energy storage." *Nano Energy* 40. Elsevier Ltd: 418–31. doi:10.1016/j.nanoen.2017.08.037.

Zhu, Cheng, Tianyu Liu, Fang Qian, Wen Chen, Swetha Chandrasekaran, Bin Yao, Yu Song, et al. 2017. "3D printed functional nanomaterials for electrochemical energy storage." *Nano Today* 15. Elsevier Ltd: 107–20. doi:10.1016/j.nantod.2017.06.007.

Index

Note: **Bold** page numbers refer to tables and *italic* page numbers refer to figures.

A

ABAQUS 91, 92
Abrao, A.M. 265
absorbed energy
  experimental and predicted values of 111
  signal-to-noise ratio response for 109, **109**
acrylates, urethane 9–10
acrylonitrile butadiene styrene (ABS)/carbon
      nanotube (CNT) composites 356
additive manufacturing (AM) techniques
      15–17, **18,** 350, 351, 364, 366
adhesive bonding
  of metal–composite interface 68–9
  performance of 68
adhesive films 70, 231
adhesive materials 8–10
  epoxy resins 8
  phenolics 9
  polyester and vinyl ester resins 10
  polyurethanes 9
  toughened epoxies 8–9
  urethane acrylates 9–10
aeronautical applications, 3D printed sandwich
      composites for 351–2
Ahmadi, H. 123
Airbus A350 294
alkali-activated fly ash paste 169
Allen, H.G. 4, 305
alloys
  aluminium 9
  titanium 65
aluminium 61
aluminium hydroxide hydrate (ATH) 362
aluminum alloy 9
  Hourglass lattice sandwich structures 48
  tetrahedral lattice truss cores 47
  fabrication process 47, *47*
aluminum face-sheet and core 121
Aly, N.M. 339
Ammar, M.M. 66
AM techniques *see* additive manufacturing
      (AM) techniques
analysis of variance (ANOVA) and regression
      approach 103
Anderson, T. 101
aniline tetramer-grafted polyethylenimine
      (AT-PEI) 353
Anjang, A. 280

ANOVA, for energy absorption and peak
      force 110
Antolak-Dudka, A. 352
aramid fibre-reinforced aluminium laminates
      (ARALL) 62, *62,* 63, 65
  grades of **64**
aramid fibre-reinforced polymer (ARFP) 32
Ashby, M.F. 207, 336
ASTM D297M-16 test method 235
ASTM D365M test method 236
ASTM D393M test method 237
ASTM 364M test method 236
asymmetrical sandwich 215, *216*
asymmetric sandwich panels 300
Atkins, C. 358
autoclave moulding process 13–14, **17**
automotive applications, 3D printed sandwich
      composites for 351–2
automotive sandwich composite structures, for
      crashworthiness 339–42
auxetic fabrics 360
auxetic foam structures 360
auxetic materials 360
axial compression behavior–experimental
      investigation 172–5
axial load–moment interaction 176, 186

B

BAG *see* bioactive glass (BAG)
Bai, R. 92
Banhart, J. 37
Baral, N. 283, 342
barely visible (BV) 77
barely visible impact damage (BVID) 77, 241
basalt FRPs (BFRPs) *172*, 180, 181
  bar/grid 169–71, *171*
  reinforced geopolymer concrete one-way
      slabs 151
Batdorf, S.B. 4
Benayoune, A. 174, 175, 184, 188
bending forces, sandwich composite structures
      laminates under 2, *3*
Bert, C.W. 4
BFRPs *see* basalt FRPs (BFRPs)
bioactive glass (BAG) 355
bio-based core sandwich panels 122
bio-inspired core designs 360
biomaterials, bioprinting of advanced 353

biomedical applications, 3D printed sandwich
        composites in 352, *353, 354*
biomimicking design approach 360
bioprinting, of advanced biomaterials 353
Blok, L.G. 336
Boeing 787 jetliners 294, *295*
Bosco, M.A.J. 263
Boutar, Y. 69
brittle fiber-reinforced composites 324
brittle fracturing crushing mode 325, *325*
Bulla, P.H. 342
Burton, W.S. 4
BV *see* barely visible (BV)
BVID *see* barely visible impact damage (BVID)

C
Cabrera, N.O. 282
CAI *see* compression after impact (CAI)
Campbell, F.C. 262
cantilever wing 298–9
Cantwell, W.J. 67, 123, 218
Caprino, G. 101
CARALL *see* carbon fibre-reinforced
        aluminium laminates (CARALL)
carbide drills 264
carbon-based materials 356
carbon–epoxy (C-Ep) 294
carbon/epoxy face sheets 199
carbon/epoxy/wire mesh face sheets
    evaluation of the flexural characteristics of
        developed sandwiches with 198–206
    load-carrying capacities *205,* 205–6, *206*
    load–deflection curves 199–204, *201–4*
carbon fiber-reinforced plastic (CFRP)
        composite 76
carbon fiber-reinforced polymer (CFRP) 32, 294
    composite sandwich panels 228
    laminates, monolithic 300
carbon fibre-reinforced aluminium laminates
        (CARALL) 62, 64, 65
carbon fibre-reinforced composites 279
Cartesian coordinate system 133
CDM *see* continuum damage model (CDM)
cemented diamond drills 265
ceramic composites 356–7
CFCs *see* chlorofluorocarbons (CFCs)
CFRP *see* carbon fiber-reinforced polymer
        (CFRP)
Chatterjee, V.A. 124
chlorofluorocarbons (CFCs) 6
CLF *see* continuous lattice fabrication (CLF)
closed-cell foams, disadvantage of 282
composite laminates, LVI and CAI behavior
        of 92
composite lattice sandwich panels, preparation
        flowchart of multilayer gradient 47, *48*

composite materials 76
    causes for the impact damage of 241
    development of 279
    man-made fibers in 100
    with particles 101
    sandwich structure of 55
composite sandwich laminates 66–7
    fabrication methods for 10–17
        additive manufacturing processes
            15–17, **18**
        autoclave moulding process 14, **17**
        compression moulding process 11, *12,* **17**
        filament winding process 11–13, *13,* **18**
        hand layup process 10–11, *11,* **17**
        pultrusion process 14, *15,* **18**
        resin transfer moulding (RTM) 15, *16,* **18**
        vacuum bagging process *13,* 13–14, **17**
composite sandwich materials 70
composite sandwich panels
    boundary conditions 135
        blast load 135
        simulation modeling 136, *137–9*
    characterization 234–3
        different impact tests 238–41, *239, 240*
        edgewise compressive strength of
            sandwich panels 236
        flatwise compressive properties of cores
            in a sandwich panel 236, *237*
        flatwise tensile strength of sandwich
            panels 235
        in-plane tensile properties 235
        standard test method for core shear
            properties of sandwich panel
            237–8
    classification of 116
    factors affecting impact in 243–6
        effect of impactor 244–5
        effect of the target structure 245–6
        external factors 246
    mechanical properties of 220
    metallic core-based
        fabrication approach and performance
            characterization of *46–51,* 46–52
        theoretical and geometrical development
            133, 133–4, **134**
    for transient analysis 137
    truss core-based, fabrication approach and
        performance characterization 52–8,
        54–7
composite sandwich panels, drilling and repair
        of 261–73
    damage 265–6
    damage sources
        environmental damages 266
        in-service damages 266–7
        processing irregularities 266

damage types 267
  combinations of damages 267
  cracks 267
  dents 267
  fiber breakage 267
  punctures 267
impact damage 267, 267–8
  sandwich damage 268
input parameters 261–73
  drill material 264–5
  drill point geometry 264, 264
  operating variables 263, 263
output characteristics
  drilling force 265
  drilling-induced damage 265
repair 268–9, 269
  chopped fiber 269
  external patch repair 270–1, 271
  plug repair 269, 269–70
  procedure 271–2
  resin infusion or injection repair 269
  scarf repair patch 270, 270
  structural mechanically fastened
    repair 271
self-healing of polymer materials 272
composite sandwich panels, numerical studies
    on failure of 85–90
  damage zone modeled as equivalent hole
    86–7
  intralaminar and interlaminar damages
    modeling 87
  continuum damage model (CDM) 87–9
  interlaminar damage (delamination)
    modeling for composites 89, 89–90
  matrix cracking and interface
    delamination modeling 90, 90
composite sandwich structures (CSS)
  advantages of 350
  blast-resistant capacity of 287
  core materials 281–4
    foam core sandwich 282–3
    honeycomb core 283–4
    skin–core adhesive materials 284–5
  crashworthiness applications of 321–45
  design of 279
  energy absorption of 344
  in marine applications 277–88
    challenges 288
    effects of seawater 280–1
    effect of temperature 285–6
    effects of underwater blast loading 287–8
  materials for skins 280–1
  mechanical properties of 279
  requirements for the skins of 280
  rigidity of 279
  use of 278

composite sandwich structures (CSS), in
    aerospace applications 293–318
  advantages 294
  analogy with composites 295–6
  cellular solids 297
  classification of 296, 297
  maintenance 316, 317, 318
  mechanical behavior
    core shear failure 305–6
    equivalent flexural stiffness 302
    equivalent flexural strength 304
    equivalent mechanical properties 301,
      301–2
    facings buckling 305
    facings yielding 304
    failure modes of sandwich panels
      subjected to flexural and
      compressive loading 302–4, 303, 304
    index lines 306
    mechanical properties and cost 300–1
  primary structures 297
    cantilever wing 298–9
    jet engines 300
    ultralight monoplane 299, 299–300
  sandwich materials employed in spacecraft
      structures, case study
    alternative materials 307, 307–8, 308
    buckling 314–15
    equivalent flexural stiffness 309, 310
    equivalent flexural strength 309, 311
    mechanical properties and cost 309,
      312, 313, 313
    mechanical properties evaluation
      308, 309
    microsatellite 306–7, 307
compression after impact (CAI)
  after full-scale blast experiments 83
  behavior
    of sandwich composite panels,
      numerical modeling 85
    shape-memory alloy (SMA)–GFRP
      hybrid face sheets 85
    of stitched composites 93
    of titanium honeycomb sandwich
      structures 93
  behavior of thin laminates 80–1
  by column compression 81–2
  experimental investigations on 79–85
  foam core sandwich panel under 92
  of honeycomb core structures 82
  influence of core density increment on 84
  influence of skin thickness increment on 83
  issue in expendable launch vehicles 83
  issue of core crushing damage 81
  issue of hybrid face sheet effect on 84
  issue of low temperature 82–3

compression after impact (CAI) (*cont.*)
  problems, numerical modeling 85
  reinforcement effect on 84
  single FE model for impact damage and 93
  stitched composite under 93
  test 77, 79
    methods and standards 78, **78**
    test model of damaged woven fabric
      sandwich composite 92–3
    thickness and ply-stacking sequence effect
      on 84–5
compression modulus 236
compression moulding process 11, *12*, **17**
compression strength 267
connectors, role 184
construction applications, 3D printed sandwich
  composites for 362–3, *363*
continuous lattice fabrication (CLF) 350
continuum damage model (CDM) 87–9
conventional FMLs 63
conventional metals, failure mechanism of 265
core materials 5, 66; *see also specific types*
core of particulate composite materials
  (PCMs) 223
core panels, corrugated 28
core shear test 238
core–skin delamination 284
core structures, for sandwich composite panels
  5, *5*
corrugated composite sandwich panels 7, *7*
corrugated cores 7, 28
  applications 38
  designs and structures 28, *29*
  manufacturing methods 36–7
  materials 33, *33*
  mechanical properties 35, **35**
  metallic 7
  panels 28
corrugated sandwich cylindrical panels,
  preparation of bidirectional 55, *55*
corrugated sandwich panels 28, 38
corrugated structures **31**
crashworthiness 321
  automotive sandwich composite structures
    for *339*, 339–42, *340*, *341*
  marine sandwich composite structures for
    342–3, *343*
crashworthiness studies, metrics used in
  mean crushing force (MCF) 324
  specific energy absorption (SEA) 323
Crupi, V. 283
crushing mode 324–6
  brittle fracturing 325, *325*
  first crushing mode 327–8, *328*
  lamina bending 325, *326*
  second crushing mode 329, *329*

third crushing mode 329–31, *330*, *331*
  transverse shearing 324, *325*
CSS *see* composite sandwich structures (CSS)
curved layer fused deposition modelling
  (CLFDM) of polymer
  composites 362
cyclic creep test 286–7
cylinder-ligament honeycomb auxetic
  structures 360
cylindrical shells 56

D
Dai, J.G. 149, 151, 156, 171, 184
Daniel, I.M. 206
Davies, P. 287
Davim, J.P. 263
DED *see* directed energy deposition (DED)
defence industry 360
density of core, on structural response 104–5
Devold DBLT600-E10 glass-fiber non-crimp
  fabric 81
Ding, Y. 169
directed energy deposition (DED) 362
direct metal laser sintering (DMLS) 351
displacement field model, three dimensional
  (3D) 132
distance fabrics 232
DMLS *see* direct metal laser sintering (DMLS)
Donga, A. 336, 339, 340
drop ball test 343, *343*
drop weight testers 238, *239*
dual-core panels 32
ductile fiber-reinforced composite 324
dynamic stress 352

E
Eberle, S. 359
eccentric compression behavior–experimental
  investigation 176–84
edgewise compressive test 234, *234*
Edgren, F. 342
E-glass fibre 280, 285
Ei-Sonbaty, I. 265
electric transport optimization 365
electrospinning techniques 354
empirical design formulae, prediction of load
  capacity from existing 187
energy absorption *336*, 339
  in composites 220
energy balance model 102
energy-based continuum damage model
  (CDM) 91
energy dissipation mechanisms 327
energy response, force and 105–7
energy storage, 3D printed sandwich
  composites for *356*, 356–8

engineering plastics 5
engine thermal barriers 352
epoxies (EPs) 281, 284
epoxy resin 8
    matrix, glass in 327
ethylene–propylene copolymer (EPC) 285
expanded imide-modified polyacrylates 7
expanded polystyrene (EPS) 172
extruded polystyrene (XPS) 172

F
fabrication methods
    for composite sandwich laminates 10–17
        additive manufacturing processes
            15–17, **18**
        autoclave moulding process 14, **17**
        compression moulding process 11, *12*, **17**
        filament winding process 11–13, *13*, **18**
        hand layup process 10–11, *11*, **17**
        pultrusion process 14, *15*, **18**
        resin transfer moulding (RTM) 15, *16*, **18**
        vacuum bagging process *13*, 13–14, **17**
face materials 4–5
faceplate laminate material, effects of 335
face sheets 2
    hybridization 199
    material, effect of 242
    materials 66
    metallic 61–71
    optimized thickness of 132
    thickness of 66
failure behaviors 206–7
failure criteria 87, 88, 90
    Hashin 93
Fam, A. 182
Fan, H.L. 55
Fan, J. 67
FDM *see* fused deposition modelling (FDM)
Feng, L.J. 47, 49
Feng, N.L. 67
FE tools 132–3
fiber/epoxy laminates, hybridization of 194
fiber metal laminates (FMLs) 117, 192
fiber-reinforced composites (FRPs) 76
fiber-reinforced polymer (FRP)
    composite face sheets 192
    integrity of 192
    materials 148
    nonmetallic 168
fibre-based FMLs 65–6
fibre composites 221
fibreglass-reinforced polyester 4
fibre metal laminates (FMLs) 61–2
    bridging mechanism in 68
    classification of 63–6
    grades of 65

lightweight characteristic of 62
    magnesium alloys in 65
    magnesium-based 66
    magnesium/glass fibre-based 65–6
    manufacturing process of 70
    mechanical performance of 70
    structural integrity of 68, 71
    surface pre-treatment, classification *69*
    tensile properties of 67
    thermoplastic-based 63, 70
    thermoset-based 70
    titanium-based 65
    types of 65
fibre-reinforced composites 350
fibre-reinforced plastic (FRP)
    advantages of 279
    anisotropic nature of laminates of 246
    feature of 279
fibres
    natural 67
    stiffness of 62
filament winding process 11–13, *13*, **18,** 19
finite element (FE)
    model 91, 93, 132
        higher-order shear deformation theory
            based 132
        simulation model for laminated panel's
            transient responses 136
Finnegan, K. 47, 53
fire growth rate (FIGRA) 363
first-order shear deformation theory (FSDT) 133
flame-retardant properties 362–3
flat plate connector 149
    GFRP laminate 150, *151*
flat sandwich panel's model 133
flat sandwich plate, time-dependent deflection
    behavior of 144
flatwise compressive test 236, *237*
Fleck, N.A. 207
flexural modulus 302, 309
flexural strength 193, 195, 262, 265, 301
    equivalent 304, 309, *311*
floor panel demonstrator 341
Florence, A. 283
FMLs *see* fibre metal laminates (FMLs)
foam-based core 119
foam cores 5, 6–7
    composite sandwich structure (CSS) 286
    materials, mechanical properties **332, 341**
    sandwich structure 45
    foam sandwich structures, macroscopic
        strength of 45
fold cores
    applications 38
    designs and structures 30–2, *32*
    geometries 31, *32*

fold cores (*cont.*)
  manufacturing methods 37–8
    continuous manufacturing processes
      37, 38
    discontinuous manufacturing processes
      37, 37–8
    gradual folding manufacturing
      methods 38
    pre-gathering methods 38
    synchronous manufacturing methods 38
  materials 34, *34*
  mechanical properties 36, **36**
  sandwich panels 32
force and energy response 105–7
forced convectional heat transfer study 352
Found, M.S. 322
4D printing 362
friction effect 88
Friedlander exponential decay equation 135
FSDT *see* first-order shear deformation theory
  (FSDT)
fused deposition modelling (FDM) 350
fusion bonding processes 232–3

G
Galerkin method 132
Gara et al. (2011) 183
Gary, L.F. 324
generalized laminated plate theory (GLPT),
  Reddy's 132
geopolymer concrete 148, 168–9
  fire and post-fire residual performance of
    sandwich panel with 160–3
    test procedure *161,* 161–2
    test results *162,* 162–3, *163*
    test specimens 161
  flexural performance of the sandwich panel
    with 156–60
    degree of composite action 157–8, **158**
    stiffness prediction of PCS WP *158,*
      158–60, *160*
    test results 157, *157*
    test specimens *156,* 156–60
  mix design of **170**
  structural performance of RC one-way slabs
    with 151–4
    code of practice and guidelines 154, **155**
    test results *153,* 153–4, *154*
    test specimens 152, *152*
Gerard, G.: *Minimum Weight Analysis of*
  Compression Structures 4
Geren, N. 196, 197
GFRPs *see* glass fibre-reinforced plastics
  (GFRPs)
Ghalami-Choobar, M. 124

Gibson, L.J. 336
GLARE *see* glass fibre-reinforced aluminium
  laminates (GLARE)
glass/epoxy polymer composites 195
glass/epoxy–wire mesh sheet sandwich
  194, *194*
glass fibre-reinforced aluminium laminates
  (GLARE) 62, 63, 65, 66
  grades of **64**
  impact resistance of 64
glass fibre-reinforced composites 362
glass fibre-reinforced plastics (GFRPs) 279
  laminate, flat plate connector 150, *151*
  trepanning tool for 265
  tubular connectors 171–2, *173*
glass fibre-reinforced plastics (GFRPs)
  connector 148
  development and performance
    characterization of 149–51
    proposed connectors 149, *149*
    test results 150–1
    test specimens 149–50, *150*
glass-reinforced plastics (GRPs) 8–9
GLPT *see* generalized laminated plate theory
  (GLPT)
Goldsmith, W. 322
Golestanipour, M. 121
Gong, X. 225
Gonzalez-Canche, N.G. 69
Griskevicius, P. 66
Guan, Z. 67

H
Hahnel, F. 30
Ha, K.H. 4
Ha, N.S. 322
hand layup process 10–11, *11,* **17,** 103
Hashin failure criteria 93
Hassanpour Roudbeneh, F. 118, 122
Hasselbruch, H.H.A. 193, 194, 197
Haug, E. 322
HDPE/elastomer films 285
Herrmann, A.S. 294
Hertzian contact law 248
hexagonal honeycomb cores 28
hexagonal honeycomb structure 8
hexagonal tube connector 149
high-density core materials 101
high-density polyethylene (HDPE) 355, *355*
higher-order shear deformation theory
  based finite element (FE)
  model 132
high-speed steel (HSS) drill 265
high-velocity impact test 116
Hocheng, H. 263

Hoff, N.J. 4
homogenous core materials 5
honeycomb core 76
  composite sandwich structures (CSS) core
    materials 283–4
  hexagonal 28
  sandwiches 192
  structures, compression after impact (CAI) 82
honeycomb sandwich structure, impact
    behavior of 93
honeycomb structures 8
hot-press method, secondary molding 54, *54*
hot-wet compression test geometries **79**
Hourglass lattice sandwich structure 50
Hourglass truss core, pyramidal truss core to
    51, 51–2
Hou, W. 119
Huang, J.-Q. 149, 151, 156, 161, 171, 184
Hu, B. 50
hybrid composites, significance of wire mesh
    sheets for 192–4
hybrid cores 117
  sandwich panels 122–3
hybridization of wire mesh sheets 193
Hyperelastic Bone® sandwich composites 353
hypervelocity impact tests 359

I
I-beam 2, *3*, 99–100, *100, 262*
IDEs *see* inter-digitated silver electrodes (IDEs)
impact damage
  causes of 241
  energy absorption during impact 242–3, *243*
  mechanism of 241–2, *242*
inkjet printing 358
in-plane tensile testing 233, *233*
insulated concrete walls *see* precast concrete
    sandwich wall panels (SWPs)
Integrated Technology Mast (ITM) 279
integrated woven textile sandwich composite
    (IWTSC) 230
inter-digitated silver electrodes (IDEs) 358
International Maritime Organizations' Safety
    of Life at Sea Convention (SOLAS)
    279–80

J
Jaisingh, S.J 195
jet engines 300
Ji, G. 322
Jing, L.B. 322
Johnson, A.F. 230
Joseph, J.D.R. 196
Jover, N. 122
Jung, A. 123

K
Kagome cores 46, *46*
Kalhori, H. 322
Kang, K.J. 46
Kanny, K. 285
Karunagaran, N. 196
Kazemahvazi, S. 33
Kazemi, M. 118
Kenaf fibre *67*
Kevlar-epoxy/STF/3D braided fibers sandwich
    panel 124
Khaire, N. 119
Khaliulin VI 30
Khan, M.A. 339
Kim, M. 280
kinetic energy (KE) 217
Kooistra, G. 47
Kumar, S. 174
Kumar, S.J.A. 282
Kurian, R. 194

L
lab-on-a-chip (LOC) systems, sandwich
    polymer actuators for 358
lamina bending crushing mode 325, *326*
laminated composite, utilization of 131
Laser Engineered Net Shaping (LENS™)
    Technology 352
lay-up sequences 66, 67, 79–80
LBPF *see* load-bearing ply failure (LBPF)
LDPE *see* low-density polyethylene (LDPE)
lead zirconate titanate (PZT) inks 358
Lee, H.K. 169
Lee, N.K. 169
Liang, X. 121
Libove, C. 4
lightweight composite materials 191
lightweight core materials 279
linear variable differential transformers
    (LVDTs) 150, 156
liquid-moulding processes 15
Li, S. 47, 55, 56
Liu, C. 124
Li, X. 286
load-bearing ply failure (LBPF) 86
load-carrying sandwich structures 216
load transfer 214
low-density polyethylene (LDPE) 118
low-density solid materials 101
low-velocity impact (LVI)
    boundary condition for 92
    effect of 77
    modes of failure and failure load in 246–7
LS-DYNA FE software 339
Lu, G. 322

Lukaszewicz, D. 339, 341
LVDTs *see* linear variable differential
transformers (LVDTs)
Lystrup, A. 262

M
Madenci, E. 101
magnesium alloys, in FMLs 65
magnesium-based FMLs 66
Mamalis, A.G. 322, 327, 331, 332, 336
man-made fibers, in composites 100
marine sandwich composite structures, for
crashworthiness 342–3, *343*
Matalkah, F. 196
mean crushing force (MCF) 324
mechanical properties
composite sandwich panel 220
composite sandwich structures (CSS) 279
corrugated cores 35, **35**
foam core materials **332, 341**
fold cores 36, **36**
sandwich composites 76, 279
medium-impact test 116
Meng, Q. 196
metal–composite interface
adhesion level of 69
adhesive bonding of 68–9
structural integrity of 68
metal–composite sandwich materials 63
metal face sheets, with textile-based core 247
metal layers, surface pre-treatment of 71
metallic core-based composite sandwich panels
fabrication approach and performance
characterization of *46–51, 46–52*
metallic cores 117
sandwich panels 119, 121
high-velocity impact properties **120**
metallic corrugated cores 7
metallic facesheets 61–71
metallic materials 4
metallic truss sandwich structure 52
metallic wire mesh sheets 192
plastic deformation behavior of 192
methacrylates (MAs) 284
Meyer's contact law 102
microtrusses, fatigue studies of 363
military applications, 3D printed sandwich
composites for 360–2
*Minimum Weight Analysis of Compression
Structures* (Gerard) 4
Miura, K. 28
modus operandi of sandwich 215
Mohamed, A. 352
monolithic CFRP laminates 300
monolithic composites 216
monolithic laminated composites 101

monolithic materials 131
Morton 218
Mousavi, S. 357
multilayered laminated sandwich composite
plate 133
Murphy, C. 265

N
Naito et al. (2012) 171, 172
Najafi, M. 122
natural fibres (flax fibres) 67, 100, 287
NCF sandwich panels *see* non-crimp fabric
(NCF) sandwich panels
negative Poisson's ratio (NPR) 352
Ngo, T.D. 361
Nguyen, M.Q. 322
Nguyen, Q.T. 362
Nia, A. 118
Nomex 8
non-composite face materials 222
non-crimp fabric (NCF) sandwich panels
79–80
nonmetallic fiber-reinforced polymer (FRP) 168
non-metallic materials 5
Noor, A.K. 4
novel AM techniques 363
NPR *see* negative Poisson's ratio (NPR)
nylon–Kevlar composites 352

O
optimization techniques 358
ordinary Portland cement (OPC) concrete 148

P
Park, S.Y. 69
Parnanen, T. 65
Pascoe, J.A. 68
PBF *see* powder bed fusion (PBF)
PCSWP *see* precast concrete sandwich wall
panel (PCSWP)
PEEK *see* polyether ether ketone (PEEK)
PEI *see* polyetherimide (PEI)
pendulum-type systems 241
personal protection equipment (PPE) 360
phenolics 9
piezoelectric transducers, 3D printed
sandwich 358
plain weave carbon fabric/epoxy (PWCF)
prepreg 82
plane woven fabric 225
Plantema, F.J. 4
plastic deformation behavior, of metallic wire
mesh sheets 192
plate-type GFRP connector 156
ply-stacking sequence effect, on CAI 84–5
polyacrylates, expanded imide-modified 7

polyester, fibreglass-reinforced 4
polyester resins 10
polyether ether ketone (PEEK) 5
polyetherimide (PEI) 5
polyethylene terephthalate (PET) 101
polylactic acid (PLA) 362
polymer composites, curved layer fused
        deposition modelling (CLFDM)
        of 362
polymer foam 192
    cores 192
    core sandwiches 207
polymeric core 117
    sandwich panels 117–19
        high-velocity impact properties of **120**
polymer materials, self-healing of 272
polymer matrix composites (PMCs), drilling of
        263, 264
polymers
    thermoplastic 63
    thermoset 63
polymethacrylimide (PMI)
    foams 7
        advantages 7
        core 81
        filled corrugated sheets 33
polystyrene foams 6
polyurethane (PU) 9, 101, 284
    foams 6, 117, 118, 122
polyvinyl chloride (PVC) 6–7, 101, 102
    compressive properties of 105
    foam 118
        compressive properties 282
        core 101
        core sandwich beams, flexural
            characteristics 198
    score, stress–strain behavior of 104, *104*
polyvinylidene fluoride (PVDF)
    transducers 121
porous 3D printed thermoplastics 351
Portland cement-based concrete 169
potential energy (PE) 217
powder bed fusion (PBF) 362
Prakash, V.R.A. 195
precast concrete sandwich wall panel (PCSWP)
        148, 156, 157
    stiffness prediction of *158*, 158–60, *160*
precast concrete sandwich wall panels
        (SWPs) 168
Puhui, C. 86
pultrusion process 14, *15*, **18**
PVC *see* polyvinyl chloride (PVC)
pyramidal composite lattice structure 53
pyramidal lattice cores 228
pyramidal truss core, to Hourglass truss core
        51, 51–2

Q
Qeshta, I.M.I. 193
Qi, J. 194, 197
quadratic traction–separation criteria 93
quasi-static three-point flexural tests 199

R
Rajadurai, A. 194, 196
RC one-way slabs with geopolymer concrete,
        structural performance of 151–4
Reddy's generalized laminated plate theory
        (GLPT) 132
reinforced concrete (RC) wythes 148
Reis, P. 263
resin transfer moulding (RTM) 15, *16*, **18**
Robert, M.J. 324
Rolfe, E. 118
Romanoa, F. 87
Rozant, O. 282
RTM *see* resin transfer moulding (RTM)
rubber compounds, synthetic 8

S
Sackman, G.L. 322
Sakthivel, M. 195
sandwich composite panels 262
    experimental investigations on CAI for
        79–85
    3D printing of 351
sandwich composites **105**, *106*
    compression test of 78
    at different impact energies *107*
    intralaminar failure and delamination of
        90–3
        CAI test model of damaged woven
            fabric sandwich composite 92–3
        composite panels and laminates 90–2
        foam core sandwich panel under CAI 92
        stitched composite under CAI 93
        three-dimensional parametric study 93
    mechanical properties of 76, 279
sandwich composites manufacturing 220–34
    core material 222, *222*
        balsa wood 223
        metallic 223, *224*
        polymeric 223, *224*
    different core orientations in sandwich
        panel 234
    face material 221, *221*
    manufacturing 222
    honeycomb 225, *225*
        fold cores 227, *228*
        integrated woven cores 228–30, *229*, *230*
        manufacturing of core 230
        3D woven honeycomb structures *225*,
            225–6

sandwich composites manufacturing (*cont.*)
  3D woven hollow spacer structures *226,* 226–7, *227*
  truss 227–8, *229*
manufacturing of sandwich panel 231
skin–core joining methods *231,* 231–3
  adhesive bonding 231–2
  fusion bonding processes 232–3
sandwich composite structures *2, 2–3, 350*
  adhesive materials 8–10
    epoxy resins 8
    phenolics 9
    polyester and vinyl ester resins 10
    polyurethanes 9
    toughened epoxies 8–9
    urethane acrylates 9–10
  common materials used in 4–10
  core materials 5, *5*
    corrugated or truss cores 7
    foams or solid cores 6–7
    honeycomb structures 8
  face materials 4–5
  history of 4
  laminates under bending forces 2, *3*
sandwich composite tubes
  axial compression of 327–31
  three-point bending of **332,** *332,* 332–9, *333–8*
sandwich cores
  shear mechanisms **209**
  wedge-shaped folded 30
sandwich laminates composite *see* composite sandwich laminates
sandwich materials, composite 70
sandwich panels 28, 76, 77
  bio-based core 122
  corrugated 28, 38
  folded core 32
  with geopolymer concrete
    fire and post-fire residual performance 160–3
    flexural performance 156–60
  high-velocity impact properties of 117–27
  hybrid core 122–3
  impact behavior of 117–18
  impact performance of 124, *125, 126*
  impact resistance of 127
  metallic core 119, **120,** 121
  new advanced 123–4
  polymeric core 117–19, **120**
  transient behavior of 132
sandwich polymer actuators, for lab-on-a-chip (LOC) systems 358
sandwich structure 131
  damage analysis of 107–8, *108*

sandwich wall panels (SWPs) 172, 174, 176, 178, *180,* 182, *185*
  failure modes of *176*
  load capacity of 186
  N–M curve of 176
  precast concrete 168
  subjected to eccentric axial load 186–7
  traditional form of 174
Sarasini, F. 122
El-Sayed, M.A. 365
scanning electron microscope (SEM) 117
Schapery theory-based constitutive relations 92
Schenk, M. 37
SEA *see* specific energy absorption (SEA)
sealant tapes 13
selective laser melting (SLM) 351, 354
Shao, R. 196
shape-memory alloy (SMA)–GFRP hybrid face sheets and CAI behavior 85
shear-thickening fluids (STFs) 124
Shiah, Y.C. 230
Shyha, I.S. 265
signal-to-noise (SN) ratio 103, 108
  effect plot for 109, *109*
  response for absorbed energy 109, **109**
simplified theoretical investigation 186, *187*
Singh, I. 285, 286
Siriruk, A. 286
Sjoblom, P.O. 218
skin–core adhesive materials 284–5
skin–core joining methods *231,* 231–3
  adhesive bonding 231–2
  fusion bonding processes 232–3
SLM *see* selective laser melting (SLM)
soft robotics, 3D printed sandwich composites for *356,* 356–8
solid cores 6–7
solid/sandwich RC wall panels, empirical design formulae for **188**
spacers 232
space structures, 3D printed sandwich composites for 358–9
space vehicles, thermal protection systems (TPS) in 359
specific energy absorption (SEA) 323
sports applications, 3D printed sandwich composites 359–60, *361*
spring elements 93
stainless steel (SS)
  Hourglass lattice sandwich structure 49
  tetrahedral lattice sandwich structure 47
  wire mesh layers 194
  wire mesh sheets
    hybridization of fiber/polymer matrix face sheets 194–7
steel alloys, in sandwich panels 121

Steeves, C.A. 207
STFs *see* shear-thickening fluids (STFs)
strength-to-weight ratio 28
stress–strain behavior of PVC core 104, *104*
structural deformation behavior 132
Sun, G. 322
surface pre-treatment techniques 69
SWPs *see* sandwich wall panels (SWPs)
symmetrical sandwiches 215, *215*
syntactic foam 282
synthetic fibres 221
synthetic rubber compounds 8
Sypeck, D.J. 46

T
tactile sensors 357
Tagarielli, V.L. 342
Taghizadeh, S.A. 33, *33*
Taguchi analysis 108–10
Taguchi's design of experiments (DOE) 103,
    104, **104**
Tang, E. 121
Tan, K.T. 93
Tan, W. 92
Tarlochan, F. 322, 324
test geometries, hot-wet compression **79**
Teti, R. 101
tetrahedral lattice truss cores, aluminum
    alloy 47
textile-based core, metal face sheets with 247
textile-reinforced composite sandwich panels
    analytical modelling 248–9
    applications 249
        aeronautics 249, *251*
        civil infrastructure 249
        marine structures 249, *250*
        railway 250, *252*
    low-velocity impact behaviour of 214–16
        concept 218–20
        importance of impact in
            composites 220
        manufacturing of sandwich composites
            220–34; *see also* sandwich
            composites manufacturing
        mechanics of impact *217*, 217–18
    numerical modelling 249
textile structure-based composite sandwich
    panels
    structure and properties of 247–8
        auxetic structures 247
        dual-core-based structure 248
        incorporation of nano-structures 248
        integrated structure 248
        metal face sheets with textile-based
            core 247
thermal damage 263

thermal protection systems (TPS), in space
    vehicles 359
thermoforming 232, *232*
thermoplastic-based FMLs 63, 70, 71
thermoplastic hot melt film 284
thermoplastic polymers 63
thermoplastic polyurethane (TPU) 357
thermoplastic PVC 6
thermoplastics, porous 3D printed 351
thermoset-based composites 281
thermoset-based FMLs 70
thermoset polymers 63
THR *see* total heat release (THR)
three dimensional (3D) digital image
    correlation (3D-DIC) method 118
three dimensional (3D) displacement field
    model 132
three dimensional (3D) printed sandwich
    composites
    for aeronautical and automotive
        applications 351–2
    in biomedical applications 352, *353, 354*
    for construction applications 362–3, *363*
    for energy storage and soft robotics *356,*
        356–8
    for military applications 360–2
    for space structures 358–9
    sports applications 359–60, *361*
three dimensional (3D) printed sandwich
    piezoelectric transducers 358
three dimensional (3D) printing
    composite sandwich structures, challenges
        363–5
    of sandwich composite panels 351
    techniques 354, 356
titanium alloys 65, 121
titanium-based FMLs 65
titanium honeycomb sandwich structures, CAI
    behavior of 93
Tiwari, G. 119
Tomlinson, D.G. 182
topology optimization 355, 358
total heat release (THR) 362–3
toughened epoxies 8–9
toughened resin test **79**
transducers
    polyvinylidene fluoride (PVDF) 121
    3D printed sandwich piezoelectric 358
transient behavior, of sandwich panels 132
transverse shearing crushing mode 324, *325*
transverse shear tests 36
triply periodic minimal surface (TPMS)
    sandwich beam structures 359
truss cores 7
Tsao, C.C. 263
tubular connectors, GFRP 171–2

U

ultrahigh molecular weight polyethylene
(UHMWPE) 123
ultralight monoplane *299,* 299–300
United States Forest Products Laboratory
(USFPL) 4
unmanned aerial vehicles (UAVs) 352
urethane acrylates 9–10
Uzay, C. 196, 197

V

vacuum-assisted resin transfer moulding
(VARTM) 15, 19, 70, 71
vacuum bagging
process *13,* 13–14, **17**
technique 271
vacuum brazing approach 48, *49*
VARTM *see* vacuum-assisted resin transfer
moulding (VARTM)
Vijayakumar, S. 195
Villanueva 123
vinyl ester 281
Norpol Dion 9500 81
resins 10
Visby class of stealth corvettes 278
viscoelastic layers 55
Voigt notation 88
Vorm, T. 262

W

Wadley, H.N.G. 47
Wang, B. 119

Wang, J. 92
wavefront error (WFE) 359
wedge-shaped folded sandwich cores 30
WFE *see* wavefront error (WFE)
wire EDM-interlocking assembly-vacuum
brazing approach 48, *49*
wire mesh sheets
in composite materials 192
for hybrid composites 192–4
integration of 193
use of **197**
work–energy principle 218
woven fabric sandwich composite, CAI test
model of damaged 92–3
Wright-Patterson Air Force 4

X

Xiong, J. 53
Yang, J.S. 47, 48, 50, 53–6, 58

Y

Young's modulus 283
Yungwirth, C.J. 123

Z

Zhang, Y.X. 119
Zhu, X. 322
Zitoune, R. 263

For Product Safety Concerns and Information please contact our EU
representative GPSR@taylorandfrancis.com
Taylor & Francis Verlag GmbH, Kaufingerstraße 24, 80331 München, Germany

www.ingramcontent.com/pod-product-compliance
Lightning Source LLC
Chambersburg PA
CBHW060751220326
41598CB00022B/2397

9 7 8 0 3 6 7 6 9 7 3 2 7